ECOLOGY ENVIRONMENT AND POLITICS

ECOLOGY ENVIRONMENT AND POLITICS

BY

C.B.S.R. SHARMA

KALPAZ

First Published, 2018

Published by

Kalpaz Publications
C-30, Satyawati Nagar,
Delhi – 110052
E-mail: kalpaz@hotmail.com
Ph.:9212142040

Printed at: G. Print Process, Delhi

Cataloging in Publication Data—DK
Courtesy: D.K. Agencies (P) Ltd. <docinfo@dkagencies.com>

Sharma, C. B. S. R., author.
Ecology, environment and politics / by C.B.S.R. Sharma.
pages cm
ISBN 9789353240264

1. Ecology. 2. Ecology—India. 3. Environmental policy. 4. Environmental policy—India. 5. Genetic toxicology. 6. Genetic toxicology—India. I. Title.

LCC QH541.S53 2018 | DDC 577 23

To

Prof. Paul Appasamy and Dr. Jeevanandam

Yoganand, Manimekhalan, Niroj Joshi, Raghavendra Rao, Anil, Amrut, Rona Wilson, Rajesh Nelson, Debasish Patnaik, Dorai Raj, Vadivelu, Samir Nazreth, Ravichandran

POSITIVE FORCES

RARE AND UNPREJUDICED

DECLARATION

Most articles are lectures delivered in 35 institutions. They are at: Aizawl, Anantpur, Annamalai, Aurangabad, Bangalore, Bhopal, Bombay, Calcutta, Coimbatore, Calicut, Delhi, Dharwar, Darbhanga, Erode, Gangtok, Gauhati, Hissar, Hyderabad, Jammu, Madurai, Meerut, Madras, Ooty, Pondicherry, Patiala, Patna, Poona, Rohtak, Shillong, Silchar, Srinagar, Tirunelveli, Trivandrum, Tirupati, Topslip, Vellore, and Vijayawada.

They are in a popular vein.
Some were given in several places, and are therefore repetitive.
There are no plagiarisms or cuts and pastes

If ecology is safe,
people are safe

Contents

Preface 9
Acknowledgement 13

GLOBAL PERSPECTIVES

1. Ecology and Resources Development and Environment 17
2. Science and Art of Ecology 27
3. Continuing Ecospasms the Perennial Concerns 37
4. Ecological World Views on a Triaging World 49
5. Globalization and the Indian Society One Step Forward, Two Steps Backward; Both Awkward 77
6. Development on Ecological Bases 89
7. Biodiversity: Perceptions and Proactions 99
8. Chemicals Worse than Radiations 107
9. Chemical Warfare 111

CONSCIENTIOUS RESPONSES

10. Environmental Movements 123
11. NGOs: Altruistic Organizations Voluntary Agencies are the only hope for ecological security and political stability 147
12. Cataclysmic Kalki in India Ecological Degeneration is Harakiric to Indian Polity 153
13. Feminisim in an Ecological Age 161
14. Ecology and Foreign Policy 179
15. Ecological Future of India 191

REGIONAL CONCERNS

16. The Kaveri Quagmire Guidance from Ecological Wisdoms 201
17. Ecography of Coastal Tamilnadu 205
18. Ecological Perils of the Sethusamudram Project 215

19. Ecological Solutions for Telangana Ecosystem - Centric Development and Administration 219
20. Ecological Revival of Tanks 223
21. Resurrecting a Tradition 235
22. Temples, Coasts, Tanks and Ecology Hybridize Them and Wait For the Harvest 237
23. One Step Forward and Two Steps Backward 243

DESCENDING PLAGUES

24. Genetic Toxicology and Genepidemics 249
25. Genotoxic Ill - Health from Pollutions 257
26. Genetic Toxicology of Pesticides in Higher Plant Systems 273
27. Genetic Toxicology in India 283
28. Urban Genetic Hazards 289
29. Genepidemics, Genepandemics and Darwin An Ominous Possibility Emerging from a Reconsideration of Darwinism in its 150th Yar 295

THEY ALSO SERVE

30. Phytomass for Energy 301
31. Ramar Pillai's Petroplants The Tragedy of Indian Science 315
32. Botanical Pesticides 321
33. Ecology Burns to Make Way for Energy 331

ETERNAL HOPES IN THE HEART

34. Higher Education in Ecology and Environment With Special Reference to Natural Sciences 339
35. Higher Education in Ecology and Environment With special reference to Social Sciences and History 351

APPENDICES

36. Forebodings of A Native American Chief "End of Living and the 'Beginning of Survival 369
37. Concens of the United Nations Taking a Stand 377
38. Vintage Wisdoms: A Bibliography 1990 383

Index 401

Preface

Only in 1969, we viewed the Earth as a planet, but the same was visualized by the wisdom of experiences by the Chief Seattle of the Suquamish and Duwamish Indian tribes in today's Seattle Metropolis, in 1854. However, their wisdom was ignored and ridiculed in practice and used for migrations, exploitation, mass murders, colonialism, and imperialism. When imperial powers could not use their usual techniques of oppression, they resorted to trade and politics, buttressed by science, as in the case of Peruvian Guano. Thus emerged the adage that politics is war without weapons and war is politics with weapons, and trade is an integral part of these two.

There were wars waged for supremacy over resources in order to empower the industrial revolution and to develop it into the next stage. This scenario continued until after the Second World War, when both the victors and the vanquished suffered irrevocably. A series of crises, unfathomable in the beginning, gradually alarmed the intelligent community everywhere. Such people gathered to form a Club of Rome and brought the Limits to Growth Perspective in 1972. Agreeing with these views, the United Nations made a major effort to sensitize and organize positive efforts globally. Together, a consensus on eleven directions emerged. The Red Indian chief is implicit in these, without which resources can neither be identified nor explored and exploited.

Two imperatives emerged from the lessons learnt by the above historical developments. Chief Seattle's concerns form the baseline. Since these were disregarded for over a century, it is imperative that the eleven directions are urgently addressed by man. Thus, a self-education, a re-education, with changed priorities became imminent. Holistic studies on ecology and resources and similar perspectives on the world history became necessary. Eventually, the western countries changed their focuses on education systems substantially, incorporating the Red Indian's Ecology and the priorities of 1972. The former is called ecology, while the latter is christened, environmental science.

It is useful to recollect the chief and his instinctual ecology, which consists of - SWAMPEA - Soil, Water, Air, Minerals, Plants, Energy, and Animals. Of these, energy is everywhere; air is above us and the minerals below; the rest exist interconnected and are therefore interdependent, man being one of the animals. As man's imperialism grew on all these, there was realization on the connections with the other three elements of Earth. Thus, this completed his hegemony over Earth, and so now we see the backlashes for the deeds performed in the past. Limits to Growth Scenario of Meadows of Club of Rome list five crises: Over Population, Pollutions, Depletion of Resources, Energy Crunches, and Hygiene and Sanitation. The UN recognizes six crises: Climate change, disasters and conflicts, ecosystem management, environmental governance, chemicals and waste, and resource efficiency. The last two are common for both, while the others too either overlap or are totally new. Consequently, a new conglomeration called Environmental Science has been created by itself. This does not exist without its mother Ecology, which also has its goal in the formation of balances, but Environmental Science became more fashionable.

What about India?

In a hurry to catch up with the advanced countries, India has ignored historical events, developments from concerned groups, and the Think Tanks of the United Nations. as well as mechanisms. Here our government and intellectuals miserably failed for themselves and for India. As a result, education was provided as free for all, imparting only partial knowledge, which was unreviewed and unfocussed. Naturally, it grew as an infantile that was paralytic. Governmental and nongovernmental activities may be complementary, but educational training for leaderships and involvement were intolerably substandard. Indian education understood the ramifications of ecology into human systems, human ecology, but ended in individual identities instead of collective ecosystems.

In the field of higher education, no two curricula are same or similar; syllabi are various; bandwagon culture or placebo culture is the norm. With every task in the university being performed internally, the faculty is the ruler. Good people continue to do good things and the bad people indulge in many vices in the name of the environment. But all are deficient in the foundations of nature history, human species and science. Today, we are badly in need of a cultured mix involving all these.

However, the thought just mentioned does not result in good public relations. There are no takers for this idea - it is all whims and fancies. The student community, however, has acknowledged this approach and in fact wants more of it. Personally speaking, more number of students from other departments opted for my elective papers. Courses from several departments in the Academic Staff College used to have my lecture on the environment. Year after year, students adjudged me as the best teacher on a 20-point comparative grading.

This experience has spread to other institutions, both far and near, to around 30-40 places where I have given more than hundred lectures on various topics on the environment. Selection of topics was based on my limited knowledge and the host's preferences.

First, the conceptual basics, components of history, philosophy, and development were focused, as these were lacking in most places. From a global perspective, we then shifted to regional and local issues but with a holistic mindset. People resisted plunder of resources sometimes with weapons. There is one area that is of great relevance to human condition-mutagens in the environment. It is not generally known that mutagens are basically responsible for cancers, birth defects, and heritable diseases, deserving wider coverage. Plants, as it is my basic field of training, cannot escape attention. The lecture collection climaxed with roles of environmental education in natural and social sciences.

For the advantage of the uninitiated in the philosophy, history, economics, and sociology of ecological ramifications in human societies, a bibliography of early literature, subjective in selection, is appended. All illustrations and tables are mine.

If this collection has tried to fill the blanks and gaps in the understanding of ecology and environmental sciences in many quarters, it is due to the innumerable students of all hues in India. To be a teacher to them has been a great fortune.

Acknowledgement

I was thrust into the field of Environment, mid way in my career. Otherwise my research field was Genetic Toxicology. At Bharathiar University, Coimbatore, I became the Professor of Environmental Science. And fate pushed me to Pondichery University. Pondicherry is a modern wilderness with unsound ecology, closing my laboratory and stifling my scientific urges and efforts. Yet I managed to survive, with goodness still intact, in teaching and societal activities.

Over hundred lectures in thirty-five institutions on about thirty topics, mostly during 1990–2004, were converted into essays, on students' requests. I am thankful to all who invited me, notably Punjabi University, Patiala and Assam University, Silchar. I owe a debt of gratitude to Mahendralal Trivedi, TA Sharma, Amarjeet Singh, and Debasish Kar. Such involvements were not possible without considerable sacrifices by my wife, Ramadevi and daughter, Swathi. In preparing the scripts, Drs..Dorairaj, Vadivelu and Ilangovan of Pondicherry University and Mr. Murugaian gave great timely help.

GLOBAL PERSPECTIVES

Chapter - 1

Ecology and Resources Development and Environment

The constant yearning of every individual is always to improve his living conditions. This can be accomplished only from the resources, which have an evolutionary history. Some of these can be directly used by the human society for its welfare; but most have to be transformed or modified before use and distributed among the people. This is called economic growth or social development, which has been happening for centuries and millennia in different parts of the world, sometimes progressively and often retrogressively. The historical inference from this long story of man is that there are limits to social and economic development in the form of insufficient resources, environmental degradation, and ecological imbalances, both in the ecosystems as well as social systems. Briefly, the issues involved are described here for the benefit of people uninitiated in these areas.

A. Resources

Water, soil and land, plants, animals, microbes, minerals, and atmosphere constitute the resource base of humankind, besides themselves. These have a variety of distributional patterns and productivity profiles, both in quality and quantity.

B. Systems

The above resources are distributed in the planet in such a way that scientists recognize about more than forty ecosystems accommodating different types of the above and developing distinguishing features of themselves in due course of evolution.

Examples are forests, deserts, wetlands, seas, islands, rivers, grasslands, and Savannahs in equatorial, tropical, subtropical, temperate, and other climatic and geographic zones and influenced by longitudes and altitudes.

C. Principles of Exploiting Biosphere for Development

Carrying Capacity: Since resources are limited population can not grow beyond the size established by the resource limits. Malthusian logic.

Stability and Diversity: More genetic diversity, more ecological stability. Agriculture: reduction of crop losses due to modernization may never materialize. "Ecological Nightmare"

Ecological Buffering: Biological matter buffers air and the geologic surface. If this matter is not in sufficient quantity, Carrying Capacity will be reduced in warmer climates.

All the above may be site specific

D. Development

Resources present in various systems are exploited for human consumption and development by the use of empirical knowledge, experiences and experimental knowledge. In other words, Science and Technology are used for development. This usage is done by a planning process by the political authorities following a political process using political mechanisms. Everything that is politically based takes into consideration the social and economic wellbeing of people under consideration. Consequently, the distribution of resources as well as the products of development constitute a basic ingredient of all these processes for equity and justice to be ensured. Trade, commerce and markets, therefore, are inseparable, including industry. During these processes, several consequences have resulted and are well documented.

E. Consequences

(a) The first consequence is always a change in the existing traditional scenarios-depletion of resources, degradation of the systems, and ecological imbalances, where the ownerships of resources and/or development change hands, sowing the seeds of injustice, inequalities, discontent and ferment.

(b) The second consequence is economic-reorientation in the internal production and distribution, external trade deficits, internal and external loans and investments, debt burdens, and the economic backlash on certain sections of people who have nothing to fall back on.

(c) The third consequence is social-inequalities, hopes and despairs, divisions, conflicts, hatred, and violence. As the traditional values disappear along with the traditional resources, according to usage patterns the world gets triaged into rich nations, poor nations, rich people, and poor people, with buffers everywhere becoming fixed stars, while hopes and despairs alternate periodically as per the whims of people in power. Goodness and other human values continuously vane inexorably making most lives worthless. This is the inherent meaning of Dandekar's Poverty Line.

(d) The fourth type of consequence is the backlash to the environmental and ecological processes, which often defy even comprehension leave alone alleviations. Once these are set in, it is difficult to reverse and restore normalcy. Examples are acid rain, ozone depletion, etc.

E. Silver Lines

But there were silver linings. The above were foreseen, feared, and sought to be studied and investigated as a beginning to avert such a calamitous fates befalling many nations and people in many countries. One such way was to attempt modeling futures based on resources, agriculture, economics, and politics to integrate all these for common good. The pioneer crusader was the Club of Rome organized by Aurelio Peccei bringing together economists, sociologists, political thinkers, jurists, and others during the late 1960s and came into reckoning by June 1970 due to the cooperation extended by King, Pestel, and Volkswagen Foundation to Peccei.

F. Futurology

Models predicting consequences and advocating rapatterning of growth have been developed by concerned scientists. These are broadly six in addition to the Global 2000 Report, which was the most comprehensive. All these today are merely of historical value because we have now routine annual reviews by the World Bank, World Resources Institute, etc. Even the First India Development Report has appeared.

(a) **World I Model: (worldwide:** 1. Forrester: Population, Capital, Resources, Pollution, Food 2. Forrester, Pestel: World Dynamics 1971.

(b) **World II Model: (worldwide:** 1. Meadows & Meadows (MIT): Limits to Growth (1972). 2. Meadows & Meadows (MIT): Dynamics of growth in a Finite world (1974).

(c) **World Integrated Model (WIM):** 1. Mesarovic & Pestel (1974): Mankind at the Turning Point (Hanover & Cleveland) 2. Interactions among Many aspects 3. Global Cooperation to avoid Catastrophies.

(d) **Model of Integrated Relations in Agriculture (MOIRA):** 1. Bruckmann (ed): "MOIRA: Food and agriculture Model" 2. Linnemann, (Amsterdam) Model of Integrated Relations in.

(e) **Latin American World Model:** 1. Herrera et al (Rio de Janeiro) 1976 2. Catastrophe or new Society 3. Socio-Political power Exploitation & Destructive Values in society 4. Poor people & poor nations (PPPN) vs Rich People & Rich Nations (RPRN].

(f) **UN World Model\:** 1. W1, W2, W3 & WIM are frightening and unpalatable to the poor nations 2. Strategies to improve PPPN through economy 3. Leontief (Brandeis U. & Harvard U.) 1973 4. Carter, Petri, Stern: Physical limitations to Economic growth.

(g) **Limits to Growth Model (Meadows & Meadows, 1972:** Of the above **six** Models the Second one popularly known after Meado- ws has attracted imagination of most people. This postulates that growth everywhere will be limited by five factors. 1. Over population, 2 Insufficient Energy 3. Depletion of Resources 4. Hygiene and Sanitation 5. Pollution.

(h) **Government Global Model (1982):** Resource depletion and Efforts needed by countries has been focused here. This contention received partial support from Global 2000 Report commissioned by Carter in 1970s and submitted in 1980. This work summarized the statuses of development in all countries on fourteen important parameters. Also it has given probable projections on all these by the turn of century.

G. Human Resource Development

Limits to growth paradigm was not taken kindly by the Third World. The Marxist thinkers felt that it was an intellectual ploy to subvert the aspirations of the developing countries. They contended that if the humans are resourceful all problems can be faced -population contained, resources conserved, pollution averted, energy crunches solved, and hygiene/sanitation ensured. The accent therefore shifted to Human Resource Development, which can not only ensure the above but also see that products of development are properly distributed.

Challenges: Unfortunately, there were two problems in improving the humankind's abilities to help themselves. Firstly, there are sections in society handicapped socially and/or economically for reasons of history. Secondly, for the same reasons of history, the world is divided into many systems of political, social, and economic orders.

Affirmative Action and Transaction principles: How to reconcile the diversities in accomplishing just development, sane conservation of ecological resources, and maintaining congenial environments? The first can be met with by a policy of affirmative actions within countries and between developed North and aspiring South. All the diversities can be reconciled to function efficiently by what is known as Transaction Principle -managing apparently uncompromising contradictions by human ingenuity of combining realities on the ground with the arts of the possible.

H. Perspectives

There is considerable variation in the outlooks and thoughts of various sections even among the intellectuals and other agents of change. While the hardcore academics talk of accounting and valuing, business peers think of markets and technologies. The altruists, on the other hand, are concerned with equity, social justice, empathy and human rights. However, it should be conceded that in the process towards the mad rush for economic development symbolized by consumerism and hedonism, human values, principles of morality, and humanism have taken a thorough beating in the hands of corruption and many forms of violence defending it. As far as India is concerned age old values of *Sanatana Dharma*, Jainism, Buddhism, Hinduism, Islam, Christianity and Universal Values of Brotherhood have given way to number games, million localisms and divisive forces in the name of development. The ecological balances that sustained all of us for millennia and congenial environment that we all aspire as a future are nowhere to be seen. The politics of selfishness and exclusion principle are taking us inexorably towards that Hindu myth of *Kalki* of Destruction-The Ecospastic Kalki of self-destruction.

Table 1.1: Resource Scenario by 2000 Ad

1.	Population	6.5 billion (2000), more effects on LDCs
2.	Arable land	Increase by 4%
3.	Food	Increased yields, only source
4.	Higher yields	Fertilizer, Pesticides, power for irrigation, fuel
5.	Wood Energy	25% increase only possible, bleak prospects. Needs 25% more than supplies
6.	Fossil Energy	Coal, Gas, Oil, oil shale, tar sands, uranium- finite -economic disparities among countries-environmental impacts everywhere
7.	Minerals	¼ of the DCs absorbs ¾ of the total resources of the world
8.	Water	Severe in many regions. 50% the world already needs 100% increases -deforestation is the major culprit
9.	Forests	Timber sources decline by 50%,-20 million hectares disappear annually mostly in Africa, Asia, South America 40% will totally go
10.	Agricultural soils	Deterioration serious -erosion, organic matter, Salinization, water logging,, alklization, desert-like conditions
11.	Air	Co_2 increased, Ozone depletion, acid rain
12.	Critical geologic	Cu, Fe, Ni, Mn, Coal, Petroleum, Pb may be used up by 2000 Zn 70% only sources remain in 2000 A.D. Most critical by 2020 A.D
13.	Wastes	Radioactive, and nonradioactive
14.	Extinction of Germplasm	20% of the existing tropical forests irretrievably lost
15.	Future of the world Economy	Growth rate of agriculture now less than 200% in the highincome countries, requirements in the low- income regions
	(a) 500% (b) 4-500% (c) 950%	- Latin America - Africa, China, Asia - Middle East

From Global 2000 Report.

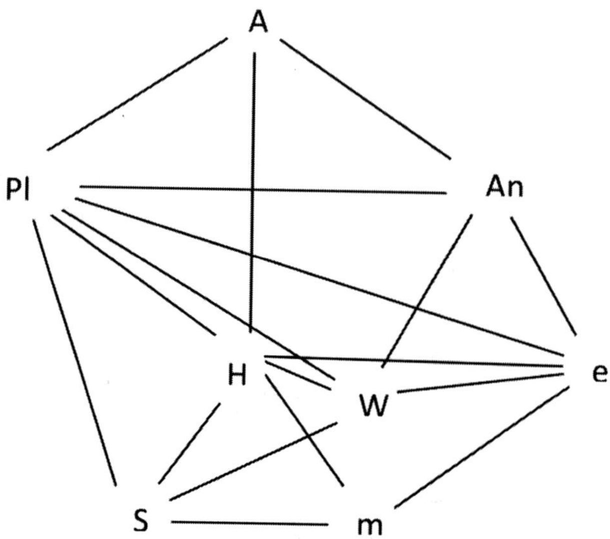

Fig. 1.1A. Ecology: Evolutionary: Balances.

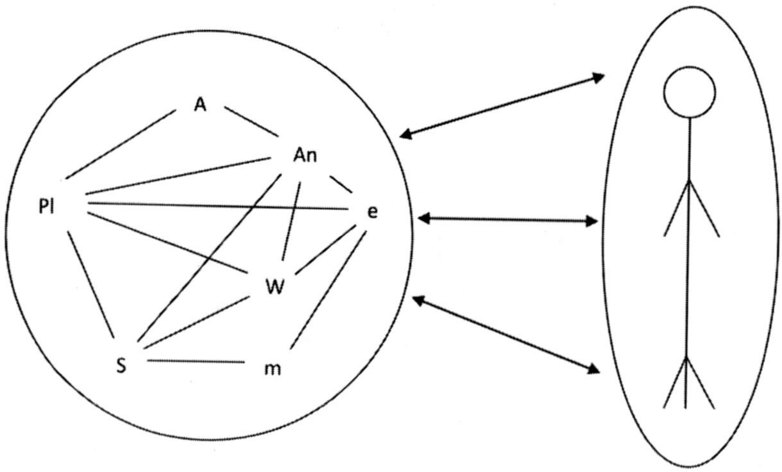

Fig. 1.1B. Ecology : Anthropocentric: Resources.

THINK GLOBALLY AND ACT LOCALLY

LIMITS TO GROWTH

1. Depletion of Resources
2. Excessive Population
3. Pollutions
4. Inadequate Energy
5. Lowly Hygeine and Sanitation

LEADERSHIPS – HUMAN RESOURCES

1. Ignorant – Avaricious
2. Opaque – Unaccountable
3. S&T – R&D, Econ-Socio
4. Ethics – Equity – Empathy – Ethos

CONSEQUENCES

1. Ecospasms
2. Inequalities – Poverty
3. Divisons and Conflicts
4. Hatreds and Violence

Fig. 1.2. Central Dogma of Ecological Crises.

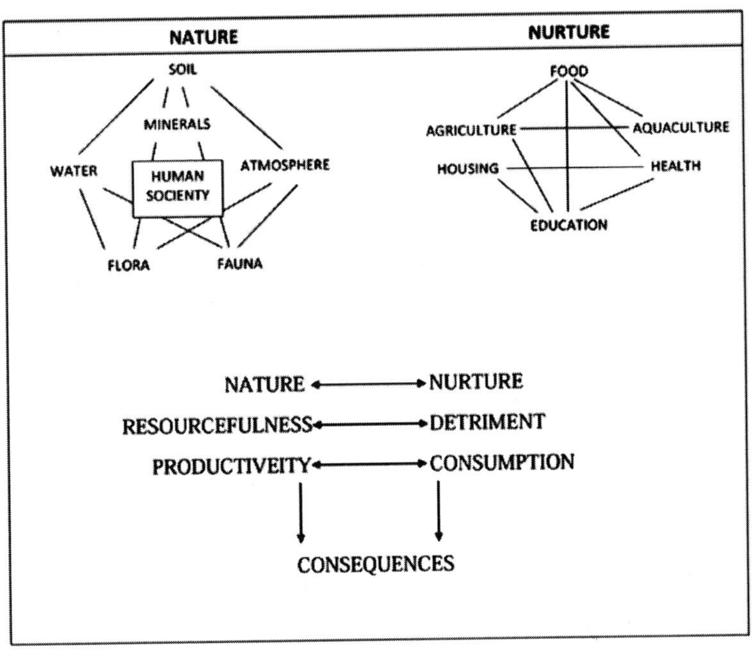

Fig. 1.3. Central Dogma of Ecology and environment.

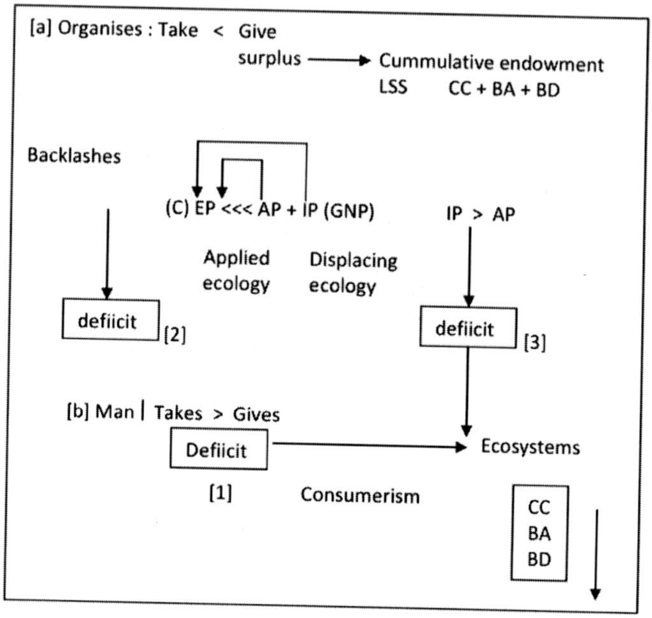

Fig. 1.4. Central Dogma of applied Ecology.

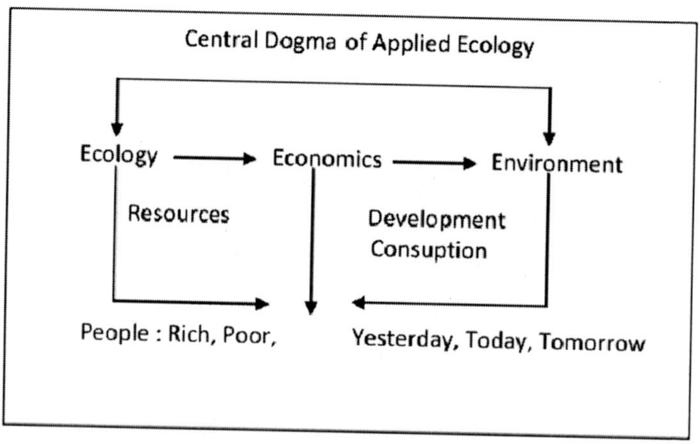

Fig. 1.5. Determinants of Social Welfare.

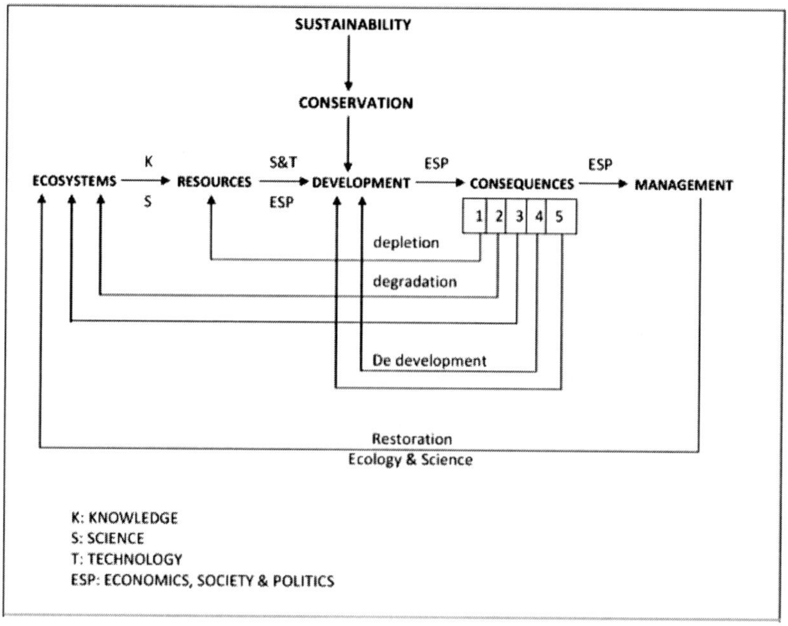

Fig. 1.6. Central Dogma of Resources and Environment.

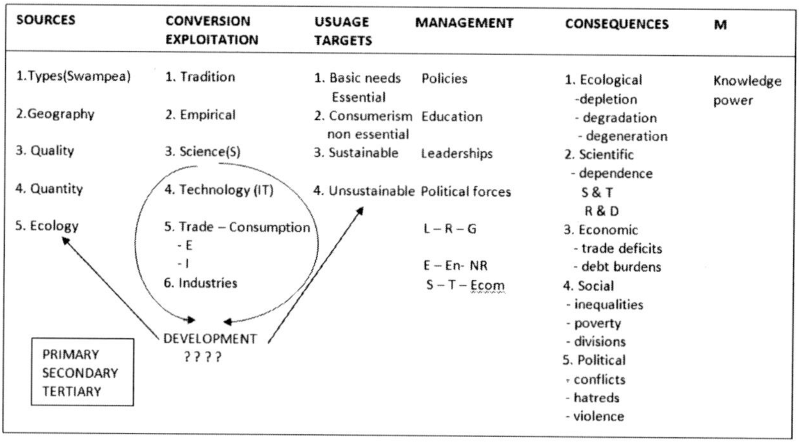

Fig. 1.7. Conceptual Expanse of [Re] Sources, Development and Envirionment Knowledge is Natural Power.

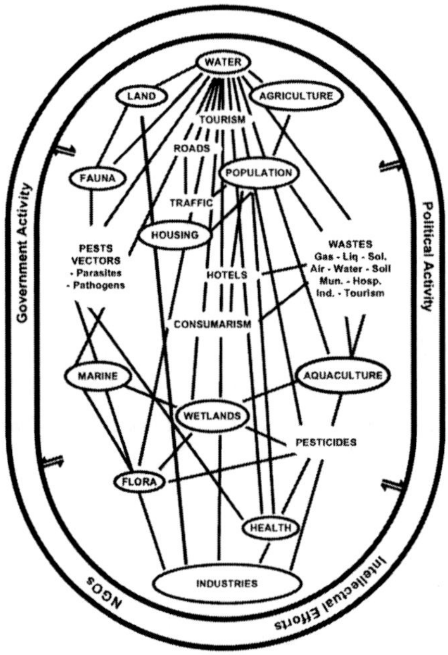

Fig. 1.8. Central Dogma pf Ecological Crises.

Everything is connected to Every Thing else in the Human Environment. Therefore the arms of ecology are too long and many to be easily cliscernable, but maintainable only by the four in the ring.

Chapter - 2

Science and Art of Ecology

A. Introduction

1. What is science?: To some, science is knowledge. To others, it needs experimental verification. For all, it includes logic and reason, inference, and derivation. It may be experiential or experimental. But more important than these, is the innocence in the motive, the query, the doubt, and directness in the approach towards answers.

(a) **Emperor's New Clothes: Innocence:** Anderson's story of Emperor's new clothes is well known. A gullible but vainglorious emperor fell to the sycophancy of the master weaver. Wearing the invisible clothes (visible only to the wise?), he went on parading. It was left to a young, innocent mind to call out that the emperor was naked! I wish this happens in our educational institutions. But here, the students praise the cheaters. Either they are born with old-aged wisdom or are wizened by parents or conscious that in an age of name plates, awards, letter heads, and publicity, Grade cards are very important. Hans Christian Anderson had not visualized this Indian anomaly of a cheating teacher and acquiescing student! Yet science motivates innocent enquiries into the truth, however bitter it may be and regardless of whether it may be hurtful.

(b) **Counting Elephant's Teeth: Directness :** The parents of a newborn child were once waiting to contact their pastor in the church for the baptism of their child. The elder child, a boy, too accompanied them. A group of priests went into the church in the morning with grave faces with the holy cross turning in their fingers. They returned by evening. None looked at the waiting family. Days went by. The impatient boy, one day, waylaid the priest on the steps of the church and asked, "Father why are you all so grim and serious. What has befallen you? Since many days we are waiting for our little brother to be baptized by you." The whole priesthood was so shocked at the audacity of the little one. Eyebrows were fiercely raised. Yet a more elderly one said, "My son, you will not understand our problem until you have grown up. We are debating, discussing, and verifying treatises as to how many teeth an elephant has?" The child mockingly shouted."It is so easy-Open its mouth and count them." Needless to imagine is the consternation of the men in robes at the simple commonsense of the child.

Ecology, as a science, is thus mostly plain commonsense, unmotivated except towards total truth in nature-call it ecosystem, planet Earth, etc.

2. Ecology Questions: The ecology question today is what Hitler once called "Lebensraum". Humankind needs breathing space. It is choked by the luckier part of humankind. The question of nature, its components, its utilities, interconnections, and characteristics have come to us tellingly when ecological limits to industrial growth have been realized. Thus, it was resource crunch that caused focused attention on the modern polity to ecological questions. Otherwise, it was the most singularly dominant feature of all ancient thought processes, philosophies, literatures, religions, and even society and its advancement. In our blind attachment to the ethos of industrial revolution in the name of science, growth, and wealth of people, i.e., welfare, we have endangered our tomorrow.

Thus, we are plagued by several problems concerning ecology, resources, and their sequels.

(a) **Depletion of resources:** Almost all resources to humankind-plants, animals, soil, water, minerals, air, and energy-have been depleted. Only humankind has not. But it is in a Dickensonian state of predicament as stated in the opening para of "The tale of two cities" Why? Excessive use-more consumers as well as more consumption. This is true all over the world.

(b) **Degradation of Ecosystems:** This has happened in most parts of the world. What is degradation? The status of the systems-structure and function-are not as good and useful as before. In other words, (a) the biodiversity-the partaking or constituting organisms of the biosphere-are not in full strength and (b) ecological productivity is not that high to buffer/buffet and carry the depending humankind. Why did this happen? It is because of overexploitation beyond limits of repair. Consequently, both habitats as well as the subsistence systems of humankind have come under great stress for survival. Imbalances in the ecosystem components are: the chains and webs among plants and animals seem to have snapped in many ecosystems of the world, especially those under human interventions; the interaction between organic and inorganic (or biotic and nonionic) components of our planetary system have come under great stress during recent times; the industrialization of agriculture has posed serious challenge to human harmony despite great productivity, for it is artificial! Why do these imbalances occur? It is either due to depletion of some resources or their total absence, or extinction of a few of the biologicals. Not merely would the productivity be further reduced. it may altogether disappear gradually if the ecological links are irrevocably banished.

3. Ultimate Consequence: We have seen before that consumer and overconsumption lead to depletion of resources. Depletion leads to degradation of the overall system and its productivity. Both depletion and degradation definitely lead to imbalances, because all constituents are not merely interconnected, but are interdependent. Imbalances lead to consequences that humankind cannot tolerate. Since we are experiencing these already, the question of ecology has become the singlemost important issue facing humankind, viewing from a holistic point of view. But are we holistic or sectorial? Were we not holistic? Were not most philosophies and religions holistic? Then why this sectarianism with its attendant consequences. This question, of where from and where to, will be examined later. Now let us understand these imbalances further.

B. Imbalances

1. Undried Fish and the Ant Hill: Once upon a time, there was a king. He had seven sons. They went for hunting, brought seven fishes and left them in the sun for drying. On returning, they found them undried and became angry. On enquiry, they realized that the haystack did not allow the sunrays to fall on them. The haystack said it was not its fault for the cattle did not eat it. The cattle could not come because the shepherd did not bring them. The shepherd failed in his routine since his wife did not feed him. She excused herself saying that she had to attend to her crying child. The child accused that the ants have bitten him. When the princes charged on the anthill, the ants protested on to why they should not bite in protection, when a foreigner pokes into our habitat. So back to square peg-the fish did not dry for consumption, and so, the resource is lost. The hunting effort was a waste, all because of effects on a niche. This is the ecology-interrelationships, interactions, and interdependences-for harmony to be there for all.

2. Honey Drop and the Hunter: The Pandavas were in the forest enduring hardships. One day, the sage Vyasa visited them, and to him they wailed their woes. He then brought a beehive and pressed it. The honey fell on the ground. After a while, a host of ants flocked around. Some amphibians were jumping to eat the ants. A little far away, Vyasa Maharishi showed the Pandavas reptiles awaiting the amphibians' return for predation. On the trees slowly a few birds were gathering. At a distance, there was flutter of activity in the hunter's den to prepare for the arrowing down the birds for food. Thus, the cycle, which is an everlasting one, continues. It can now be said further that the dead bodies of hunters and all humankind mingle in the elements of the earth and atmosphere and become the nutrients for the plants. The plants grow, flower, and secrete nectar. The nectar is collected by the bees to transform it into honey!

3. Spinster and the Naval Might: The English spinsters were considered to be responsible for the British overlordship of the seas! How? Because of ecology. The story may be summed up this way-The British navy mostly feeds on beef. The cattle feed on clover. The clover needs bumble bees for pollination and reproduction. More bees are found in and around human settlements than in far away agricultural fields. The field mice, which normally eat away honey combs and larvae, are controlled by the predation of cats. Thus cats allow bumble bees to survive and pollinate clover, which become the feed of the cattle, on which the navy depends. These patriotic cats are reared and cared as pets by the spinsters. The first link of clover, bee and rat and the second link of bee, rat and cat are harmonized by the cattle and navy on the oneside and the women on the other side.

These three instances of interrelationships underscore a philosophy, a morality constituting a reality for living in balances and harmony, i.e., ecology.

C. Invisible Roots

1. Brazilian Groundnut and British Poultry: Once upon a time, there were mysterious deaths of poultry in England. The business was affected, supplies were affected, and costs increased. Why were there mortalities? Investigations revealed that the reason could be their feed. The feed was based on groundnut. The nuts came by a shipment from Brazil. The scientists embarked on to the actual fields from which the nuts grew; they found that the crop was infected by a green mould. The green mould and the nuts were analyzed and they were found to possess some common chemicals. The extracted chemicals were tested and were found to be toxic to rats, producing cancers. They performed the same experiment in poultry and turkey. Thus, pharmacologists, veterinarians, nutritionists, biochemists, toxicologists, agronomists, botanists, and administrators finally established the link between a South American fungus and Britain's pricing of poultry, its eggs and meat. The story did not end here.

The imports were ordered to be stopped unless the fungus was controlled. Part of what was earned by Brazil, they had to spend it in buying pesticides. When the costs were found to become inconvenient to their local economy, they were forced to develop them locally instead of buying. A branch of a foreign pesticide company cropped up in Brazil. In due course, it was realized that lots of money was going out of the country than it was earning for itself. Therefore, ownership became joint ownership. Yet, the knowledge system necessary to understand the links was still in the hands of the British. Technology transfer was not adequate. It then dawned on the government and the ruling party that the knowledge system itself has to be developed indigenously, technology by local talents and locally popular control measures, if Brazil has to really benefit from its agro resources, i.e., develop genuinely, where the balance sheet of the total cycling leaves net gains for the society and the country.

This long road from poultry death to political realization and to self-sustainability took 25 years. By this time, Brazil's population increased by 50%, slums by 100%, and desperation yet further. All these social pressures-increasing numbers, illiteracy, and rising expectations from communications blitz-lead to divisions, conflicts, and violence impeding the sustainability efforts and confusing even the leaders of knowledge systems. Governments were run using the military and having no other alternative but to sell their evolutionary wealth-Natural Resources-through unequal trade practices. On most issues, this was the case in most South American countries.

The root cause of all these, thus, are the natural resources, their interlinks, balances if broken strifes result-plain Ecology ! This is what happens in most countries of the colonial world. Natural libraries, i.e., natural wealth, is burning in the quest of computers and electronics!

2. Hamburger Connection: Hamburger is a popular delicacy that has traveled from the USA even to Russia and India. Like Coca Cola, this has become available even in unthinkable places! The meat for it has to be obtained from a specially raised breed of cattle. The places in the USA are either unsuitable or exhausted to meet the demands. Therefore, the meat is imported from Brazil. Or could it be Brazil's policy to earn precious dollars! The cattle are raised in ranches maintained on lands cleared of forests in Brazil. In other words, Brazilian deforestation is the necessary prelude to hamburgers in North America. Along with deforestation, there is impact on the biodiversity, surface soil, water holding, and some unknown ecological chains, webs and cycles. The consequences of these can best be realized only after some decades, perhaps. Now, we can calculate the cost of each hamburger pleasurably consumed in terms of biomass lost and the time it takes to form. The treasure accrued for some centuries is destroyed to raise ranches over years for consumption in minutes! What a price? Several, similar stories can be found where they clear the forests and use the land for mining, ranching, agriculture, or as human settlements. Yet Brazil alone can boast of a Sao Paulo, and satisfying consumer appetites, yet the temptress of development never satisfies.

D. Cause here Effect there: Ecology at Play

Earth's systems are interlinked, ignoring political boundaries and sometimes even geographical barriers, where many mysterious things happen, for which the reason is not apparent immediately. One can give dozens of examples to illustrate this, but two would suffice-representing air and water. We can also see here how good neighborly relations can be affected or how a technically strong country can hoodwink neighbors.

1. Acid Rain: The rain falling on Canadian soil, lakes, and forests was found to be acidic. The acidic rain was affecting biota, their productivity, and ecological conduct. But, from where did the acid originate? It came by the dissolution of the oxides of carbon, nitrogen, and sulfur suspended in the air. The question is: links From where did the air pollutants come? They came from the industrial complexes of North America because of the wind direction. Similarly, the acid rain of Norway and Sweden came due to the pollutants from Poland, East Germany, and Baltic States. Thus the cause is in one place and effects elsewhere, and it is difficult to pinpoint sources when the whole geographical system is made up by several polities. International relationships get soured and ecological skirmishes may develop from ignorance. Some nations can resort to ecological mischief with border countries. The case of pollution of the North Sea and Rhine river by West German industries is a telling instance regarding the futurology of ecological cunning and deviousness.

2. Oil Pollution: Whenever oil spills on sea either due to tanker mishaps or accidents, in productions and transmission, or war as in the Persian Gulf, the movement of the hydrocarbons knows no boundaries in spread and effects. The consequences are suffered by a large number of innocents. This is an ecological crime on the ecological innocents. And the planet has witnessed more than twenty episodes during the last 20 years. The total impact of any one of these have never been gauged in full. What an ecological unaccountability.

E. Growth Threatens Subsistence: Tragedy of Ecological Wrongs

When Science and Technology is applied to convert natural resources into better products of social utility through policy interventions, a betterment or development of social life is expected. Sometimes, this leads to a threat to subsistence of people for whom the development was meant. We will take just two examples-one from the US history and another from our recent past.

1. Excessive production-Retrenchment-coal at Uncle Tom's Cabin: A famed novel based on the US depression during the 1930s tells us that there was no coal to cook, for there was no money with Tom. He had no money because he lost his job in the colliery because there were surplus hands, because coal production overreached the target of need! Yet Tom, who produced surplus of coal, had no coal for himself. The question is, "for whom is the development, in the present case of coal production?"

2. Triple troubles in Indian Rural Ecosyste:ms: The village producer is a tenant or a small land holder. He loses it, becomes a coolie or an urban migrant, loosing his habitat too. The products of development do not reach him. On the contrary, he is bestowed with the byproducts of rural land seize and urban development-migration to a subhuman urban slum and exposure to all kinds of pollution. This is the story of Indian rural artisans. To my mind, these are a quintuplet of troubles: (1) loss of land, (2) loss of habitat, (3) no growth/development, (4) degradation of living habitat, and (5) degradation of water, food, air and health environments.Thus, a series of ecological wrongs turn the life of a traditional producer into a modern urban parasite

F. Culprit and Victim Under Boomerang

Ecology unfortunately does not distinguish the causative forces and the victims and threatens both. Ecology is blind like the proverbial justice, quite often.

1. Ozone depletion : Excessive industrialization, while converting many into fugitives for a subsistence living, unleashed a great variety of pollutants. Some of these deplete the ozone in the stratosphere allowing the passage of UV rays into the Earth's biosphere. The entire biosphere is threatened with mutagens, some of which, especially to humans, are hazardous, while in others, a rise in spontaneous mutations is expected.

2. Green house effect: Similarly, many oxides of carbon, nitrogen, and sulphur entrap the heat from Earth's surface rendering the entire biosphere as if it is a glasshouse. The resultant warming may lead to a series of consequences, such as rise of sea level and inundation of coastal ecosystems.

Both the above were caused by the developed North-the rich countries. But the suffering has to be borne rather stoically by the undeveloped south as well, the traditional victims.

G. Community Justice

The foregone amply suggests the need for justice-justice not in accordance with the perceptions of any section of humankind but from an ecological wisdom as seen in the community life in any ecosystem. A few ecological morals for guiding sociopolitical life may be mentioned here: (1) every organism gives to the community what it can and takes from it what it needs. (2) all organisms lead a free life of choice in their characteristic niches, i.e., with equity (3) nature does not need man, but man needs nature (4) there are no gender inequalities in any organism, but a mere division of labor. (5) everything is interconnected with everything else-all are interdependent, where a disturbance to any one in the web disturbs everything else-checks and balances are for common good.

Thus equity, ethics, individuality, humanism, reverence for fellow life (Schweitzer), feminism, and the like are ingredients of the community justice or Ecoism.

H. Mother Earth Concept

An understanding of ecology as a science, natural, social, and humanistic, assuring harmony among all Earth's creatures through justice is possible by regarding Earth as the Mother of all Mothers. This is by no means a new concept.

Earth was considered as a Mother, first by Indians. Ancient wisdom regarded it as Bhoomata/Bhoodevi and was extolled in Prithvi Suktam of Rigveda. In fact, the entire ecological ethos was contained in Vedas, as suggested by the signature lyric of the recent TV serial of Chanakya on Doordarshan.

Woman, like Earth, is an embodiment of all virtues, hardwork and patience, hope and inspiration, endurance and sacrifice, love and peace, justice, and compassion, through all her seven stages as daughter, wife, mother, sister, teacher, grandmother, and finally, Shakthi, the goddess. Thus, a man trying to become a woman becomes a Mahatma. But a woman trying to become a man becomes a devil. It was often said that calamities fall on humankind when the tree, the land, and the mother were hurt and made angry.

I. Conclusion

Thus, ecology would mean perpetual motherhood, justice, righteousness, and humanism. The inscrutability of its power to backlash and its basically everlasting resourcefulness adds a dimension of spirituality and divinity. Ecology thus becomes the new god: All rules to maintain this harmony though balances constitute the new religion, Ecoism. When humankind, in its awe and reverence to nature's power and resourcefulness, created gods and religions, why cannot these be redefined and substituted once the bases of these resourcefulness, checks, balances, and the resultant power have been discovered to be mere ecological and evolutionary forces?

Each organism
Gives what it can
And takes what it needs

Patience is extreme,
When lost anger irreversible

Cause here effect there
Cause now, effect later,
Everything is connected,
It is powerful, it is fragile,

Someone prepares food,
Someone consumes,
Someone else nourishes

It is useful, it is harmful
Thus commons are strengthened
For common good

Chapter - 3

Continuing Ecospasms the Perennial Concerns

A. Introduction: Our Life Today

An aging man travels without ticket in the rear of a dilapidated bus on a pitiful road with heavy traffic, accompanied with the noise of honking horns on one side and the fury of the exhaust fumes on the other. In such a situation, he is expected to be at peace with himself and life around. This is the life of an average human today. Something is seriously wrong somewhere and therefore everywhere. No one is sure where to begin for the cause or a remedy. In such a scenario, ecological balances are said to have vanished. This is the ecological question in a nutshell.

The words "Ecology" and "Environment" are very common, as they are perceived to affect human life everywhere. And the effects are likely to increase. Let us examine these questions by trying to understand the major hazards threatening the social continuity of humankind.

B. Ecological Balances

We shall recapitulate three examples from the earlie chapter - [1] Undried Fish and the Ant Hill fable from children's literature, to understand ecology prima facie, [2].Honey and the Hunter episode in the Mahabharata, a philosophical taleon the laws of nature, and [3] Spinster and the Navy connection, a historical reality of our times. These three instances of interrelationships underscore a philosophy, a morality constituting a reality for living in balances and harmony, i.e., ecology.

C. Balance To Backslashes

These certain links established during evolution are difficult to comprehend at any given point of time or space. They announce themselves prominently only when they breakdown somewhere threatening the destinies of the people. Such breakdowns, as sequels to meddling with links, chains, webs, cycles, and thus balances, have degraded this planet into several worlds and subworlds. A quick look at this canvas shows a panorama of hazards, crises, backslashes, and afflictions that will aid us to understand the great ecological question on human survival. However, an understanding of the most basic words "Ecology" and "Environment" are necessary.

D. Ecology and Environment

From a human angle, the environment of man consists of natural resources, their converted versions through science and technology, and balances among these natural resources, such as water, soil, air, plants and animals. These balances-links, webs, chains, and cycles-constitute ecology. From an evolutionary point of view, the planet has biotic and nonionic components. They are all mutually connected because they constitute each others environment.

E. Gathering Storms: A Checklist

What happened to the ecology, the environment, and the biosphere on the planet Earth-all are threatening man's future? This was the concern of the recent UNCED in Brazil and its precedents at Nairobi in 1982 and Stockholm in 1972. They all have warned of the dark clouds on the horizon of human destiny. The gathering storms today are perplexing indeed, as the solutions are not in sight; if in sight, they are not approachable; and if approachable, the efforts are daunting or wanting. Some of these hazards are truly global. Others are regional or zone specific. Let us briefly try to understand them.

1. Depletion of Ozone layer: A thick layer of ozone intercepts the solar radiation from totally reaching Earth's atmosphere. It prevents ultraviolet (UV) radiation from passing through; the biota on Earth, including humans, is spared from the ill effects of UV rays, which are known to mutate genes. In humans, the mutations lead to cancers (among others), especially of skin, as it is the most exposed recipient organ system. In other biota, mutations lead to more variations, the materials to the evolutionary grill.

There are reports that ozone layer has been thinning for sometime leading to a hole geographically or cartographically, just above Antarctica. The fallout was also supposed to be high in South Africa, Northern Queensland, and Southern India. The skin cancers are also of higher magnitude in some of the above regions. What causes ozone depletion? Can it be contained? It is believed that chlorofluro carbons, released by the cooling and refrigerating technologies, are the chief causes. It is also agreed that substituting them is the only way to control the problem. This subject was greatly dealt with at the recent Rio Conference and at Montreal before.

2. Green house effect: Green house is a glass house where plants are grown. The carbon dioxide that is released is trapped within the house because of the blockade by the glass top. Consequently, the house becomes hotter. Similarly, if the atmosphere of the planet Earth becomes warmer due to various gases, these could be serious consequences. Ice caps may melt, sea levels may raise, coastal regions may submerge, and other related effects on a global scale.

The warming up of the planet is due to the concentration of oxides of carbon, nitrogen, and Sulphur, and other gases in the Earth's atmosphere. These gases are popularly called green house gases. These are produced during the combustion of fuels for various purposes-most of which go with the generic name, development. There are only two ways of alleviating the impacts of this phenomenon. First is to improve the technology to reduce the emissions. Second is to increase the green cover, especially the forests, so that many of these will be absorbed into plant-productivity processes. International understanding and conventions are fortunately very clear on these matters.

3. Mutations in the Biosphere: Developmental processes not only release green house gases and ozone depleting CFCs but also a myriad more chemicals. It appears that there are about 5 million chemicals, of which nearly 100,000 are in regular use and 10,000 are added annually. Some of these enter the biological systems, especially to man through direct consumption, usage, occupation, food chain, and accidents, causing many varieties of toxic effects, sooner or later.

Some effects are seen very late, sometimes generations after. They appear without warning signals. These are the effects on the genetic material, whose consequences are not predictable and causes are untraceable. Thus the genetic diseases, birth defects and cancers in human populations are on a rising course for the past two decades. These lead to personal misery, detriment in human performance, and heavy social costs.

The genetic diseases in man, numbering about 3000 syndromes, afflict about 10% of the population in the developed countries. Cancers appear to account for 25% of the population. Having adopted the same development model, the poorer countries also will demonstrate similar trends. With the addition of ignorance, inferior science and technology, and teaming millions, the problem may be far worse when it becomes more evident. In fact, the gene-based diseases may assume an epidemic form in the developing countries sooner.

4. Germplasm Depletion: A great part of the planet supports microbes, plants, and animals, and of course humans. Origin and evolution of these took millions of years. It is only during the last 10,000 years that man has learned to use plants and animals as his resources. The number of consumers and the level of consumption were so low until a couple of centuries ago that the resources appeared everlasting. Also, most resources were not even seen to be explored for their utility. However, due to science, technology, population rise and consumption, the scenario has changed. Resources have been dwindling, even the reproducible ones-plants and animals. Since their characteristics were genetic and heritable, once they are lost, they are gone forever. This is the problem on our hands-the germplasm loss.

The plants and animals whose knowledge is known to us are very few. Before we could know, hundreds of them perished permanently. Most of these were from pristine natural ecosystems of the developing countries. As the vegetation is sheared, forests cut and degraded,the biodiversity is lost, germplasm is lost. Man has killed what may be useful to him in the future for food, medicine, etc. The plants and animals existing for millennia vanish without man ever knowing them or their utility. Poignant indeed! Premature biological murder leads to genetic death of organisms, otherwise everlasting.

Furthermore, as they cease to exist, becoming extinct, some links in ecology are snapped, with some consequences somewhere, which will be discernible only when the impact is evident. In other words, there is an ecological death following genetic and biological deaths. Can humankind survive if ecology dies?

5. Acid precipitation: The greenhouse gases mentioned earlier are sometimes intercepted by rain. As a sequel, the rainfall and snowfall becomes acidic. Soil, lakes, ponds, and biota receiving this acidic precipitation are affected both on a short-term and long-term basis. Since all these affect not only productivity but also interrelationships among plants, animals, soil and water, this has become a serious problem to contend with in the northern hemisphere.

The causes and effects of acid rain were well documented from Canadian experiences of American emission and Scandinavian experiences of European emissions. The only way to deal with this hazard is to minimize emissions of combustions through technology or increase their absorption through forestry.

6. Drinking water scarcities and crises: In his long history, humankind never found scarcity for the most important need of life, the elixir of existence of all life. Despite rivers, rains, and an underground sea of water, half of the world's population experiences water shortages. Why? For two reasons-depletions of underground water and pollution of the overground resources. Adding to this is a cumulative problem of population migration to cities sprawling them beyond the limits of substantiality, vis-à-vis water. Due to overconfidence, the world has never tried to harness all the rain water from running into the seas. Surface water resources, rivers, are dammed and the pooled water distributed for irrigation and other purposes. Unfortunately, the quality has been becoming gradually poorer; this is because of locating industries on river fronts to release effluents for dilution and passage.

Underground resources became dearer due to overexploitation for drinking, irrigation and industrial uses. Recharging is impeded by its inherent slowness and seasonality. A clear case of use beyond renewal, i.e., sustainability. In fact, there are several causes of underground water pollution from seepage from surface pollution.

Rainfall has not changed much globally-only its periodicity and relative fall. Recharging the water table or impounding through tanks has been a time-honored destiny of such rainfall. Both these are in jeopardy now. More tanks are dry for longer periods of time. Some tanks get filled up for housing purposes. Water tables in most places consequently go lower and lower.

The poignancy of water problem is that while population increases, availability of water decreases; while science and technology making strides and development is everyone's urge, sustainability of life itself is under jeopardy by the nonavailability, chemicalization, biological pollution, etc. of water, the very elixir of life.

7. Sea water intrusion: The depletion of potable water around the coastlines assumes a greater consequential destiny to humankind than anywhere else. Here, the riverine water is saline and ground water fairly rich. For various reasons, extensively along the beach and river mouths, many industries took up positions. Even agriculture and agricultural practices have accelerated the use of underground water resources through wells and tube wells. Consequently, the usage of underground water is perhaps greater in the coastal ecosystems.

As a sequel to the depletion, spaces appear in the geological systems making the water table. Instead of getting recharged by rainfall, it gets the nearest marine water. Once salinity enters, there is no going back-the system is lost perhaps forever. This is the problem facing most of the coastal ecosystems where underground water is extensively used by agriculture, industries, human settlements, etc. This has become a reality in many places, and most places supporting rich human populations will have to contend with this crisis sooner or later. Migrations perhaps may be warranted.

Notwithstanding, the legal limit of 2-6 kms for human settlements, other activities and underground water extraction, the exploitation continues. The law is observed in its breach just as the common sense is most uncommon. Therefore, the only solution is that the users limit consumption in their own interests.

8. Soil Erosion: The importance of soil for plant life and through it to animal life and human societies is well known. From a rock of any type to oil of very small particles, nutrients, microbes, and humans retaining and releasing water as a result, to difficult extents, take centuries and millennia. Soil-water-plant-air continuum is the very premise of biosphere. The soil erosion threatens the existence of everything else.

The first type of soil eroded is the forest soil. Deforestation exposes it to the direct impact of rain. After one or two seasons of such exposures, alternated by dry season, the soil will be washed down into streams, rivulets, and rivers. The alluvium may settle in the reservoirs, silting them, or form into deltas. The reduction in plant cover renders rain more furious, floods more violent and loss of soil more ominous in the future. In the process, dams get silted and deltas become unstable.

The second type also results from loss of plant cover in semiarid regions, i.e., with low precipitation. Grazing practices gradually reduce plant cover exposing the soil to the fury of sun and continuous periods of dryness. Such a soil can easily be blown away by wind baring the substratum of lower layers, which will not be able to support vegetation. The blown-off soil will have to get deposited somewhere along the wind direction. Being unstable, it will result in the recipient habitat too becoming unstable. Desertification thus continues, spreading and never ending, aided and abetted by man's activities.

Whenever the above phenomenon is intercepted by periodical rains and low temperature, one gets the waste lands, or wasted lands, which may perhaps, at best, support herbaceous cover. This happens in many verdant lands-Savannahs, scrub jungles, deciduous forests-resulting in a highly reduced biodiversity and community interaction.

The only way to deal with all these is to maintain as much dense and diverse vegetation as possible. Thus deforestation, silting of dams, desertification, soil erosion by wind or water, excessive grazing, and floods have all only one connotation-loss of soil within 1/1000 to 1/10000 of time that it takes to form. This is caused by plants and, therefore, preventable by plants through human intervention.

9. Chemicalization of the Habitats: Much before people became aware of the words "Environment" and "Ecology", pollution was heard everywhere-air pollution, water pollution, and soil pollution by pesticides, odious compounds, smoke, heavy metals, etc. Now, it is well established as to the causes of pollution, sources, kinds, nature, and effects. Equal to these industrial waste products, human life absorbs many more chemicals through consumables-products made by nature (plants) or man-made drugs artificial food, etc. Many nonconsumable products also possess chemicals to which man is exposed.

In other words, the society appears highly chemicalized-air, water, soil, household consumables, food chains, etc. The impact is very obvious-chemically induced diseases or toxic symptoms. If microbes cause a disease, one can resort to chemicals to control them. If chemicals cause diseases, what is the remedy?-removal of chemicals from the environment. How is this possible-when on one hand we are introducing more chemicals for human comforts. Condemning them on the other hand-is self contradictory. A conscious decision has to be taken.

The toxicities of chemicals may be to the skin, eyes, liver, kidneys, or respiratory, nervous, immunological, and reproductive systems. In extreme causes, the genetic material in the cells of any tissue may be altered leading to cancers, birth defects, and heritable diseases in the forthcoming generations.

10. Hygiene, Sanitation, and Diseases: Chemicalization of air and water and its consequent toxicity is only one side of the problem. Whether or not water is polluted, its physical status can have some effects. In other words, water that is stagnant and receiving domestic wastes has been a long-standing source of scourges to man. Agricultural run off, industrial effluents, and domestic effluents alone or mixed with unpolluted water are These three instances of interrelationships underscore a philosophy, a morality constituting a reality for living in balances and harmony, i.e., ecology. often found to be stagnant for want of drainage.

Such habitats become a good breeding ground for vectors, parasites, and pathogens, which cause diseases in the human populations. Stagnant water as a source of ill health was known for centuries. When it is known to have more chemicals, questions arise whether these are mutating the pathogens, parasites, and the vectors as well. If so, the problem of control becomes all the more complex and in fact uncontrollable. Mutation could be one of the reasons behind the reported resistance of many vectors to pesticides and drugs. Protozoan, bacterial, fungal, insectal, and helminthal origins of many human diseases is well established. Mutations in these hamper the total elimination of any disease. While malaria has made its reappearance, many others have hardly been eliminated. The solution lies in hygiene and sanitation through proper drainage and decreased use of chemicals for various purposes.

11. Human Populations: It is generally agreed that human populations are on a phenomenal rise. Such a rise reduces per capita share of natures' resources. The resources naturally tend to decrease. The products of science and technology and governmental policies of improving the standards of life through health, education, housing, etc. also do not reach most people.

More importantly, the people cannot be employed and become unemployable. This leads to enormous inequalities. As the resources deplete, costs of living increase and standards decrease, and welfare measures become unreachable and unemployment everlasting-all these cumulatively increase divisions and conflicts. People suspect each other, with distrust in an atmosphere of constant struggle and hopelessness. Violence will be everywhere-with finding a scapegoat becoming a pastime!

The population problem is different in different countries and regions. Some are very rich, with small populations, growth rate zero or near it (Europe and North America). Some have large areas but small populations (Russia). They encourage high growth rate. Some, though they have large areas, they cannot support even the existing small populations, either because of poverty or lack of development despite resources (South America) while a few others are living in degraded environments with depleted resources in addition to poverty, illiteracy, and low level of development (Africa). In contrast to all these, there are many countries rich in resources, culture and heritage, and area development but huge in populations. Here poverty and affluence coexist (Asia).

The scenarios change from country to country because of complications due to factors of religions, cultural heritage, electoral arithmetic, group security, etc. While some countries maintained 0% growth norm, others one child or two children per couple norm, some societies registered 3-5% growth rate. While some counties depend on immigrant labor force, others are choked with excessive numbers.

Migrations from most poor to less poor countries and poor to richer countries are the order today. Such migrations necessarily lead to social tensions, conflicts, and violence and make military strength a necessity despite the end of cold war politics!

Beleaguered by problems of resources and development, the psyche of every man is besieged by the fear of security tomorrow. Appropriation and exploitation increases and support this clannishness. Added by helplessness and hopelessness, people resort to anything, not even short of murder and mayhem. Most violence today can be traced to these clichés, schisms, breaches, and hiatuses.

The solution lies in equity and empathy. There must be reduction of a market place morality on one side that includes less profiteering and less consumerism, and reduction of birth rate at all costs and by all means. The crowded Earth is crumbling by its own weight.

12. Degradation of Ecosystems/Habitats: Most ecosystems have degraded. What do we mean by degradation? Overall productivity is reduced, i.e., its capacity to carry the depending animals and humans is reduced and may eventually vanish with increasing pressures. This can happen in two ways.

Firstly, as the resources are exploited, the diversity and stability are reduced. The interlinks are snapped. This will have a chain reaction of further snaps. Ultimately, the balances are endangered yielding place to a free-for-all situation, an uncontrollable anarchy.

Secondly, extraneous chemicals may enter the system through the phenomenon popularly known as pollution. Pollution causes toxicity to the body functions of all plants and animals-changing their health, behavior, productivity and finally the interacting abilities and interdependencies. Once the diversity, stability and productivity are threatened, the ability of the system to withstand any serious climatic or environmental threats will be reduced.

Furthermore, its ability to buffer the impacts on substratum, for example, soil, will also be reduced.

Thus degradation of a habitat results from overexploitation and/or pollution and manifests in the loss of productivity, carrying capacity, diversity, and buffering ability-and ultimately ecological death.

Forests have degraded-diversity reduced and productivity reduced, endangering the soil as well as rivers ultimately. Fresh water ecosystems have degraded chiefly due to pollution. Estuaries and coastal ecosystems have degraded due to overexploitation of resources and pollution. Agroecosystems have degraded in the past in some states of Northern America due to intensive culturing. Yet the same practice of more fertilizer and more pesticides inputs on crops is followed in the population-cum-agriculture intensive regions of Asia, Africa and South America. Degradation here also is imminent. Semi-arid tropics have degraded, due to grazing and suicidal human activity, into deserts. Ponds are getting filled up for land use. Mountain ecosystems have been under degradation due to deforestation and plantation crops.

Human habitats too are a picture of ill health. Urban migrations render rural ecosystems in jalopies. Urban immigrations checked all systems to the limits of capacity, sustainability and endurance; thus, there are slums, insanitation, polluted air and water, nonformal labor, inequalities, crime, and violence in most urban societies. Human life itself is degraded-families are under stress, breakdown, psychic problems, and what not. Thirdly, this has lead to the sporadic episodes of droughts, floods and accidents.

The problems we have surveyed so far were global or regional, encompassing fairly large sections of global populations. In contrast to these, there are some sudden happenings or those with reasonably brief warning and lasting for a very brief spell or long spells. These are industrial accidents-sudden, unexpected and decisively hostile for human existence. The best example is provided by the Bhopal Gas leak accident of India. All others are of lesser magnitude, but occurred almost in all countries. The next category are the floods, which can be anticipated in all rivers during the rainy season-as is happening form times immemorial. Then what is new about these-it is its suddenness, intensity, repetitiveness, and the devastation. As the forest cover is lost, the rainfall comes down heavily into the plains pushing soil, sometimes whole land masses.

Almost all rivers, the source of civilizations, brought immense sorrow to people occasionally in recent times. A more recent feature, especially in India, is the flash floods in the ephemeral rivulets. A sudden downpour in the catchment area, a dry bed of river/rivulet/stream without hurdles to the gushing waters, with absence of an embankment, and washing down human settlements, bridges, etc., leaving an aftermath of devastation and loss of life and property has been the scenario in recent years.

Droughts are the opposite of flood-the difference being in the duration - waterlessness continues for long spells. The misery of droughts need not be elaborated, for every one knows the hardships of life without water. The issues here are: How are they caused? Can they not be prevented? Is there a nexus between floods and droughts? India is a perfect case in this regard. There is no year without drought or flood in one or the other region of India. The great sub-Saharan drought that was combined with over grazing, involving Somalia, Ethiopia, Chad, etc., is a recent memory and a current experience.

F. Causative Factors: Development or Consumption

Why are so many problems chasing humankind? They are environmental problems-happening in the human's environment. Basically, they are ecological, as they are caused by the broken interrelationships among the components of the environment. Why are these broken? Due to pressures from human populations brought forth in the quest of their own development. Development, in its amplest definition, is to improve the lot of people gradually in all ways possible-one route is to convert natural resources through science and technology to nearer forms of consumption. In this process, people get employment and obtain returns in the form of better-quality products, or new products altogether, which are bought by the wages from employment. Thus, the quality of life is bettered. Unfortunately, sometimes, more products and of better quality are demanded by some sections. As they have more purchasing ability, supplies are commanded. Consequently, the supplies, trade, science, and technology are geared to exploit the natural resources further. This so-called development initially increases employment and quality of life but actually caters to the desires of the employing classes. A situation arises when natural resources deplete, become scarcer and wanting altogether. This means development has taken place beyond the limits of supply of resources. Why? Due to the excessive demands of people with purchasing power. What are the demands? More products and better products. Are these necessary for a good survival? No, then why produce these and endanger Natural Resources? In other words, the consumerism of the purchasing power is the destructive force of resources.

These resources are in the ecosystems in the form of plants, animals, minerals, water, etc.; once these are depleted, the ecological links, webs, chains and cycles are affected leading to various types of hazards that we surveyed earlier. Thus, for all ecological or environmental problems, it is the consumerism of the employing class of humankind, which they euphemistically call development. Development is not the villain, it is the consumerism-oriented development [and not basic needs supplying one] that is responsible for all ecological crises.

G. Solutions: Balances Lead to Harmony

Having established the root cause of the current crisis, is it really possible to solve; for a solution everywhere affects the employing classes-the leaders of society invariably. Increasing the resources or developing alternate resources, therefore, is attempted without curbing the consumerist passions. The real solution lies in protecting the existing resources and curb development to that extent, rehabilitate the lost resources everywhere, and stagger development to an extent, until then curb consumerism and invest the saved purchasing ability in increasing the productivity of the biosphere or controlling the degrading factors. In other words, human society itself has to behave as a part of the ecological web in nature-***Sarve Jana Sukhino Bhavantu***-the greatest of all ecological principles contributed by the Hindu thought-'***Jana***' meaning not people alone, but all creatures on the planet.

H. Hopes: The Efforts are the Reasons

The situation in the world vis-à-vis the realization of ecological problems and translating the responsibility into action is not hopeless. But, it is always "*too little and too late*" scenario. Science and Technology does its best. Trading pattern is less hostile to the underdogs. Governments across the world announced their own agenda of priorities and collecting together on global foray like UNCHE at Stockholm in 1972 and at Nairobi in 1982 and UNCED at Rio de Janeiro in 1992. In response to the inadequacies of governmental actions, both in intent and content, precept, and practice-voluntary organizations sprang up in most countries, some of them accepting no assistance from governments. Education and awareness on ecological crisis have been taken up by specialized institutions and media beside the NGOs. In rare cases, people have taken up political actions outside the usually accepted political foray-popular movements and agitations by affected people. In a few countries, political parties with ecological manifestoes originated, while in many others, the conventional political parties assumed ecological postures.

I. Despair: Escaping From Problems

Instead of averting the ecospasms, most people who could do something indulge in Nimbyism [***not in my backyard***] and Nodmyitism [***not my tenure***] and escape from the issues. Simple problems become spasms owing to this selfish outlook. The common-scenario of prosperity and welfare worsens into tragedies. That is why perhaps Russell lamented that a great many people enjoy wars if they are not around their doorsteps! It is only through introspection, and a work-word-thought ethic of equity and empathy that the society can be redeemed from ecological hazards-nay tragedies, for, we are already in a world of ectoplasms! What is an ECOSPASM is explained in the conclusion.

J. Conclusions

The interrelationships among the physical elements on earth were established through a trial and error revolutions across a bilinear situation. Humankind meddled with these in its urge to make life pleasurable, mistaking it for peace and happiness. Therefore, the subtle interlinks that render a "spinster determining the might of a country" are unfortunately broken forever. Many more may breakdown with the passage of time. We know something is wrong. Out of confusion, we make everything wrong, converting tomorrow to hopelessness, instead of repairing whatever little could be.

This is the message from ecology for mending the damages we are causing to our society, and to ourselves. The cycle is truly vicious. Unless the ecological balances are maintained, social problems will always arise and become intensive. Unless social equities are established, ecological balances cannot be maintained. In between and in due course, there will be more divisions and conflicts among people leading to cumulative hatreds and violence. All the putting together constitute ectoplasms when no one others are there to see reason and trace causes and their roots. A situation envisaged in Hindu mythology is that ***"at the end of it all is Kalki to destroy the sinning world."*** To my mind, Kalki has already entered several societies in the form of **ECOSPASMS.**

WHAT IS IN THE PENCIL IS GREATER THAN WHAT IS AROUND IT. THE TALENTS IN YOU ARE GREATER THAN THE ENVIRONMENT SURROUNDING YOU. YOUR POTENTIALS WILL CHANGE YOUR ENVIRONMENT.

ISRAELMORE AYIVOR

Chapter - 4

Ecological World Views on a Triaging World

Preamble: Earth summit has come and gone. Hardly any problem was left unmentioned. The official congress was attended by a maximum number of Heads of State as no other conference has ever attracted, save one of the UN sessions during Nehruvian era. The dissenters had their parallel sessions, having, no faith in governmental activities and, an explicit concern for human ecology. How did it happen? What are the portents? What are the forebodings and fears? Is there scope for hope; are all these identical for the whole humankind and if they differ, how and why do they? What is the range of possibilities to different sections of the world? What about India -shall the whole of India have a single bad future or some parts can hope for ecologically viable future. Politics, development, natural resources, and interrelationships among these and with human societies form such highly intricate labyrinth that all the proverbial mazes put together are rendered a minor challenge. A perception of this labyrinth inspired by the Earth summit is sought to be shared with concerned students regarding this great issue of humans future on planet Earth.

A. Introduction

The most common words nowadays seem to be Environment and Ecology. However, these two words seem to be different things to different people. All shades of opinion covering these two magic words can be summed up in one word, "Environmentalism". How this consciousness is developed. From where did it start? Where is it likely to take us? What are its roots, vis-à-vis the man's origins, ancestry, and relationships with nature?. What has all this to do in the future of humankind of our planet?

Today, environmental consciousness pervades most of the talks in the world. Today, intellectual man is after the four P's: power, profit, pleasure, and popularity. Today, India is in the age of a package, "***I, Me- Mine and Greed- Crime- Cover- up- Innocent talk***". Yet today, the prosperity of humankind and its ability to help the less fortunate is the greatest in history with war being pushed to the ken. Today, all existing prophecies are tempered by pragmatism. Indeed we are in the Dickensonian times as described in the opening paras of the Tale of Two Cities, Abel and Cain coexisting. Yet the pilgrim has to progress with challenges more exacting than what Bunyan visualized. Consequently, ethics and empathy became the greatest necessity as never before in human history. As more realistic solutions happen to be the idealistic ones, we are indeed at the "***hour of the unexpected***."

Everything has a history. Therefore, Man's predicament today must have roots deep in the uncharted beginnings of its emergence and growth in space and time. It is therefore appropriate to begin with the interactions between man and nature from the earliest times.

Scores of nations assembled. Their leaders discussed problems and detailed panaceas on common causes were sought to be made. This was in Rio de Janeiro, the capital of Brazil during June 1992. The subject was Environment and Development. Though development of humankind was happening for centuries and millennia, for the first time, it is perceived that their environment is affected. In other words, development cannot continue anymore because the natural resources and their ecology are adversely denatured due to development. Shall then development cease? No. It should go on - the show must go on with dividends. How it can happen in an age of scarcities, more consumers, more consumption, more hoarders, and more backlashes? This is the million dollar question. The United Nations conference on Environment and Development concluded that it can happen provided the development is truly ecological- interdependent, interactive within a global common of the entire planet and its resources. The age of ecology at last arrived. Everything else, ethnicity, culture, religions, science, and technology, all have to work within ecological rules, norms, and matrices. The ages of Religion and Materialism- Science and Technology- gave way to an age of Mutualism and Interdependencies- Ecological Age.

This age has not descended all of a sudden. It was not a magical revelation. It has its own story of trials and errors, pitfalls, and close calls, ups and downs. Viewing the world from an ecological sense began a few decades ago. The views were from different angles. They constituted different perceptions, often different radically. But the common concern was the same- the denatured environment of man and the possibility of its retrieval for human welfare, with fears lasting and hopes lingering.

1. Is A World View Necessary?: To understand this ecological world view and its journey in time, one may have to go to the beginning. The first question- Is there a necessity of a world view? Yes. All of us chose our own ways of life with some perceptions on what we want, i.e., motivations. This is at an individual level. Families and groups sections similarly have their perceptions, motives, and objectives, on which are based the actions and pursuits, successes, failures, pleasures, pains, and peace. The mind is at the root. In a higher plane, it can be said that philosophies are at the root, paving routes of actions and changes. The history of the entire world and its humankind is cluttered with myriad philosophies and many more actors and actions, ultimately, with a few moving with an aura of permanence. Synthesizing them, a broad perspective emerges, a world view. This World view always emerges from the experiences of the past, and hopes for the future. Inevitable. Therefore, developing a world view consciously is like heralding a dawn before sunrise especially when a change is imminent as a necessity.

2. Viewlessness: Before the early ideas on the world at large were expressed and recorded, there was a vacuum "***Viewlessness***", as per today's description. In other words, the views could never be recorded for posterity. But the study of history and evolution tell us that local views were the norm. And these local views centered around the local environment and people-living with nature, using nature's resources, domesticating animals and plants, making a few tools, and revering and worshiping nature in awe and astonishment of its beauty, splendor, might, and power. And there were only a few human settlements and civilization and they were far apart in the vast expanse of the globe.

B. Ancient Times

It is well established that life originated on Earth about a thousand million years ago; and that the once united continents have drifted to give the present geography and that the life has evolved during all these years giving rise to a complex of fauna, flora, habitats, and climatic conditions. It is noteworthy that all these still exist all over the globe. Despite the variations in environmental conditions, no part of the Earth is really devoid of interacting life. Into this scenario came man when continents were still together or had land bridges. The erect man perhaps migrated far and wide across the planet inhabiting every biogeographical zone.

The humans were first nomadic. Gradually, they formed into herds. The Hunter-Gatherer, Stone Age man, gradually organized into groups and domesticated plants and animals. Thus, the roots of civilizations were laid. Though these processes have taken a few thousands of years, human settlements got organized in all the continents. The levels of organization differed reflecting their levels of intelligence. Thus, the human societies developed in the forest, on the mountain slopes, and in the river valleys. Gradually, civilizations evolved from these. Thus human populations existed in South America, North America, Europe, Asia, Australia, Africa, and in several islands. Consequently, the human species has become the animal widest in distribution covering all continents, climates, and ecosystems. As the environmental conditions vary greatly in these regions, these human settlements dotting the entire globe developed into distinct civilizations, all of them in harmony with nature, without which the people would not have been there.

1. Human Settlements: The humans needed water, plants, soil, and animals for their food, shelter, and clothing. And nature provided all these. Resources coming from nature, without which man cannot survive, assumed the status of a great force of life. Therefore, nature's components are to be worshiped, i.e., plants and animals. Fire and wind also were realized as great forces, influencing the life of man. Consequently, all these forces became objects of worship and gradually became gods.

Thus human populations, living on nature's bounty, gradually flourished in numbers and began to occupy considerable territories denuding the nature for habitat, agriculture and transport. Yet, the nature was hardly affected owing to its immensity. Every settlement must have learnt of other settlements and perhaps discovered certain resources with them that they themselves did not possess. Desire to acquire them must have initiated the practice of exchange, and ultimately, trading practices.

2. Gods and Faiths: Concurrently, the infinite and awesome resourcefulness of nature has created in man a need for knowing its causes and developed a healthy respect and fear of this powerful causal forces, which were collectively named GOD, subsequently. This concept of causal force naturally was used for managing the settlement. Thus a creed, a philosophy, a system, and ultimately, a 'religion' developed resulting in the dictum, "if a religion is strong, the population is strong." Therefore, conquests through might and emergence of societies with considerable cultural and material development happened. Creeds or faiths to keep people together, which are basically indicative of a psyche of fear and respect towards the awesome nature, have also taken roots. Thus an ancient creed of values, a ***Sanatana Dharma*****,** existed a few millennia before Christ, as the mother of some modern religions - Buddhism, Jainism, and Hinduism. And it is within the realm of human imagination that systems, faiths, or creeds have bound human settlements in different parts of the world, at least for different durations. All said, humankind, whether in nomadic stage or in highly civilized stage, was always looking towards nature for its resources and GOD as the causative force.

3. Religious View of Life: With the increase of settlements, both in numbers and sizes, interactions grew, management necessities grew; a meaning in the nature -cyclical and continuing-was discerned. These were thought to be guidelines for man to live his life with fellow beings and norms, rules, commonalities, restrictions, and justice, all within the frame work of nature, emerged. These were to be used both in exploiting nature as well as running the societal lives. But, nature remained an all powerful enigma owing to its unfathomable complexity. This almighty nature was converted into a concept, and personified sectorially and holistically. The entire complexity and enviable power were attributed to them. Rules to manage humans in society were also attributed to these Gods, lest they may be disobeyed. These principles were to be the guiding spirit for human society forever. Thus the God and Religion were created by man. But God was called the creator. The man's principles for himself were considered as orders of creation, by God, so that other men will obey and observe. We shall see if this happens.

(a) **Pre-Christian:** Hinduism and Buddhism were the major pre-Christian religions, the date of Hinduism being controversial. Whether there was a religion like that was in question (it was only a ***Sanatana Dharma***, a conceptualization in the times of yore of righteousness). But the scriptures had telltale contents. ***Vedas*** constitute the first world view based on nature and righteousness of man in society- the religious way of life. Buddhism, coming several centuries later, constituted a high print in respecting all living creatures including fellow men through simplicity, compassion, and service. It is remarkable that Buddhism and Hinduism lived in comparative nonfriction for centuries. All perceptions and concepts, all emanate from the same source and aim towards the same goal was perhaps realized by both these even in those times! What a contrast with today? Both religions considered that man is a part of nature and nature has to be treated as such.

(b) **Christianity:** Christianity with its own roots, apostles, and holy book preached a reverence for life through love. The nature was supposed to be a bounty created by God for his creation endowed with the power of knowledge from the apples, for the Man; here, Nature for Man concept originated. Incidentally, this is the only religion appropriating nature for man.

This world view has changed the global environment today.

(c) **Providential Ecology:** All these dogmas created 'God' and norms for righteousness, the religion, and credited to "Him" that all animals and plants were created with specific duties and rights. These have to be under interactions, a course of life, a cycle of death and birth. All these suggest that nothing is created new and nothing is dead permanently. The balances among them run the course of life around man, as ordained by God, and therefore to be revered. This is ecology of the Providence -Providential Ecology. It is significant that ecological understanding existed even in ***Vedic*** times and in their scriptures!

Hinduism, Buddhism, Christianity, and Islam occupied significant areas of the planet. Kingdoms, empires, trades, wars, campaigns, victories, defeats, expansions, amalgamations, and conversions happened, yet belief in providential ecology, a God, and a religion continued.

It was suddenly proved by Darwin that providential ecology does not operate. There were struggles for existence, survival of the fittest, and everything evolved leaving a history behind Thus everything created by 'God' does not remain for eternity! Darwinism was based on observation of life around, and its scientific analysis based on logic and reason cause and effects.With this, religious world view took a backseat, yielding place to scientific world view that was waiting in the wings. The sobering fact was that science,tools, and technologies came from centuries before and the Industrial Revolution occurred only during the late 18th century, generally credited to Bacon and James Watt. But the view of Science did not emerge to an extent to replace religious world view. These were mere forerunners. It has to wait until the enigmas of natural forces, ecology, were explained by Darwin and dethroning the until then ruling creed of providential ecology! As ecology was too realized as the prime force, providence was marginalized.Thus scientific views of 16th and 17th centuries had to wait until mid 19th century to become a world force, as alternatives to God and Religion.

The ancient philosophers of Greece visualized the planet Earth and its life as a creation of the Providence where all the life was designed to live in peace and harmony through interactions and balances. But the bubble of this providential ecology was pricked when Darwin demonstrated a struggle for existence and survival of the fittest and by the discoveries of fossilized life. All religions also more or less visualized life as a one-stroke generation by the creator to live in everlasting harmony by following the precepts of the religion. Perhaps Hinduism alone visualized the ultimate demise due to the avaricious psyche of man.

During these times of direct dependence on natural resources, humankind developed insights into the structures, functions, and behavior of natural systems. He has put this knowledge into use. Consequently, he has improved or modified the resourcefulness of nature and methods of exploring and tapping them. These have constituted the science and technology of the ancient times. A little of it is documented, but lot of it is in evidence. This trend continued until the advent of industrial revolution, the giant leap into the modern times. However, intervening these two periods, there is a period of preparation for the modern times, which is labeled as medieval times.

In summary, it can be said that human settlements started and flourished wherever natural resources were bountiful, especially water and soil. They grew in area and population size depending on the resources and environment. Trading started among these settlements for exchange of resources; trade routes became the channels of information about the relative prosperities of human settlements. Attempts to annex such settlements along with their resources and further such advances have been a legion until the middle of the

15th century. Enslavement of human resources has been very common during these times.

C. Medieval Times

The onset of industrial revolution galvanized the human ingenuity in resource appropriation. This period is supposed to be the forerunner of all that is modern, which can be broadly considered under three heads: trade routes, resources of others' lands, science and technology for efficient exploitation of resources, colonialism and its mechanisms, and social Darwinism.

This is further characterized by the migrations of people along with the enslavement of some people, creation of first world and third world, and the beginnings of transition from agriculture-oriented feudalism to industry-oriented capitalism, whether or not the political forces were democratic or imperial.

1. Trade Routes, Resources of Other Lands: As time passed, the human settlements gradually came to know the resources of other settlements. It may be noted that these settlements were in the form of kingdoms, empires, nations, or tribal formations. The common feature of all these people was the desire to improve their standard of living or quality of life. Each one, therefore, desired to obtain the resources of others. This was possible either by trading or by conquests. Thus, there were trade routes from Europe to Asia and Africa and vice versa and tradesmen moved freely. However, some of these trade routes were blocked by the conquests of enemy kingdoms, e.g., Turkey. Consequently, new trade routes had to be established. This is how that the Atlantic countries had to resort to seafaring.

The Atlantic countries like England, France, Holland, Belgium, Spain, and Portugal launched new trade routes towards the lands of plenty, i.e., Asia. In the process of establishing new routes to old countries in Asia, they discovered new lands like South America, North America, and Africa. Not content with trading, many of these lands were dominated or owned up by these maritime powers due to the allurement to their resources.

2. Colonialism and Social Darwinism: As stated above, trading of sea routes had also resulted in finding new lands. This has lead to migration of people from Europe and domination of the local natives. This domination and ownership had different mechanisms depending on the state of development of the local societies. For example, in Africa, the people could be easily enslaved due to their very low level of development. In North America, patent genocide had to be resorted to. In South America, the tribal were deep in their own tractable forests, leaving the shore lines to the colonizers. In Asia, where societies were fairly advanced, they had to employ a mixture of market, military, and missionary techniques. Thus the European nations had taken over a large part of the planet for resource exploitation. This was justified even by religious sanctions invoking the Darwinian philosophy of survival of the fittest. All these had lead to the formation of the First World (the dominant) and the Second World (the dominated). Further, the principle of self-reliance (autarkism) and barter was eclipsed by the rise of capitalism in the European imperial -colonial nations.

Thus the maritime advancements and the advent of industrial revolution bestowed tremendous advantages on the human resources in Europe. These advantages were used to colonies Africa, Asia, and the Americas. Not only were the lands taken over, but native human populations were also exterminated (North America, Australia) or enslaved (Africa). In Asia, neither could be attempted due to population size and cultural status, therefore, cunning had to be resorted. Through these ingenious ways, the resources of three continents were appropriated by the European Nations, leading to conflicts among them in sharing the spoils. Consequently, Europe determined the destinies of the World's nations.

3. Scientific World View: That every thing has a history and that every action has a source or cause is the substance of science. Its method is logic and reason; the goal is the truth - a question answered and several raised. To understand these, techniques were often developed by the ingenuity of man; tools too were invented for the same purpose. Thus, a philosophy to start with went through the processes of logic, and reason discovered truth in nature, and to know more invented tools. Science and technology thus made the entire universe as their domain, a subject for onslaught by its philosophy, probing methods, and end products. The end products were materials that had to be for the use of man for they were generated from previously existing materials in nature. Just as nature was used, by man before, the products from nature through the intervention of science and technology too had to be used. Technology became a byword for materialism. As man became enslaved to materialism, technology became the neo-God. Religious righteousness was forgotten. Nature was considered endless in resources. Materialism grew into consumeristic hedonism, thought to be as endless as resources, and technology allowed it.

Alas, this dream had a black hole in the form of ultimate technology, pyrotechnology-atomic energy and its destructive forces. Humankind was brought perilously to the brink of annihilation by a nuclear holocaust. While everyone was concerned with consumerism on one hand, on the other hand, nuclear threats, as a slow poison, spread, and slow rot was set in, with slow denaturation of habitat occurring, whose consequences may be as far reaching as the nuclear catastrophe could be! Besides, it cannot be prevented, regenerated, or remedied by a mere policy decision!

4. Scientific Vs. Religious World Views: While scientific world view was ruling the world at least for two centuries, religious world view persisted for reasons I do not wish to elaborate here. While the latter, despite its hold on the world for full eighteent centuries and by corrupting humankind, did not endanger the continuity of life. It only divided them. But scientific view not only divided the world but also threatened its continuity as a resourceful biosphere! Though this was suspected from 1950s, it was recognized during 1960s, and it became an accepted eventuality only recently when time seems to have run out for many remedies.

It must be realized, therefore, that religious world view preceded for 15-50 centuries before the advent of the age of religions, which continued for eighteen centuries. Scientific world view existed for -three to four centuries before the advent of the age of science in the 19th century, which continues still. The religious world view continued in the age of science. Similarly, both religious and scientific world views will continue to have sway in the age of ecology. And the age of ecology, starting in 1992, was preceded by fragmented world views on ecology for nearly three decades. Let us examine a synthesis of these.

5. Science, Technology And Exploiting Natural Resources: Development of new routes of trade, new trade objects, and maritime power could not have taken place without advancements in science and technology in the colonial countries. Most of these included methods to modify the raw resources into products. This process of technology was given an organizational set-up so as to convert resources into industrial products and subject them to economic and political mechanisms of distribution. This is the beginning of exploitation of global resources by a few European countries and a few sections in them at the expense of the rest of the world. The class conflicts among haves and have-nots in these countries and impoverishment of many nations that the world has been witnessing have roots here.

D. World War I

The conflicts for resources lead to the most major war in human history among European countries- some land-locked countries longing for colonial resources while others holding a monopoly over them. The hopes of the former were foiled.

We have seen that disparities among nations have already originated and disparities among people have already been accentuated. But there is one more. There are countries in Europe without colonies resulting in disparities among the ruling classes of these countries. It is believed that this simmering discontent has lead to political strifes and ultimately to the First World War. This war, therefore, was essentially between colonialist vs. colonialist spatially limited to Europe but motivated for control over the world's resources. In this process, however, some of the colonialists even dragged gullible colonies.

When the war was over, the victors and vanquished came to some understanding on the colonies in Africa and the exploitation of the Arctic, but the oppressed classes in the European countries turned against their own masters, so that they could develop their resources for all. The victims of capitalism, imperialism, and industrial revolution in some European countries have organized into an entirely new politico-economic force. This is the origin of the Second World countries, borne shatteringly from the First World countries. This Second World country, popularly known as the communist world, has provided for nearly 70 years an alternate form of development where man does not exploit man. Man exploits nature for man. From this signal, it is not far to assert the beginnings of stirrings from the colonies to assert themselves to become masters of their own resources and their exploitation and for their own people.

E. World War II

Germany, frustrated in the First World War, teamed up with two other nations similarly longing for the wealth of other nations and brought on the Second World War. Once again, the desire to share the spoils of the colonial appropriation having failed, Antarctica, high seas, sea beds, and outer space were internationalized as a probable compensation for the likely loss of colonies.

That the next war started in two decades speaks of the shocking stupidity of humankind. This war, the Second World War, was also fought among the colonialists. The theatre of war has slightly spread over to other continents - North Africa and Asia. New Asiatic colonial power is the new addition. Many of the former colonies fought with their masters. At stake was the same, the global resources, and thus the Second World War was a repetition of the First World War, except for the magnitude and spread.

Why did this war come about? This is chiefly because of progress in science and technology converting the resources into commodities and the consequent accumulated economic wealth. This happened not only in the two groups of colonialists but also in the allied economies, such as Canada, USA, and Australia. The same has happened in the Second World countries also -the communist world. It is indeed characteristic of these societies that considerable wealth developed from nature's resources has been diverted for building military machine to destroy the humankind for whom the wealth is supposed to be built. This is social Darwinism at its worst. This is exclusion principle of Population Ecology in action. This is the beginning of life boat ethics at the level of conscience.

The consequences of World War II were catastrophic for all involved. The Second World War unleashed two forces. Firstly, there was the realization on the part of the victors as well as the vanquished that they have to share rather than fight. Secondly, the colonies began to assert that they must be the masters of their own resources. But the most dominant development was the realization by the colonial powers that they can no longer control the resources of their colonies because of their thorough beating in the war.

1. As a sequel to the sharing principle, there were some treaties covering international seas and Antarctica. The USA aided both the warring groups in the form of the Marshall plan. The rebuilding of these countries, which took nearly a decade, could be accomplished chiefly due to rapid strides of progress in science and technology and the support from the USA, which was hardly affected by the war. In fact, this is the beginning of neocolonialism.
2. The colonies in Africa and Asia realized by now that the natural wealth of their nations has been siphoned to the European masters and this can no longer be tolerated. The struggles for political freedom, therefore, cropped up or were accelerated. In these efforts, the example of Soviet Russia, as a Second World country offering alternate mechanisms of social and economic growth, became very attractive. Thus, most of the Third World countries looked towards the Second world countries for guidance and help in building their own nations after obtaining political freedom from the colonialists. Consequently, a competition between the First and Second World countries to influence the Third World countries began and thus the seeds of future wars were sown.
3. At last, the world was divided into the First World comprising of ex-Colonialists and their capitalistic friends, the second world of communist countries, and the Third World of former colonies or vassals of the First World. The First World has greatly progressed economically and militarily. The Second World countries were lagging behind. The Third World countries were looking for help and obtained them through scientific, technological, and trade exchanges with the developed countries. This initiation of exchanges, related to resources ultimately, is the beginning of a protracted war of neocolonialism, the Third World War. A cold war between the First and Second World countries has therefore become the necessary political ingredient of this Third World War because of their desires to wean the developing countries into their spectrum of influence.

   First and Second World Wars were thus waged between two groups of developed countries for hegemony over the world's resources, i.e., between colony holders and nonholders.

F. World War III (1950s -1970s)

World War III is a euphemism for the events that have followed in the

World War II right from 1945. This Third Global War was unleashed by the reentry of the First World into the Third World for resource draining and political domination, i.e., neocolonialism and neo-imperialism. This initiates a global action for global retrieval, which will be dealt with under the heading World War IV. For the time being, let us examine what has happened between 1945 and 1970. These collectively earned the name World War III for various reasons covering the three worlds.

This war was basically a war of consumerism. This was between the ex-colonialists and their allies on one side and the ex-colonies and other poor countries on the other side. In other words, this was a war between urges of consumerism (aided by science and technology) and the urge for development. The European and North America countries had developed very fast from 1945, in science, technology, industry, and consumer goods. Consequentl,y their standard of life increased enormously. Correspondingly, their military growth also expanded. On the other side, the majority of the countries in the world needed basic education, science, and technology to develop their societies and improve the standards of life of these people. They only had resources and the urge, but they were totally dependent on the rich or advanced countries.

The advanced countries can be grouped into the capitalistic block (20 countries) and the socialistic block (10 countries), named as the First World and the Second World. The remaining 100+ countries of different geographies, nationalities, resources, and problems, mostly in Africa, Asia, and South America, constitute the Third World. The advanced countries are mostly to the north of poor countries, except Australia and New Zealand. Therefore, these two groups-First and Second Worlds-are called the developed north and the Third World is called the developing south.

1. First World: This Third Global War was unleashed by the reentry of the First World onto the Third World for resource draining and political domination, i.e., Neocolonialism and Neoimperialism. This has happened for three reasons.

(a) Marshall plan was initiated by US due to which enormous development took place in Europe by the military-industrial complexes both for war purposes and rapid consumerism. This has raised standards and costs of life of people together with the national wealth in all the developed countries. All this could happen chiefly from the resource appropriation from the Third World carried out for three centuries-which functioned like a fixed deposit-capital from the rest of the world.

(b) Most of the decolonized countries unfortunately opted for the kind of development their colonial masters had in their own countries little realizing that they had neither the "colonial capital" nor the scientific and technological bases. They had to therefore depend upon their former colonial rulers of Europe and their allies- USA, Canada, and Australia for almost everything.

(c) The developed countries began to assist the countries, which have chosen their developmental models with capital and manpower for science, technology, aid for specific projects, loans, and trade concessions. Wherever the above could not happen to their satisfaction (or not in their own interests) regional wars have been engineered. These embers are still burning around a dozen places, which will be dealt with later.

2. Second World: The Second World countries are also developed. But in the international politics of cold war with the First World it has to grow in military strength. In addition, it also has to help some developing countries either for spreading their philosophy of communism or for the purposes of political power with the first world, the cold war. Consequently, the entire resources, manpower, science, and technology of the Second World countries,were geared only for the above purposes. This has resulted in the nongrowth of living standards of their own people -a kind of self-sacrifice just to survive as socialistic states and help other people emerge with similar goals. In the face of growing wealth and geographical spread of and opposition from market economies, the science and technology of the Second World countries could not be used to satisfy the aspirations and demands for higher standards of life of their own people. This is, but expected, because, the First World countries had the foundations of colonial wealth accumulated for over a century and did not have the burden of helping the Third World countries. Unfortunately, the Second World countries were often plagued and beleaguered by the Third World countries on several other fronts.

3. Third World: The Third World countries needed science, technology, and education to improve their own manpower to utilize their natural resources. They further needed financial inputs to initiate infra-structures. These could come only from the First and Second World countries. As a result, considerable memoranda of understanding have been signed and implemented between the North and the South. Money flowed-freely, conditionally, with other unspecified riders. Needless to say, most of these were initially advantageous to the recipient countries, but ultimately beneficial to the donor countries. As a result, the developing countries developed a bit accumulating a lot of debt. The developed First World gave anything primitive and dispensable and accrued considerable wealth from the Third World through various agreements. This scenario continued until most of the third world became irrevocably indebted to the first world, in various degrees. A new group of countries emerged from the Third World whose entire national income was not enough even to pay back the interest on the principal debt. This group, about seventeen in number, can be hereafter called as the Fourth World.

The Third World demonstrated a lack of dedication to improve their own lot with the help the First and Second world countries. Most of these were not only debt burdened but also lost resources due to population explosion and inefficiency. Consequently, rivers were polluted, forests depleted, and human habitats degenerated. People from rural ecosystems migrated to form urban slums and the health education being traditionally poor became monopolized. The little wealth that has accumulated also became monopolized. Consequently, fissures, frictions, and violence became the norm.

Thus the rich became richer and poor poorer both among people of a nation, as well as among the nations of the world. Meanwhile, the nature got plundered irrevocably due to the irrational greed of man. But no one has realized that this will have global impact, i.e., which ever part of the body is affected the whole body has to suffer.

4. Regional Wars: We are already concerned with the competition between the two worlds for global influence. But neither of these was interested in direct confrontation. They would not risk another war due to the lessons of Second World War where the victor and the vanquished were equally affected. Furthermore, the addition of nuclear arsenals made another war a military suicide. Consequently, proxy warring was resorted to and the whole Third World was beset with internecine conflicts-separatist movements within countries, revolutions by the oppressed, border conflicts between neighboring states, and regional conflicts of historical nature. In many cases, they were historical legacies. In a few cases, they were insurrections. In a few cases, the ideology was involved. In some cases, ethnicity of people was the concern. In others, it was regional domination. In many countries, it is an unabashed seizure of power by the Military. All these either directly or indirectly were fueled by the First World and the Second World. And in all the cases, the resources of the land in question were involved. Thus, all these regional wars started after the Second World War continued for nearly 20 years in South America, Africa, East and West Asia; examples are Korea, Vietnam, Laos, Cambodia, Malaysia, Bangladesh, Kenya, Ethiopia, Nigeria, Cuba, Equador, Columbia, Peru, Bolivia, Chile, and the West Asia conflict between Israel and the Arab countries.

The history of Arab-Israel conflicts was well-known. Whatever may be one's perceptions, the consequence of the third conflict during 1967 are very far reaching from an environmental and ecological point of view. It made the entire world realize that what is true with biological units is equally true for the socio-economic life as well among the political units of the world, i.e., interdependent balancing. More details are given later.

5. Environmental Ethics From The Regional Wars: Every catastrophe has its lessons to offer provided there are takers. During the Third World War or war for consumption and development, there were universal loss of resources and human refugees from degraded environment. These happened to different degrees in Africa, in Asia, and in South America. But the most telling effect came from West Asia. For the first time, a resource was used as a political weapon, that is the fossil oil. Following yet another defeat in the hands of Israel, the Arab countries stopped selling oil to the West. The prices were raised astronomically. Most of Europe reeled under shortage of petroleum. The difficulties of different countries though were enormous; one of the most affected was Italy. Almost at the same time acid rain was bursting on Canada and Scandinavia from the industries situated in other countries. These developments have shown not only the interdependence of countries on one another, but also demonstrated how natural resources can be used in politics. Some of the plagues unleashed by this Third World War are listed below.

Enormous pollution of the entire globe was unleashed by the fierce development of the European nations-a kind of war on the environment.

The Natural Resources of the developing countries systematically depleted at a rate fastest in the history.

Development in the Third World was not satisfactory due to outmoded technology, lack of scientific manpower, pressures of fiscal deficits, and finally the development model (disregarding land/man ratio) unsuitable to the local people.

Balance of trade, aid with strings, balance of payments of loans, and migrations/sucking of trained manpower resulted in the further impoverishment of these nations despite the apparent facade of growth. Many countries reel under debt burdens. While others have an ever increasing population below poverty line. Divisions and conflicts, and hatred and violence is the order of the day everywhere. The entire world got entangled in the strangle hold of their former colonial powers.

Poverty, pollution, depletion of resources, starvation, insanitation, energy crises, appropriation of other's wealths, and over population resulted in conflicts and violence all over the world. Environment on the inexorable march of degradation threatens globally.

Some of the above have been done in an indirect way through the auspices of international organizations.

6. Club of Rome Principles: Thinkers become vocal on the growth of societies at the expense of resources and their consequences to the world at large. However, the most notable among these voices of concern is the Club of Rome organized by an industrialist Pecked. This Club is essentially composed of industrialists, economists, sociologists, managers, and the like. One of their initial activities was to commission an analysis of the global development and environment. This has resulted in the Report called the Limits to Growth by Meadows et al. The summary of this is as follows:

1. Over Population: This is mostly relevant to the Third World, countries of Africa, Asia, and South America.
2. Depletion of natural resources: Mostly relevant to Third World, but also to the First and Second World countries to a lesser extent.
3. Pollution: Relevant to First, Second, and Third world countries.
4. Energy: Nonavailability of conventional energy in due course of time. Is it a problem of the entire world?
5. Hygiene and Sanitation: This is a problem of the Third World only

Having summarized the milestone report of limits to growth of nations, it is necessary to summarize the other developments elsewhere. The efforts of Club of Rome were emulated by other intellectual groups in thinking loudly, warning societies, offer alternatives of development, as well as conservation. There were also movements by concerned people for saving the Earth and its environment in various parts of the world. These two developments forced many governments to initiate actions, to restore sanity in the exploitation of their regional resources for future. All these together made an international effort necessary.

The intellectual exercises, people's activities, governmental actions, and international efforts heralded the beginning of the Fourth World war-a war by the entire humankind against itself to save, protect, maintain, restore, manage, and improve resources of the planet bequeathed to them through history. Soul searching, making amends for past wrongs, and opposing any new blunders are the watch words of the philosophers of this war so that planetary resources and their ecology continue to be everlasting and healthy.

G. World War IV

The activities during the post-war period were affecting the entire globe with only regional variations of intensity. This could, however, be delayed and averted only by cooperative efforts by all nations thinking globally, but acting locally. The Fourth World War, therefore, essentially constitutes follow-up actions against boomerangs from the Third World War. The above realization came to intellectuals first as usual that lead to local people's actions and international activity. The governmental actions came subsequently. Finally, these resulted in governmantal cooperations between countries and groups of countries and even political parties are ecologized or ecological groups politicized.

1. North-south Divide: Unfortunately, the perceptions of developing and developed countries differ from each other on the agenda for actions. This is the Real North-South Divide-and the underdog, but why? This divide is due to the oppressed nature of the "South" and the domineering attitudes of the "North". This South and North-metaphoric/ Symbolic description of poverty line neighbors-are present in every country. Poor countries have rich people and rich countries have poor sections! The entire agenda of work, effort, mechanisms, and even motives are dictated and implemented by the North in each country-by any name, any form, and any mechanism. Does this not remind us of the global North and South in the politico-economic dispensation of resources? Therefore, the environmental crisis will remain as the everlasting burning fires of the mythical Ravanasura's corpse! The South in each country will be used as a buffer between the decaying environment and the selfish North exactly in the same way as the global North (G-7 and friends) treats the global South (G-77 and friends), i.e., a buffer between their interests and the decreasing resources of Earth. With these two sets of North and South, is not the effective population of the world less than 50% of the numerical count, the others uncountable as humans!

The Fourth World War thus is basically a war between the past and present generations, on one hand, and the future generations, on the other. Only some of the present generations bother to support the interests of the future civilization. If these forces win, there is hope for the globe. If the dark forces of the past triumph, ignoring future generations, doom is inevitable.

At this stage, perhaps a recapitulation may be appropriate. In summary, it can be said that exploiters also cannot escape environmental back lashes. Developmental models should depend on the demography, resources, productivity, and history of nations and cannot be borrowed. There cannot be open-ended growth as envisaged by both capitalism and communism: Politics cannot be right or left but front conscious of writings on the wall in the front. Values cannot be neutral but socialistic and righteous. Means are as important as the ends. Agro-ecosystems are far more determinative of future of humankind than the urban systems. The only realistic solutions appear to be the idealistic ones.

2. Providential Ecology and the Religion: The ancient philosophers of Greece visualized the planet Earth and its life as a creation of the providence where all the life was designed to live in peace and harmony through interactions and balances. But the bubble of this providential ecology was pricked when Darwin demonstrated a struggle for existence and survival of the fittest and by the discoveries of fossilized life. All religions also more or less visualized life as a one-stroke generation by the creator to live in everlasting harmony by following the precepts of the religion. Christianity maintained that man is the inheritor of Earth's resources that God had created. On the other hand, Hinduism visualized the ultimate demise due to the avaricious psyche of man. We shall delve into this at a later stage.

3. Hardinism: The thread of Darwin's struggle for existence perhaps is taken over and carried by Hardin, following the route of carrying capacity that life on Earth is like on a life boat. There can be place only for some- If all have to be in the boat he insisted that the boat shall sink. In other words, a good future for some or none at all for all is the alternatives because modern man is dominated totally by hedonism. And the struggle is just for this. The survivors take over the life boat, the planet, and the rest will be excluded from the Noah's Ark - Pessimism predetermined?

4. Spaceship Earth And Common Future: However, this gloomy fate awaiting the most of humankind is sought to be averted by Boulding, Stevenson, and Ward. This is possible perhaps according to them to think in terms of a common future to the entire humankind -a term borrowed by Bundtland commission, recently. They visualize the planet as a spaceship. Any problem anywhere will affect the whole spaceship necessitating a living together helping each other. Either it stands united in the cosmos or falls divided gradually as the frontispiece (cover page) of the first issue of Time depicted in 1990 (1989). From this came the slogan -Think globally and Act locally. In other words, the optimists hope that the doom can be averted by human action for good economics. But is this hope realistically knowing the human apathy and immorality wearing all pervading mask of management ?

5. Self Regulation Properties Of Earth: The Gaia hypothesis of Lovelock and Margulis consider Earth as an open enough ecosystem to redeem itself. It is said that the Goddess of Earth in Greek, Gaia has infinite patience to redeem itself. Having passed through enormous vicissitudes of changes, it would overcome the currently perceived crises also. This optimism comes more out of faith than reality. The reality is that Man assumed such a dominant role in Earth's history that catastrophies generated by him outnumber and outweigh the natural calamities, which brought extinction of life in the past. The camel is bearing the proverbial last straw on its back and yet astonishingly it is standing. This gait is acquired by the crutches built by science, technology, management, and propaganda. The facts of the global South (G77) and internal South in each country constitute 60% of humanity. Then optimism for whom? On what Basis? On the hopeful basis of imparting permanence to this divide? Knowledge itself was so deficient that peers did not realize what the ***Bhoomata*** or ***Bhoodevi*** concept of Vedic wisdoms professed and prophesied!

H. Ecological World View

All saints, religions leaders, philosophers, writers, thinkers, messiahs, prophets, native people, and common folks in all lands from centuries have bestowed a sanctity on nature and considered all life interdependent, which must continue so. This can be seen in Hindu scriptures, Buddhist practices, Christian preaching, of St.Francis of Assissi, Confucius, Tao, Gandhi, Mao-

Tse-Tung, et al. One can see these in the concerns of Leopold, Sears, Darwin, Vogt, Osborne, Schweitzer, Wallace, Banks, Bates, Moore, Gandhi, Theoreau, Elton, and Salisbury during late 19th century to mid 20th century. Though the views on future began to emerge from Hardin, Aldous Huxley during late 1950s. But, it is Carson's Silent Spring and the Reith Lectures of Darling on the BBC in the 1960s that alerted the conscious. The six day war between Arabs and Israel, use of oil, an ecological resource, as a retaliatory political weapon, gave origins to the Club of Rome of Aurelio Peccei and its efforts. The efforts of Margaret Mead, Barbara Ward, and Kenneth Boulding, to name a few, ultimately led to crystallization of public opinion on environment, ecology, development, and future of man. These were followed by a few global models on man's economic future under the aegis of Club of Rome. United Nations organized meetings in Mexico, Switzerland, and Sweden, which resulted in the institution of United Nations Environment Programme (UNEP). Around this time came the views and efforts of Lovelock, Paddock, Ehrlich, Commoner Kahn, and Simon, to name a few

1. One World -Several Worlds: Concurrent to the above -mentioned activities, philosophies and movements to rehabilitate nature to its resourcefulness sprang up in Europe, especially in Germany; these have a mix of Marx, Gandhi, and the latest ecological creeds. Then the time for ecological movements was heralded and through Gandhian techniques!

In some countries, these simply remained as non-governmental voluntary activist programmes while in others they assumed political overtones and even adopted political mechanisms. While the melting pot of ideas and philosophies continued during the 1980s, the robbing, fleecing, and melt down of the Planet Earth continued politically economically, and culturally with active application of science and technology and trade policies. Under these weights, the world at last broke down into several worlds- First, Second, Third, Fourth, Fifth, and Sixth-all on economic and ecological bases. What were a mere three worlds on political premises of balance of power and cold war equation are now at least Six 'worlds' on the strength or weakness of ecological productivity despite thaw in war possibility, reduction of militaries, and monopolarity and suggestions of possible peace. So complicated are issues now that one wonders how to differentiate right from wrong, and truth from falsehood. Yet we will make an attempt to organize the available information and the plethora of developments to draw the ecological world view, which is the main theme of our present concerns.

In this process of melting pot convulsions on perceptions of humankind's future, a lot of new phrases embellished English literature concerning ecology and environment-Triage principle, Spaceship Earth, Global commons, Life Boat Ethics, Sinking Boat, Doomists, Boomists, Gaia, Deep ecology, Permaculture, Ultimate Resource, Limits to Growth, North and South, Tragedy of the Commons, Presinking Ark and the prefixes of "Eco", "Environmental," and "Green" for anything conceivable as ecologically relevant under the generic name of the very topical, and fashionable "Environmentalism".

It is very remarkable that the word ethics is mentioned more often than anytime before in human history. "Green" is mentioned almost everywhere as a connotation for productivity and therefore to resourcefulness. "Environment-friendly", and Ecology-friendly as euphemisms and cover-up phrases have also littered our media! As if devils quoting scriptures, many wrong people use these right phrases!

2. Three Major Views: The world view on ecology of humankind can be broadly categorized into three. Here it must be recognized that no one has so far used the term "World View" for any ecological philosophy.

(a) Doomists: The first category can be termed as Doomists. These views predict a certain doom to humankind for various reasons.The first among these is the Life Boat Ethics of Hardin whose concepts were derived from ecological theory and practice. The tragedy of the commons-here global is a well-known adage in many languages and cultures. Only a fraction of the vast humankind can have a place in the rescue boat- rescuing from the sea of scarcities and problems. The rest must sink, therefore, lest the whole will perish; Paddock's Triage principle implies that affluent countries have to discard unsupportable nations to their own fate, if more self reliant and sustainable societies have to be helped to stand on their own. There are too many evidences in support of this pessimism based live of action!

(b) Boomists: The second school of thought is that Science and Technology has answers for all costs of production will fall. Technology will create new materials for further processing without dwindling the limited resources. The Earth is resourceful and the man has the ultimate resource, the intellect! These are evidences in favor of these views, too. But whose Science and Technology? For whom they serve? For whom is the boom? Gaia is the Greek Goddess of Earth, ever self-sustaining. Therefore, no matter what man does, Earth will not collapse from raising temperatures or sea levels and other catastrophies as it is an ecosystem, dynamic enough to buffer any eventuality and beef itself from damage into recovery and productivity. This is the Gaia hypotheses of Lovelock and Margulis.

(c) The middle pathists: The third view, akin to the Buddhist middle path, suggests a continuity of the current status of development with a few modifications here and there. Call it a state of equilibrium, steady state, or status quo, or simply the continuity philosophy.

Here, one can discern a few shades of opinion because no one wants a doom. Everyone fears it and most do not share the optimism of the boomists unless one is a Nimbyist or ***Nimyltist*** (Not in My Backyard/ Not in My Life Time). Thus the current "***Earth Managers***" have emerged!

(i) Spaceship Earth: The early thinkers consider the planet as a Spaceship- anything happens anywhere affects the whole. The imperative is that global commons must be protected and conserved globally with appropriate local actions.The slogan "Thinking Globally and Acting Locally" came from this perception. And local actions spread all over the globe. There are yet various shades of opinions as to how these actions can be done.

(ii) Gandhians : Gandhians propose zero growth like the Club of Rome pioneers, until at least the ecosystems regenerate enough. Equity is more important than productivity for concentration. Needs have to be curtailed. Basically an extension of Hindu way of traditional life and shades of Marxism, Gandhian ecology found global attention.

(iii) Greens: The Greens in Europe while supporting Gandhian simplicity do not advocate growth to be at zero level, but a redistribution of excessive productive capabilities of the North, drastic cuts in Northern consumerism. Shades of Marx are easily seen. Some of these, however even advocate the philosophy of religions for green ideologies to succeed in practice.

The predominant view however is to allow the present development model to continue. But cut down consumers, i.e., population. Modify all development by employing "environment-friendly" technologies. Conservation has to go also simultaneously. Science and technology has to be geared to greater efficiency and success so that environmental security through food systems at least is assured. These views are best articulated by Lester Brown and his group of opinion leaders.

But all are concerned about Earth, its resources, its human population, and the ecology that connects all these. Everyone knows that Earth is plundered. It can recover. But it will not give the psyche of man from history - Appropriation, ***Nimbyism*** and **Nimyltism** (Not in My Life Time-ism).

I clearly see the emergence of Kalki to destroy the world, as predicted by prehistoric Hindu thinkers, when they described what I consider as the theory of Evolution- Dashavatharas, Kalki, the last Avatar, however, will be the form of environmental backlashes and catastrophies social violence from divisions, conflicts, and hatreds. All violence in society is a Kalki manifestation. It has not yet really come. It is on the anvil. 2000 AD probably will see it in many societies despite all the pipe dreams associated with this magic date!

Globally, triage or exclusion principle will be implemented against a few scores of countries. This will be the "Kalki" in action globally. In each country, selected in by triage, there would be local North's (affluent) and South's (poor)- the latter will be kept under control as- captives. But countries will survive. Among the rest, as stated before Ecospasms -ecological and economic tragedies, and social violence will take place- Kalki in full play.

I. World War IV

Challenges from the First World, ever dependent Third World, raising expectations of people, degrading environment, dwindling resources, inefficient science and technology, and above all a weariness arising out of a state of ever readiness for war with the First World without natural allies resulted in the breakdown of the Soviet system. This system was the most major check and the balancing force against colonial exploitation. When this ceased to exist, there was change of gears among all the countries of the three worlds.

A kind of Triage Principle is being implemented in the name of globalization. The Paradigm of "***Think Globally, but Act Locally***" stands mutated to "***Think Locally and Act Globally*". **Those, we cannot comply will be left out. Marginalization of classes within the countries and among the countries has begun. The Fifth World War, a decisive one, has been unleashed. Let us briefly survey a few relevant issues.

1. Perestroika And Glasnost In The Eastern Europe: Why the Eastern Europe was thoroughly opened up? For their own citizens to think afresh? Why was restructuring encouraged? Is it recognition of the pulse of people for reforms? It is all these. Why did this come about?

The Second World countries were propping up many Third World countries, while having a running battle with the fellow North. The entire economy, resources, science and technology was geared for these purposes. Consequently, the people were able to get only a minimum/optimum standards of life-while their neighborhood counterparts in Europe outstripped them in the quality of life. The First World blared this message through the communications revolution into the yearning psyches of socialist people reminding their unending labours of Sisyphus with no end in sight. On the other hand, the developing world, propped by Socialist Bloc, never showed signs of self-reliance.

This scenario continued for two decades. A people cannot be expected to sacrifice their legitimate lot for maintaining a military machine and unimproving Third World nations, including the overtly Marxist ones. Abandoning these dead weights in favor of better living standards for their own people became a historical necessity. Therefore, there are Glasnost, Perestroika, and initiation of market economy.

Second World countries ceasing and yearning to merge with the first one has two impacts on the Third World - cessation of benevolence from the Second World and shift of First World's interest towards their erstwhile adversaries, turning them into friends. Both these left the Third World beleaguered, optionless, and totally at the receiving end of the monocentric overlordism. We have experienced this for centuries! A return to East India Companies is imminent.

2. Exclusion Principle: When the above are happening, the North will make their choices, selections, and rejections-quite naturally -the Darwinism and Hardinism come to play their ecological and political roles. Hardinism is already well known for its Noah's Ark approach.To my mind, the exclusion from the proverbial Noah's Ark, as today's world has more or less become with the Volte-face of the Socialist Second World, will affect the Third World countries in a biogeographic way. The life boat has no place for all. Exclusion principle has to be followed - ecology in its profoundest decisive role. The brunt will be on Asia. Perhaps the first one risking an eclipse may be India. Let us examine further.

South America has considerable natural resources and very low human density. The resource to man index is very high. The populations are mostly Spanish and Portuguese origin. These frontier people will not be let down by their racial relatives in the North. As a corollary, as a bonus, the native tribals also will not go down the subsistence levels.

Africa presents a different case. The north is a mix of Arabian and Negroid stock. The vast continent has not only a highly unproductive desert and degraded habitats but also yet surviving virile ecosystems. Population density is generally low, barring a few exceptions like Egypt and some West African Nations. The resource to man index is still high. Only a "few people" in the Sub-Saharan region have to be excluded to lapse into the final stages of ecological Hara-kiri! If an appropriate management through conservation is resorted, Africa can at least cling to the Noah's Ark!

The case of Asia is different with highly populated and developed societies -culturally, economically, and politically -where natural resources have been depleting irrevocably. The best human resources have been sucked into the havens/heavens of developed North. All countries have a poverty line of minimum survival standards -20-50% people in each political unit belong to the lower side of this line. The Mongoloid among these (i.e., Japan, China, and Indo-Chinese countries) are in a better position on a ***land-to-man*** comparison. Most of these countries are comparatively more homogeneous racially and religiously. Even China has a greater land mass than India giving it a far better land-to-man ratio. So where does the pendulum stop? Where does the needle point? Its points on South Asia -Pakistan, Nepal, Bangladesh, and India.

While it is agreed that Bangladesh presents a hopeless case, it may be India that will see its worst days. Why? Why our country is will be the most preferred political unit to be consigned outside the Noah's life boat? It is the diversity dividing the original unity.

3. Buffering: The strong societies of the better off 50 + countries will use other countries as buffers throwing eternal hopes and crumbs from the Caesar's tables here and there, now and then insulating themselves from violences arising out of want, poverty, helplessness, and hopelessness. Similarly, the rich sectors in the rest 120+ poorer countries will endeavor to create strong middle classes as buffers against the poorest of people, showing charity now and then emulating the rich North. No one will reduce consumerism and no one can live without hunger and thirst. The poor can no longer survive as there is no ecological commons and they cannot fight because their GOD, Marxism, has been discredited and jettisoned. This is the psyche of the mighty everywhere.

If this has to be avoided, all actions must have an ethical streak. In other words, concerns for natural resources became a kind of a religion - new religion where the new paradigm of God comes into existence in the form of Ecology -Productivity- balances for all.

J. Global Prognostics: Redeeming Features

During 1991, the world has witnessed the failure of centrally controlled economies to meet the twin challenges of arms race and welfare of the people. This was realized by capitalist world also. This world also was affected, but not to such an extent as the communist world. This is because; the former has got an historical capital property and no stakes with a people they exploited. On the contrary, the communist world had stakes with all the exploited people of the world. Without any historical capital, they have to maintain arms race, maintain welfare of its own people and maintain concerns for the exploited people of the world. Obviously, the game was fore doomed and it had to be lost one day or the other, and it happened in 1991 under the leadership of a man who perhaps realized it 5 years before. All these events have followed globally.

The First World that came out victorious is distinguishable into two groups - G7 and the rest. The Second World countries deteriorated into two classes, the latter to the level of the upper crust of the Third World. The Third World is divided into three groups. The upper with floating future, the middle uncertain and oscillating, the last sedimenting to a stage of unirascibility. Consequently, many regional groupings may be formed in the entire world, such as OAS, EEC, PEC, Eastern Europe, CIS, ASEAN, SAARC, G7, G77, etc. Regional peace becomes the most vital and the developed nations are likely to help only those countries having sustainable growth regimes, as per the philosophy of triage. The world today thus has eight sub-worlds.

Therefore, we will find World Bank and international monetary fund will dominate the entire world. Sooner or later many countries and people will be left out of reckoning in the process of recovery. In other words, if the world has to survive some part of it has to be sacrificed. Thus, the philosophy of triage and the life boat ethics or Noah's Ark altogether will come into complementary and supplementary action in the utilization of resources by people. These will run the world in one or the other form. As resources are converted into commodities, and economics alone will dominate the people lives. Public affairs of some countries govern the economics of other countries. Political economy's role will have a singularly decisive impact on the futures of all nations' decades to come. All this may ultimately turn good, because the wasteful inputs into militarism can now be diverted to peoples' welfare.

On the other hand, if consumerism dominates, there will be a North and south not merely globally, but in each country and violence will be unavoidable and this will be tellingly apparent in the former communist countries and in a majority of Third World nations.

That the Earth is our resource or its components deserve our worshipful treatment has been a running theme of all religions. But during the growth of civilization this has been obliterated by the consumeristic and indulgent passions. Consequently, the planet Earth is plunged to a situation of resource crisis and these resources were effected both in terms of productivity as well as their longevity because, if the balances are lost, productivity will be short-lived. This has been highlighted, time and again by National and International leaderships, Organizations, and Groups.

1. International activities: UNCHE-UNCED, IMF-WB, UNEP-WHO-FAO-ILO-IUCN, North-South Dialogues, South-South Co-operations, Technology Transfers, Trade Practices, Debt burdens and Reliefs, Scientific collaborations, Aid Strings, Cancun-Williamsburg-Paris. Roles of Brandt, Carter, Indira Gandhi, Palme, etc. USA, Canada, UK, Australia, China, Scandinavia, and Japan's Questionable role.

2. Peoples' Movements: Greens' Movements, Green Peace, Friends of the Earth, Local Movements in India, USA, Europe, Malaysia, South America, and Scandinavia.

3. Intellectual Activities: Club of Rome -Limits to growth -Models (1-5)- Carter-USA: Scenario exercises -Green philosophers. Individuals committed to their own special fields of investigation, but with an underlying concern for human welfare did not lag behind and in fact some led from the front. Thus, lighting a candle is better than cursing darkness. Therefore, there came about many efforts to redeem the loosing scenario. A brief account of some of these noble figures may be appropriate.

Leopolds, Sears, and Osborn were all voiced their concern about the deterioration of the habitat. But the time to listen to them had not come by then. Moore felt, there should be a natural resource conservancy because he can see the writings on the wall. A Science Congress in the USA have recorded their concern about the resource crunch and cautioned on the necessity of conservation. Such voices were raised from time to time expressing concerns about the nature's health. Carson felt that the entire Biosphere is likely to be polluted by the man-made pesticides and fears that the spring may not be musical. Mead was amazed that the plight of native people all over the world. All this lead to a great oration in the U.N.O. By Stevenson, where he likened the comity of nations to a spaceship of interlinked and interdependent parts. This was the origin of the concept of spaceship Earth elaborated by Ward and Boulding, subsequently.

These were preceded by lectures on Environment on the BBC by Darling and an article by Aldous Huxley on Ecology and Politics. After two decades have passed, only now the entire world has awakened to the reality that something is going wrong irreversibly on our planet. Where, When, and How became the concern of Club of Rome and resulted in a famous outcome, the Limits to Growth Model by Meadows et al. We shall examine these contributions by individual's stills further.

Commoner felt that technology is the chief villain in the degradation because it empowered the economic and political forces by adding efficiency to the conversion of resources into commodities. This has encouraged not only a coomodification of more resources but also scouting for new sources and creating new consumption patterns. A new deity, nay devil, Technology, began to preside over the lives of humans.

On the other hand, Ehrlich felt that it is the number of consumers, which deplete resources. In his book, Population Bomb, he went into the extent of calling pregnancy as a nine-month disease. Even when populations were low, was there justice in the distribution of natural resources among them? No. In the similar vein, it can be argued, even if the population decreased still the resources will be monopolized. Low population does not ensure justice in resource sharing. When justice does not exist anywhere why undue opposition to a natural phenomenon of births? Therefore, this alarm on population explosion should be viewed as a psychological desire of the existing populations for greater shares. Since the resources are any way not enough for all, by any method of distribution, it is better to control population, so that some justice can be done. This justice is imperative in the philosophy of Ehrlich. This is only to be expected from him because of his Jewish background and training genetics and Ecology of plants and animals.

The famous biologist, Dubois, felt that the biosphere has the in-built capacity to deal with any contingency because the properties of organisms and those of physical and chemical environments are complementary to each other. Unfortunately, these relationships are being eroded from all sides leading to an unstable planet. How can human societies be stable then?

The population Ecologist, Hardin, bluntly tells that there are not enough resources for all. Each nation is a life boat with limited rations for its members and all boats are not identical either in the resources or in the sharing humans. So some boats have got to sink. In some boats, some people have to be thrown out. This life boat ethic has mainly come from two sources -the exclusion principle in Population Ecology and the tragic ethos in the human psyche that all should be well with self and "***let the rest go to hell***". This is the terrible dispensation for humankind predicted by Hardin.

In similar way, Paddock felt that the highly developed countries cannot but discard some of the nations has hopeless. In this philosophy of triage the replenishments or medicaments should be given only to those who work for recovery to normalcy. Any inputs to others will be a mere waste and amounts to injustice to people who can really recover. Thus the weakest, the squanderer, and the poor world have to be allowed to decimate themselves - a kind of ecological euthanasia.

There are many others -Polunin, Bodyko, Mumford, Myers, Vogt, Brown, Goldsmith, Ekholm, etc, just to mention a few of the growing number of serious intellectuals concerned about the future of humankind on Earth. India was however the earliest to realize this as suggested by the Vedic wisdoms, Jainism, Buddhism, and an array of saints, savants, and reformers of the Hindu pantheon in all ages and finally Gandhi and Gandhians.

K. Perspectives: Light a Candle Than Curse Darkness

1. *The Concepts of God and Religion: The Ultimate Truth:* God is supposed to have created all organisms to live forever and in harmonious balance. Religions endeavored to maintain the above in human societies as well with sets of do's and dont's.

In the present context, ecology constitutes (not merely endeavor) harmonious balances among plants, animals and non-biological components of the globe. To see that these are ensued environmental do's and dont's are prescribed for man, as the environmental ethic. Thus Ecology becomes the new god and Environmentalism, the new religion for future. God basically being a concept and religion a set of principles to follow ecology and environmentalism replaced all concepts of God and religions as a viable synthetic alternative -The powerful and enigmatic nature from whose awe 'God' arose. From the eating of the forbidden fruit of knowledge, man has grown in knowledge until his future is threatened. Then came the wisdom that interdependences and community living are after all the God and religion!

2. *Reemergence of Marxism: Phoenix Rises from the Ashes:* Therefore, the Marxism that is being given an indecent burial becomes relevant to the depressed people and depressed nations as never before. Self-reliance and self-sufficiency within limits are set up by their natural resources. In other words, carrying capacity of the habitat may have to be sought overtly to limit development. In other words, development cannot take place beyond the net productivity of an ecosystem. I visualize, therefore, that political relevance of Marxism will be matched by the developmental relevance of ecology. Many ecologists may look towards forming political parties with left and front attitudes or Marxists may turn idealistic ecologists in many third world countries by the turn of century. Either way violence is inbuilt because of the resistance from market forces who would champion the development model of the North and build their monoliths. The excluded people will get hollow freedoms and freedom to hope against experience. Therefore, violence and divisions are inevitable.

THE WORLD IS IRREVOCABLY MARCHING TOWARDS
DIVISION AND CONFLICT HATRED AND VIOLENCE

LEONTIEF

Chapter - 5

Globalization and the Indian Society

One Step Forward, Two Steps Backward; Both Awkward

Preamble

With confusion widespread among youth, regarding Globalization, Liberalization, and Privatization, I hazard to explain them, for whatever worth it is.

1. The only global phenomenon for centuries has been knowledge; it was universal, for which reason the highest centers of ***knowledge, wisdom, and cultures are called Universities.***

2. As the history progressed through wars and exploitation, the First World War took place, a war among the European powers for Global Resources. The result was a political platform, League of Nations. It failed. Second World War took place, again for global resources; UNO, as a world Political Platform, is organized. This ***Political Globalism*** began during 1945-50

3. With the Arab-Israeli War of 1967, came the realization of Limits to Growth and human welfare due to FIVE drags. Environment and Ecology began to be considered Global. The world was exhorted to "think globally but act locally" by Club of Rome, The greens, UNEP, and its major countries. All the six world growth models focused on the centrality of environmental constraints for any growth. This ***Environmental Globalism*** began during 1968-80.

4. When growth became unsustainable, USSR had to follow capitalistic mechanisms, permit private players, for which rules needed to be liberalized. Economic growth is so vital for any country that other countries followed USSR -Eastern Europe, SE Asia, China, and India too in 1990. In the process it had to begin with Privatization through Liberalization, two global phenomena. This Capitalistic globalism began during 1984-90.

5. Resources, developed into products, shall have to move across countries, as an imperative of capitalism, for which barriers need removal and

restrictions against injustices, erection. Therefore, a General Agreement on Trade and Tariffs among countries was developed gradually culminating by Dunkel Draft and WTO. This **Economic Globalism** began in 1994.

6. In a vast country, people need to know each other as partner players. This can be possible only by our own expertise, a political will - both provided by Pitroda and Rajiv Gandhi and a social need. India ushered in **Communication Globalization.** This connectivity flooded Knowledge systems with Informations and knowledge through Internet to upgrade the quality of teaching and training of students, who in turn, as the newly empowered, transform the society in whatever sphere they are working. But this did not happen despite Policy Inputs and Technological inputs. The middle men, agents of change in the knowledge systems misused the facility for selfish purposes. Surveys also revealed that only 10% utility could be discerned. This failure of **Information Globalization** continues from 1987.

7. For our products to be competitive, **technologies** need to be bettered, **resources** need to be maintained, and **efficiencies** had to be raised. These are possible only by human resourcefulness - Education, Science, Technology and Hard work, summed up as **knowledge.** When this happens, only, we can get even with other developed countries.. Otherwise, the country will be exploited both from within and without, as in the colonial times. Crime and Profiteering, Division and Conflicts, Hatred, and Violence are already current in India. Good in concept, but bad in practice. A True Globalization indicated by the dominance of Knowledge in Society is unlikely in the foreseeable future. This foreboding began in 2004

8. *These are matters for educators to introspect and think loudly if possible.*

A. Introduction

The idea of one world is not new. It has been a part of Vedic wisdoms- Vasudhaika Kutumbakam. Karl Marx expected both capitalism and Communism to reach the same goal implying their ultimate unity. HG Wells talked about One World. Similar opinions and perceptions were shared by many thinkers. More recently, during 1970s, an ***Anthropologist*** Margaret Mead, a ***Sociologist*** Barbara Ward and a ***Political thinker*** and politician Adlai Stevenson viewed the world as a '***Space Ship Earth'.*** This was followed up by the Greens who asked to "***think Globally and Act Locally"*** then altruists and dissenters, ***now social workers and activists.*** In India, 17 years after embracing the globalization concept, it remains a challenge with Contradictions, Paradoxes, Subversions, Hypocrisies, and counter-productive measures of various provincial governments. The time is to begin actions against negative impacts of globalization and consolidate the positive ones

1. Globalization essentially consists of markets everywhere for everybody- those who cannot sell cannot buy! It is full of consumers

looking for products of human resourcefulness applied on the basic resources - SWAMPEA. These resources are intricately related to each other, called ecological. Globalization intrinsically means rapid improvements at micro level, albeit at the expense of ecosystems, to a sustainable extent.

2. Approaches towards an understanding of this recent development in world's history are different depending on the personal outlooks. My approach is Historical, Holistic, Basics First, and Introspective, i.e., against resource appropriation and ecological degradation in the Indian polity, ecosystem-vise, not provincially, regionally or sub-locally.

B. The Idea Of Globalization

Many factors and events contributed to the emergence of Globalization thought. Many opinions and interpretations are expressed. All are right in their own way. It will be a futile exercise to try to pinpoint who the pioneer was! Enough if we care to understand Tagore's imagery on child birth!

1. Some of the events considered as the beginning of globalization are:
    (a) UNCHE
    (b) UN World Model
    (c) The Greens,
    (d) Club of Rome and World Models
    (e) 1919 (LN) and 1945 (UN)
    (f) Global 2000 Report in 1980
    (g) Spaceship Earth Concept
    (h) GATT Talks-1,2,3
    (i) WTO
    (j) Earth's view from Space
    (k) Perestroika and Glasnost
    (l) Global Domination Policies from Theodore Roosevelt onwards
    (m) When USSR wanted help from USA and Germany
    (n) When Dunkel Draft was signed and
    (o) When it was heralded by Manmohan Rao Dunkel (1990).

Globalization began when populations were counted as consumers -during the GATT talks at Williamsburg, Cancun, and in Uruguay. Thus, India's **population became 500 million consumers.** However, a brief sequential account may be appropriate.

1. By 1967, it was clear that economies cannot grow anymore due to the depletion of Natural resources in the several studies of the Club of Rome.

Several debates generated new thoughts. One of them was that the growth can be sustained to some extent if the human resourcefulness is tapped and improved. Based on this, premise models were prepared on future trends wherever possible. Thus originated the concepts of secondary, tertiary, and renewable resources, sustainable growth, Human Resource Development, Computer Modeling, and Futurology.

2. From now onwards, there were many activities by UNCHE, GATT, and UNEP. Club of Rome came up with several growth models, and culminating in an exercise of global forecasting in 1977, by Carterf the then US President.. By 1980, it was evident that single country models are irrelevant and are to be replaced by regional and sectorial global models. Human resources on one side and their products on the other end have to cross barriers and transcend societies, if the world was not to march irrevocably towards '***division and conflict, hatred and violence***,' as Leontief warned. Thus, the Globalization Paradigm came into existence evolving from regional models, and the second UNCHE at Nairobi was named as UNCED, which now is called WCED. Prof Dunkel prepared a draft on trade norms, for the entire life is commerce after all; GATT transformed itself into WTO. As a sop to the underdeveloped countries, Davos made its appearance.

3. Thus the concept of human resource development, originated after the limits to growth paradigm of Club of Rome, was widely recognized. Therefore, HRD, globalization, ecology, and environment are siblings or cousins.

4. These developments influenced all countries. USSR disintegrated despite Perestroika and Glasnost, China became economically liberal, and Yugoslavia was broken up by the efforts of David Owen, UK, and Germany. The phrase Balkanization is replaced by Bosniation; from 1990, India too liberalized policies; privatization increased gradually; western forces entered on a large scale; India became ready for grabs. Globalization is visible everywhere influencing every aspect of India -both positively and negatively.

C. Some Insights and Outlooks

When events take place, we are often unable to understand them. As they are replaced by newer ones, and experiences accumulate, we understand them better. This knowledge alters the way we think, the outlook. In other words, the hindsight wisdoms build a track to anticipate forthcoming events.

1. When the Europeanization of other lands (colonization) happened, consequent to industrial revolution, women, children, peasantry, and artisans were exploited. A proletariat was created. Social Darwinism divided the society into three sections. These gave birth to Marx and Engels, who left a lineage of Lenin, Mao, Hochi Minh, Castro, and Muzaffar Ahmed, MN Roy, Sundarayya, and Singaravelu in India. Naxalism's growth is hardly surprising!

2. USA is an unavoidable bedfellow despite its perfidious conduct pertaining to oil reserves, nuclear energy, consumption levels, knowledge appropriation, trade controls, breaking of Yugoslavia, USSR, destabilizing the Middle East, and buying up the best human resources, while it continues consumption, uses up energy, and continues to divide the world. While it conserves its resources and environment, it destroys planetary ecology, and coerces other countries to fulfill its agenda, such as global warming, climate change, terrorism, etc. Naturally resistances develop, spread, and adopt newer techniques. So is terrorism, whose origin in India is wrongly associated with Muslims and Pakistan, while it was provoked in Palestine by UK and USA. The question remains -how are we protecting ourselves from western designs? Ecological nationalism is the only way left at this juncture in history.

3. Globalization, socially, is a post-Second World War Americanization and Post-Marshall Plan Europeanization (1950-70), in all its negative impacts. But India did not examine literature on those societies during that time in order to draw relevant lessons in time. Due to this failure of Indian intellectuals, India remained unprepared when the negative aspects of those times are repeating themselves in India. If we study them even this late, some rear guard action may be possible.

4. Instead of holding the feudal and capitalist forces responsible for the havoc created in several parts of the world, the developed countries have been harping on ' ***clash of civilizations***' for the last 10 years. Extant civilizations, existing only geographically, are much mutated and more dis-integrated, Unfortunately, plurality became too much for a brotherhood living. Consequently, identity crises developed everywhere, many of them within the otherwise cohesive societies. Now the issue is clashes within civilizations. ***Huntington is already replaced by Senghaas!***

D. Impacts on India

There is all-round degradation of society and growth of Bads, which may undo whatever progress was made. Most threatening is the metamorphosis of *Homo sapiens* into *Homo hypocriteus* and *Homo rakshasus*. The worst corrupted are the very agents of change, politicians and academicians -newer species out-learning the old horses in enjoying the virtues of easy money.

Some General and Pervading Impacts are as Follows:

1. Political parties are not studying history of market societies, sequels to globalization, consequences, and imperatives for action -except the Marxists and the Swadeshis. Holidays from reform processes are often forced by two -[1] the leftists of all shades opposing them or subverting them (ideological); [2] excessive subsidies by the central government and the state governments rather competitively, for preserving or building vote banks (political blackmail)

2. The central government is treated like a rich widowed mother-in-law by some states, always fleecing on some pretext. Subsidies and freebies, demands, blackmails, fascistic threats by some governments (such as TN and Pondicherry) are shamelessly blatant in undoing the reforms of the center just to build '*vote banks with ATM facilities*' at the expense of real development (political blackmail), destroying the individual in the process.
3. Criminal indeed is the conduct of those entrusted with power and responsibility. Power is exercised, but responsibility is shunned by following mechanisms starting from NIMBY of Toffler (Not In My Back Yard) through ***Nodmytism*** (NOt During MY Tenure), ***Nimyltism*** (Not In MY Life Time), but definitely following ***Nerlicofism*** (NEpotism, Regionalism, LInguism, COmmunalism, Favouritism). In these respects, India is ferociously united.
4. The benefits from globalization are considerable. But the negatives are overwhelming. Narrow identities replaced loyalties enshrined in the constitution. Cultures degrading, morals eroding, work ethic decaying, consumption spreading, crimes accelerating, trading creeping in everything, and the resultant social burdens became a drag on the polity. The worst sufferers are women, children, natives, old people, and farmers. People are marginalized or totally excluded.
5. A lot is heard about NGOs, MNCs NRI, MNC, FDI, SEZs KFC and McDonalds, travails of nuclear families (DINK, DISK, SISK), knowledge society, patent laws, Intellectual Property Rights, GM crops -all just in a generation time! This is the power money, or easy money. But each is an area of twilight or darkness. The distance between the apparent and the real is great. Unless a thorough understanding of each of these in all the dimensions is made, globalization phenomenon will remain nebulous and harmful in the long run. The Nobel Prize winning transaction principle itself stands corrected in this continent of Circe, India.

Impacts on Traditional and Developing Societies

1. Anthropology: First people, natives,aboriginals,tribals, and ethnic groups:

(a) Environmentally significant as promoters of carbon banks, and ecologically vital as conservationists and promoters of balances.

(b) Encroaching tribal lands continues for 60 years, accelerated during the last two decades. *Modus operandi* simply is to deforest and call it degraded, and try to restore it through tree *c*ulture, during which time they are relocated with ephemeral incentives!

(c) Any interference in their affairs led to resistances in the past -over 100 years in British India. Now they join the communist insurgencies, as in Chattisgarh, and continue revolting.

(d) Since globalization, mining the forests, often for selling the minerals, has greatly increased.

(e) Some SEZs are located at the edges of the tribal lands for their eventual sucking.

(f) Where is the nationalism Regional Issues -only sublocalism, a suicidally reductionist version. Internal migrations, identity crises that plague even NRIs, and political blackmails within the country are evidences for this. Ancient wisdoms, Gandhi, Patel, and Nehru bound over five hundred kingdoms to the idea of India, which their enemies are gleefully destroying under the influence of the most powerful intoxicant elixir ever created in human history, the VOTE BANK. This is a consequence to Nerlicofism practiced by the Indian National Congress after the deaths of Gandhi and Patel, and followed by virtually every party.

(g) The implementation of SEZ is criminal, for it being at the expense of established ecosystems. It reminds us the native kings doling out regions to the British Colonialists for mutual profits. Four hundred SEZs for India is a crazy concept; they should have started a couple in each state on an experimental basis, locating them only on neglected lands, such as, East-Coastal regions of TN between Adirampattinam and the south of Sayalkudi. India is truly on sale for the second time in history. Neocolonialism is back; aftermath of industrial revolution is in attendance -unfortunately both in the same country.

2. Sociology: (a) Nuclear Families develop psychological problems - DINK, DISK, DITK, SISK. (b) Excluded people -gerontology. (c) Gerontology of excluded professionals -social and economic backgrounds, discarding ageing wisdoms means discarding historical knowledge and hindsight wisdoms. (d) Educational impacts both ways are well known and we are currently experiencing them.

3. Social Work: Burdened with problems unleashed by globalization, some old, some new, some simple, and some complex, we must include social activism and resistance movements.

(a) Orphans, girl children, exclusions, disabled, aging people vs. aging professionals, affirmative action -recipients -their problems at school, college, universities, vis-à-vis teachers.

(b) Farmers: Premchand, Telangana, Bimal Roy then, Sainath Now. What else is needed to study -actions are only needed.

(c) SEZs: Neither can we emulate China nor can we innovate, these will benefit only the middle men with commissions. A natio of castes, cuts, and commissions; in Pondicherry parlance, the 10% syndrome.

E. Hopes Versus Realities

Considerable hopes for improving the country's economy and their percolation to the lowermost were generated during the early 1990s. Gradually the scenario has been changing. Some examples are as follows:

(1) Globalization envisages a world view of life. But there is neither a religious world view nor a scientific world view, and what to talk of the genuine one, the ecological world view.

(2) Globalization envisages greater integration, but gave birth to narrower identities, same cultures fighting for separation, caste and subcaste societies, treating the neighbors as foreigners, a peaking Nerlicofism, and others similar to these.

(3) Globalization envisages growth with development, but there is only growth of wealth and wealthy people but did not result in the development of societies excluded from the mainstreams-the exclusion principle of ecology was surreptitiously implemented behind a variety of smokescreens.

(4) Globalization envisages a culture of knowledge, its transmission, and more efficient work ethic. On the contrary, teachers disappeared; only examiners remain. Salaries increased! Research falsified while money is flooding. Doctorates multiplied by following myriad tricks. Theses are never rejected or asked for revision? Career Advancement Scheme promotes all; denying to none! This is our academic character during the last 20 years!

(5) Globalization envisages greater cohesion and efficient transactions. But their slots are occupied by entertainment, crime, cricket, cinema, and consumption. Poverty, poor people, migrations, and exclusions increased. But everyone is crying hoarse against Maoists including yesteryear communists. Globalization has accelerated this rebellion against injustice during the last 20 years and threatens to continue.

(6) Globalization envisages to maximize human resourcefulness at micro-, macro- and megalevels so that productivity and distribution can reach most human populations, with minimum exploitation of resources of nature.

F. Knowledge Power

Independent India's first cabinet when confronted to make a priority choice between education and land reforms, decided, quite rightly, in favor of the former. But the beneficiaries, by and large, betrayed India except the first batch of five hundred youth who went for higher studies returned and manned many positions in India with distinction. This patriotic spirit almost vanished now. On the other hand, land reforms, a must for agriculturally sustainable growth and natural justice, were never implemented in spirit except in Bengal where also it was too late and too little. Land lords became congressmen and industrialists, and farmers became coolies, riksha pullers, migrant labour, and communists. Then comes globalization accelerating crimes against their futures, driving many youth to the violent creed of Maoism generated by the decision of the first cabinet of independent India against land reforms!

(1) For many, globalization means liberalizing regulations for private players to demonstrate their entrepreneurship in improving the lot of people through economic growth. Since these need good human power, globalization would mean revolutionizing higher education to the level of the developed countries or to the level of the developed oases in India -IITs, IISc, TIFR, BITS, etc. One may wonder why so much accent on higher education. Because it is the mother of college education, which in turn is the mother of secondary and primary education. The products of these three operate everything in the country and determine everyones' futures.

(2) If the country does not run well, education imparted was bad, and the teacher community is at fault. In India, there are no teachers in universities, only faculty, Heads, Deans, Directors, etc; teaching and training merits are not attached any value either at entry level, or at any of the promotion points; feedback perceptions either are not obtained or are ignored. And students are dutifully controlled and side-tracked, instead. This is the service globalization did to education.

(3) In this regard, China and Russia have lessons for us. Russia brought reforms politically, ignoring economy, and it disintegrated. Since its knowledge systems are strong, it has recovered. China, on the other hand, did not change politically, but upgraded economy and rebuild higher education. You see the results yourself. Education means maximal use of human resources -it is on this foundation that all developed countries became prosperous and powerful.

(4) In India, on the contrary, the polity is bankrupt, economy looking up for a third of India only because growth did not result in development; education deteriorated by the leaderships. The seeds of national disintegration sown long ago have now crossed the seedling stage, and an Ecospastic Kalki is hovering over India.

(5) While the regulations are liberalized, industries are privatized, regulators remain as they were. The new blood to replace them is worse because they are trained by the present knowledge leaders and reflect their self-centered character. Do we then become a global player, no. If at all, we shall only be a twelveth man in a cricket team or an LOB in an infantry platoon! This too

because of our massive population of over 500 million consumers and some great individual achievers!

(6) Thus, globalization will fail to bring prosperity to India unless education undergoes revolution; feudal and fascist political forces are resisted, and justice begins to be fought for, everywhere. Enemy of prosperity is within us, around us, and not somewhere else. Please think over this. China, Japan, and Israel could be our role models in many things, even Malaysia. History tells us never to trust feudal money bags -they have got to perpetuate poverty and turmoil to remain in perpetual comfort!

(7) he purposes of education are three-fold: [a] to impart knowledge from information and enable gaining wisdom in time, [b] skill development for livelihood and self-employment, and [c] imbibition of values to develop the fist two objectives into a culture. All this is to empower them to take over leaderships from their forerunners. Today values that normally come from teachers and universities are nonexistent except in the form of a course named '*Value Education,*' as if values can be taught! We have also not benefited much from computers and internet -probably except in self-propulsion -the misuse quotient being about 90%. Then where are we going?

(8) Look at the current budget. Considerable funds for education; sixteen more central universities. All of us know what will be their fate. Talking of monitoring and evaluation will come to naught, if follow-up action is not taken, i.e., penalizing the defaulters and rewarding the achievers. So much for climate change-we did not cause it. We cannot predict it. We cannot alleviate it. Simply because we also will be sufferers from these changes, do we have to prioritize it?

G. Paradigm Shifts

During the last 60 years, most incentives in India have failed to build a welfare state. Making positive inputs and incentives is like flooding a stream of poison with nutrients, and expecting dilution of the poison and its gradual disappearance. But poisons react with the diluents and convert nutrients into newer versions of poisons. It is a matter of common sense that the poison should be cleaned up first.

(1) It is a paradox that optimistic talkers do not promote hopes and pessimistic critics do not spell out alternatives, and both do not resist dark deeds. If evil is prevented, good carries on, not needing incentives. Our minds must perforce, focus on crimes, their exposure, and retribution. This shift in our cultural paradigm, against dark forces, only leads globalization towards prosperity in India, not incentives.

(2) Secondly, a change in thinking will be better; from '***what happened here when***' to '***what happened there then***' and '***what may happen here.***'

This approach prevents repeating others' mistakes and bad experiences and helps planning better. These are issues of psychology and historical sociology.

H. The Foreboding

A great opportunity to work and build prosperity in India through the power of knowledge is wasted due to the selfishness of people, empowered for this task. India as a multination is disappearing. According to Kushwant Singh, it is already dead. India, as Leontief warned 30 years ago, is irrevocably marching towards "***division and conflict, hatred and violence***" despite the efforts by the Rainbow Warriors of Chief Seattle in the British history in North America

Table 5.1
Economic Growth Models Preceding Globalization

1.	Latin American Model (LAM, Mallmann, Sabato, Oteiza , Herrera , 1972)
2.	World 2 and world- 3 models of Forrester and meadows 1972 (W2 & W3)
3.	World Integrated Model of Mesarovic and Pestel > 1972 (WIM)
4.	Model of International Relations in Agriculture (MOIRA, Linnemann >1972)
5.	UN World Model (Leontief , 1973)
6.	US Government Global Models (Several , Upto 1977)
7.	Global 2000 (Barney Committee, 1977-1980)1982

Gatt-1, Gatt-2, Gatt-3, Dunkel Draft (1994)

Table 5. 2
Globalism is Good. Globalization, A Threat to the Planet

"In the middle of the 20th century, we saw our planet from space for the first time. Historians may eventually find that this vision had a greater impact on thought than the Copernican revolution of the 16th century, which upset the human self-image by revealing that the Earth is not the center of the universe.
From space we see a small and fragile ball dominated not by human activity and edifice but by a pattern of clouds, oceans, greenery, and soils. Humanity's inability to fit its activities into that pattern is changing planetary systems, fundamentally. Many such changes are accompanied by life-threatening hazards.
This new reality, from which there is no escape, must be recognized and managed." -WCED. 1987
"From One Earth to One World," Our Common Future
Therefore, the "Ultimate Resource" (Julian Simon 1981) is the human mind, from 20 years into Our Common Future, 2007

Table 5. 3
Global 2000 Report to the President, 1977-1980 (1982)

	"The most explosive and important documents of the 20th century "
	"The most detailed and authoritative review of the planet problems"
	Predictions
1.	If present trends continue, the world in 2000 will be more crowded, more polluted, less stable ecologically, and more vulnerable to disruption. Serious stresses involving population, resources, and environment are clearly visible ahead. Despite greater material output, the world's people will be poorer in many ways than they are today.
2.	Barring revolutionary advances in technology, life for most people on Earth will be more precarious in 2000 than it is now - unless the nations of the world act decisively to alter current trends.
3.	The struggle to sustain a decent life for human beings on our planet will be enormous - yet there is reason to hope
	Globalization of economies is the imperative for a proper balance among population, resources, and environment

Table 5. 4
Spirit of the Rainbow Warrior

This we know: all things are connected like the blood that unites us. We did not weave the web of life, we are merely a strand in it. Whatever we do to the web, we do to ourselves.
We love this Earth as a newborn loves its mother's heartbeat. If we sell you our land, care for it as we have cared for it. Hold in your mind the memory of the land as it is when you receive it.
Preserve the land and the air and the rivers for your children's children and love it as we have loved it.
When the EARTH is sick and the animals have disappeared, there come a tribe of people from all creeds, colours and cultures who believe in deeds not words and who will restore the EARTH to its former beauty.
This tribe will be called 'WARRIORS of RAINBOW.
Words of CHIEF SEATTLE, Leader of the Suquamish and Duwamish Indians, spoken to the U.S.Government during negotiations for land in the mid-1850s
Therefore, Good people from all walks of life have to come
Together now in India to save its polity, culture, and resources.

Chapter - 6

Development on Ecological Bases

A. Introduction: From Resources and Welfare to Illfare and Violence

Nature embodies soil, water, atmosphere, minerals, plants, and animals, which are mutually connected for their vital activities. Human society used all these and modified them through its intellectual ingenuity for further exploitation. Consequently, what we see around us are all products-selected, modified, and excluded-of natural evolutionary processes, as well as human interference. In other words, humankind drew from the natural resources both prosperity as well as penury simultaneously to different sections of people varying only in magnitude and consequences. This is the sum and substance of our continuing environmental crises generated by development ignoring ecological principles the roots and routes of which are given in a self-explanatory fashion.

B. Ecosystem Approach to Development

An approach to development following ecological principles stands on the premise that local resources are converted into products by means of local people are familiar and improved further through science for continuous utilization. People, resources, efforts, and products are all considered partners that feed each other back, thus continuing the system despite exploitation.

This approach is by no means new or novel. It has been experimented with degrees of success/failures at different places, at different times in different names.

1. Development of a region should be based on the resources available there or nearer to it. It should be limited by their longevity or renewability or recyclability, i.e., the carrying capacity. Anything beyond this will be vulnerable to back lash impacts, the first being a dent in the balance sheet- tottering balances among soil, water, plants and animals. These ecological imbalances affect humans and pave to social and political problems.

2. Ecological development therefore will be niche-dependent. The habitat has to be conserved and restored for development to continue. This should use feedback impacts and necessarily guard against backlashes. Needless to emphasize that water, the elixir of life, constitutes the nucleus of the approach.

3. The above background on the need to maintain resources and ecology for posterity may serve as a foundation bearing in dealing with the current problems of regional planning and local development focused in this conference. This may be further augmented by a few basic concepts generated by ecological thought and perhaps govern the fate of developmental efforts.

C. Ecological Concepts Relevant To Development

Of the many ecological principles, three appear to be more relevant to development in all interconnected microlevels-products reinforcing or diminishing resources, host-parasite relations where producers are dominated by consumers and physical distances, and transactions among the participants of the process of exchange. In other time, space and velocity of interactions between inputs and outputs determine the fate of any plan on natural resources.

1. Linkages, Feedbacks, and Backlashes and the Four Pillars: Humankind not only utilized soil, water, plants, animals, and minerals directly but also converted them to other forms of consumption and use-foods, medicines, transport, agriculture, aquaculture, forestry, tourism, energy technologies, irrigation, animal husbandry, rural and urban systems, industries, etc! As these are all interconnected, the end product(s) sometimes reinforces the resource and more often endangers it. Unfortunately, this basic contradiction between production and consumption was not realized until the swelling under the poverty line. This crisis from a [a] loss of resourcefulness of the ecosystems, [b] loss of these resources for the people due to [c] improper development and [d] improper distribution of products has loomed large over most of India for over 50 years.

While this was realized by all the three forces of social change-politics, bureaucracy, and intelligentia, governments jettisoned their duty of reforming through people's mobilization at grassroots level, in preference to power! As a result, a new metaphor emerged from the disgruntled elements from these very forces-the fourth force-nongovernmental organizations (NGOs).

Today, development and ecological balances depend on these four forces or pillars of this modern citadel - Political forces and mechanisms, governmental bureaucracy, non-governmental agencies, and intelligentia. Anyone wrong among these can scuttle the whole scheme of good tidings. Anyone good can resurrect hope and sustain it. This quartet holds the panacea to all problems and also the curse of passing the buck, ensuring a tragedy to the commons.

It is noteworthy that the consciousness of ecological balances in the face of materialistic pursuits continues to be our scriptural, philosophical, religious, and traditional heritage. Yet most NGOs in India are only on paper!

2. Host - Parasitic Relations: Every urban area can be considered a growth region attracting goods from the neighboring producing places for local consumers and drawing outside consumers to local producers and markets. This producer-consumer distance, direction, and interactions gradually enhances. The neighborhood totally depends on the infrastructure, amenities, and markets of the growing urban centers. Gradually nibbling away the facilities, they become parasitic, stifling its planned growth as excessive and unexpected consumers. This kind of relationship is similar to that of host and parasite in biological systems. So as to ensure that planned growth does not become a victim of such a parasitic neighborhood, it is necessary to include this aspect in the planning process of civic bodies. A balance sheet of interactions across producers, consumers, resources, inputs, and markets therefore becomes a prerequisite.

3. Ecological Distances and Boundaries: The distances and communication system between the host system and different neighborhoods will be various. Therefore, the ecosystem of the growth center is not limited to the municipal limits. These urban systems will also vary from place to place. Urban planning therefore should take into account these distances. Problems become graver if these towns are in the border areas between districts, states, and countries. The unplanned growth of urban centers, chaotic results, and varieties of problems in the border regions are all due to the neglect of ecological distances. These should have been mapped for all such regions, as a basis for civic environmentalism.

A few examples that may fall into this category among the nearly 10,000 towns are: Karimganz, Murshidabad, Malda, Cherapunji, Noida, Gurgaon, Hosur, Jhansi, Digha, Belgaum, Pondicherry, Karaikal, etc.

D. Niches of Ecological Development

1. Gandhian Autarkic Experiments: All Gandhian models of local development enjoyed feedbacks and never suffered backlashes. On the other hand, the modern consumeristic models -both communistic and capitalistic - experienced both and with time more backlashes. This is the well-known story of ecospasms and environmental crises. Many examples can be cited for models where ends strengthened the resourcefulness -Sevagram, Gandhigram, Kunrakudi, Sholavandan, Anandavan, Yelamanchili, Alwar, Chitrakoot, etc. All these followed conservation of local resources, employment of local artisans, and feedbacks. Modern methodologies were used at appropriate places to support the mainstream. These are unreliable, unless leaders are available.

2. Tank Systems-About 200 000 -system and non-system: Most of the 200 000 irrigation tanks are in the states of Maharashtra, Karnataka, Andhra Pradesh, and Tamilnadu. These water storers for irrigation of crop fields, are both system and nonsystem type, useful for up to 30 villages in rare instances. If we treat them as ecosystems, the futurology changes for the better. The system consists of catchments, commands, inlets, outlets, shorelines, beds, embankments, plants, animals, soil, water, wastes, and the human systems. A wetland forestry and through it a bird sanctuary can be maintained, improving the water and soil qualities through its guano emissions. The soil amendments and a few inputs can maintain social forestry, silviculture, sericulture, apiculture, floriculture, refugia for local plants, and aquaculture, in addition to sustaining the existing agriculture. These developments, however, need to be accompanied by diversion of rainfall into the tanks, strengthening the bunds, removal of encroachments, and empowerment of local panchayats gradually. The Tank Rehabilitation Programme financed by the European Union and being implemented in Pondicherry and Tamilnadu is therefore basically flawed as it concentrates on engineering and social work and totally ignores ecology.

This model is replicable with minor local changes. But a census of the systems, diagnosis of their health status, and estimation of their likely value is however a prerequisite. This approach thus not merely revives the age-old system but adds many scientific inputs improving about 1 million villages mostly in peninsular India.

3. Coastal Systems -A length of 3000 - 6000 km: Indian coast is about 6000 km long with ecology, resourcefulness, settlements, activities, needs, and the extent of saline environment varying widely. These areas mostly remained undeveloped in a sustainable manner to their potential. They are merely considered fish and salt producers. Potable water continues to be a perennial plague. A fresh look into this system, therefore, is warranted.

A sustainable system has to be water-centric. Consequently, the coastline needs to be divided into zones based on water availability patterns. The centric place may be on the coast (Karaikal) and even as far away as 20 kms (Pattukottai) All such units need to be conserved from many angles -rainfall harvesting and storage in nonsystem tanks, modifying soil quality following a long-range plan, stabilization, and reclamation of land by growing suitable vegetation, some of which of economic value, restoration of mangroves, salt marshes, estuaries, wetland stretches, etc. The salt pans and aquacultures need to be regularized and mandated to take part in local ecological renaissance. All coastal streams, perforce, need to be checked from entering the sea and stored in land. Extensive vegetating with grass and plants of annual and perennial herbs, shrubs, and trees is required. As stated earlier, all these need to be connected to a keystone habitat or link or arterial township of water sourcing. Such a model with well-defined sources, inputs, and distances can transform the desolate and saline dreariness into symbiotic microecosystems in a mega coastline ecosystem of India.

Water storage in the tanks and especially in the natural wetlands serve as subterranean check dams against the underground salinity intrusion due to their geological properties. This is the only insurance against the spread of salinity to the underground sources of potable water and need to be conserved and restored.

Since this model envisages simultaneous conservation and development, it is replicable in some places and unreplicable in others depending on geography, water, and other resources.

4. Small Town Ecodevelopment -About 10 000 towns: There are nearly 10 000 small towns in India and all are focusing on growth and expansion. It is in the interests of future that the growth happens in an ecologically sound manner. Here, two aspects must be borne in mind:

1. Parasitism of the neighborhood on the growing town, and
2. Dependence of the town for raw materials and markets on different places.

These are the ecological distances or boundaries of inputs and outputs. Development plans therefore have to take the parasitism and distances into cognizance. While this outlook is relevant to all towns, it is more pertinent to those in border areas -where people, resources, and outputs move more significantly and with consequences.

In this approach, both conservation and development have to take place hand in hand because both resourcefulness and consumption interact continuously. This model is unreplicable, unless geography and resources are similar.

E. Ecology in Solving Chronic Problems

Militancy in the North- East: Militancy is the term used for resorting to violence. It is a means to obtain justice perceived to be unavailable through normal, democratic, and peaceful means. Justice is synonymous with enjoying the fruits of natural resources either directly or indirectly through the development process- science, technology, economics, trade, and politics. When justice is perceived to be deprived and hope of ever getting it is lost, violence is resorted to. Once it seizes the human mind, it goes berserk, deviates, and loses direction and control.

Every militant movement in the world has the same reason -deprivation of resources and ecology in their ecosystems- Khurdistan, Palestine, Ceylon, Columbia, and Peru, Punjab, North -Eastern India, Andhra Pradesh, and other movements in the past in Malaya, Burma, Bolivia, Cuba, Nicaragua, etc. Solutions lie in addressing to the question of resources and development to the people first.

In the North East, three Hindu- Indian societies of plains and four Christian- Indochinese societies of mountains are geographically together. Peace lies in increasing the mobility of the former and reducing the nomadism of the latter by promoting enterpreneurship through roads and communications, economic inputs, and increasing scientific and educational capitals. Solutions lie in niche, specificity in policy inputs, contract farming, and corporate agriculture, processing factories, and trading the products with neighborhood, gradually increasing the area of trading.

1. Corporate agriculture: A large scale farming of locally practiced/ suitable varieties can be carried out by corporate bodies in food sectors involving mostly local people. This has a double benefit-local productions and expertise and product processing and expertise. HRD is the bonus for agricultural and consumeristic growth.

2. Contract Farming: A contract between the producers and collectors, whatever it is, wherever it is. The producers are assured of a reliable income, and collectors, a produce for marketing for consumption or processing. Both of them are ensured of a job and people's continuous supply of consumables. This is necessary in all uplands and mountains, where conventional developmental models cannot be implemented. Associating local leaderships will yield best results.

The above two approaches are expected to decrease shifting agriculture, subsidies, laziness, parasitism, and backlashes and increase work, trainees, trainers, products, quality of life, and feedbacks for rising hopes.

3. Small Town Development: The big and small towns in the North-Eastern region should be first experimented on the basis of the proposal made earlier in this script for the obvious reasons of urgent need for development of people by the people and for the people and the strategic location of the region geopolitically.

F. Higher Education in Environment

The Chaos Must End: If we agree that social ills are mostly due to bad education and educators, proper and timely inputs here may change the scenario for good. Higher education on environment was flawed from the very inception. Today, it is the most licentiously continuing activity in the universities. Contrary to the objection by the DOEF, UGC allowed Environment Departments in the Universities at New Delhi and Waltair. The objection was based on the premise that Environment deals with many disciplines -Botany, Zoology, Chemistry, Physics, Geography, Geology, Sociology, etc., necessitating serious homework leading to guidelines. But they became free-lancing extravaganzas in about hundred universities, so far. Not stemming the degradation of resources can be directly traced to this otherwise potential saving grace for a country. What needs to be done?

Our experiences with many universities tell us how not to start and run environment departments- refugia for unqualified and irrelevant people, student training, and irresponsible to social concerns. People refused as lecturers in mainstream departments became Readers; Non-doctorates became faculty and continue politics!

(a) Environmental Education should be compulsory in all the more than three hundred universities. But the existing departments should be abolished and faculty distributed in their parent disciplines. In their places, many more courses should be organized- discourses, training, and practice in semester duration involving all youngsters relevant to each course! This faculty should become in due course leaders in their fields within the university limits.

(b) The courses must deal with concepts, geography, ecosystems, and global and local issues. Further, there have to be field and lab works, societal and library works, besides contacts in the classroom, with good use of computer systems.

Perspectives

The Swadeshi Model: Considering that route maps of nature should always precede road maps of development, the present approach attempts to apply basic ecological principles to local and regional development. This mechanism involves soil, water, flora, fauna, and others for modern and traditional productive practices as demanded by the growth of consumers. It is imperative, therefore, to generate local leaderships and ensure the participation of local artisans and people. Then the outputs become inputs for future- a typical pristine ecosystem-ecological distances, ecological boundaries, natural resource inventory, existing and possible interlinks, feedbacks and backlashes, water status and transaction costs. Such a scheme verging on a dream could perhaps touch the lives of a million villages and up to 200 million people living along coasts, hilly areas, and peninsular arid zones. This could, when worked out in greater detail than presented here, emerge as a blueprint for a truly Swadeshi renaissance of both land and its people of India.

An integration of ecological principles with Gandhian means becomes a Swadeshi approach to overcome Swadeshi problems. This is a tall order if we consider the stakes involved - 6000 km long coast, 200000 tanks, 10 000 towns, and many niches in the hill areas The essential pre-requisite is the basic data for each of the above units, the Ecological Growth Centres Centers: host- parasite relations,

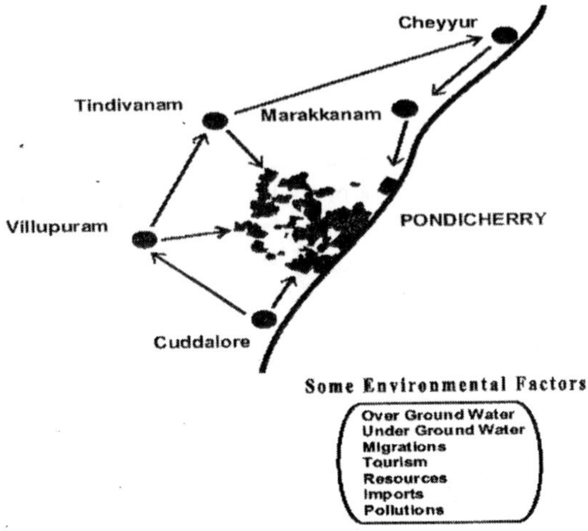

Fig.6.1 : Ecological Boundaries and Political Boundaries of Environmental scenario of Pondicherry depends on Transboundary Influences

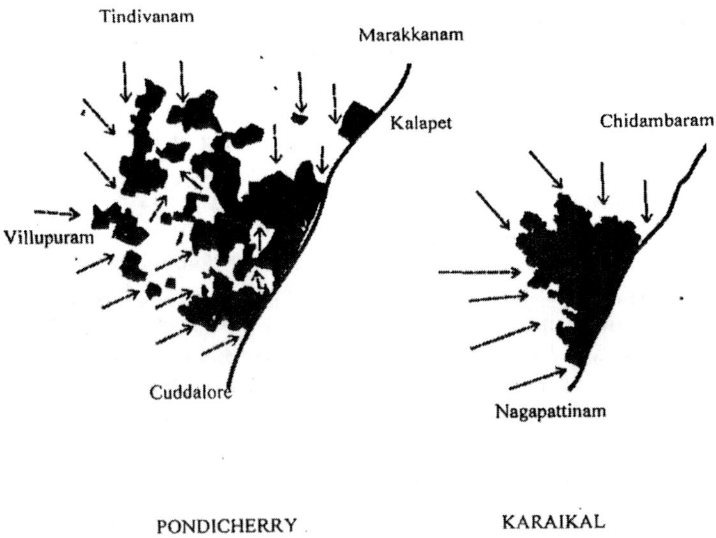

Fig.6.2. Host - Parasitic Relationships between Pondicherry and Karaikal and their surrounding areas of Tamilnadu. (Migrations® Water consumption® Land use® Work force® Infrastructure and Facilities)

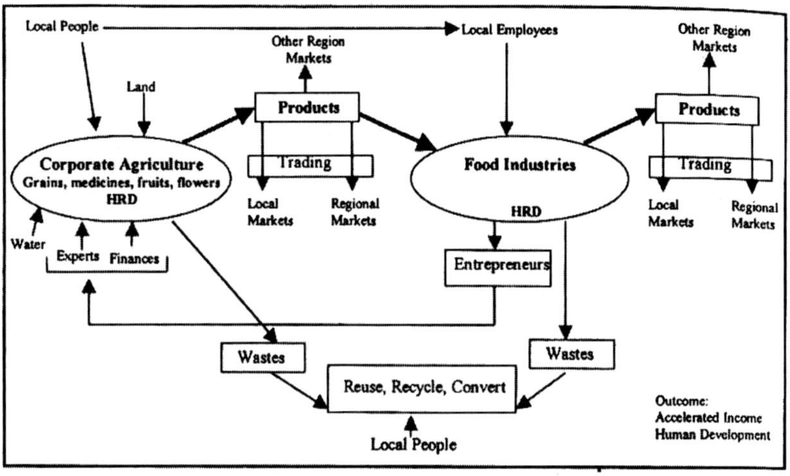

Fig.6.3. Producer - Consumer Model in Major Growth Centers.

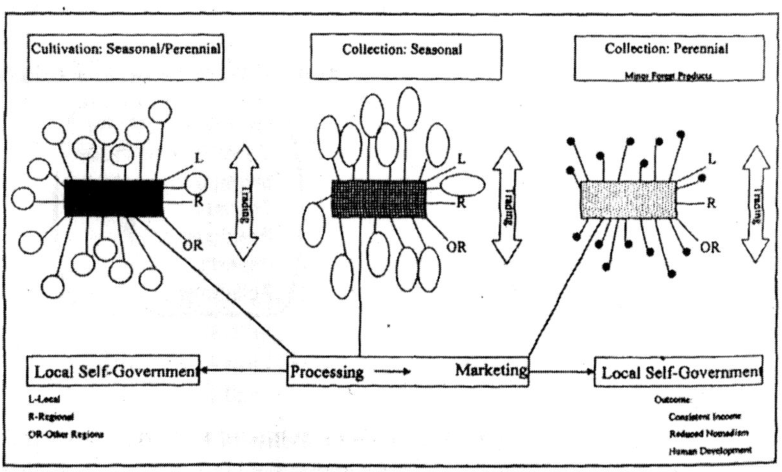

Fig.6.4. Contract Farming in Minor Growth Centers.

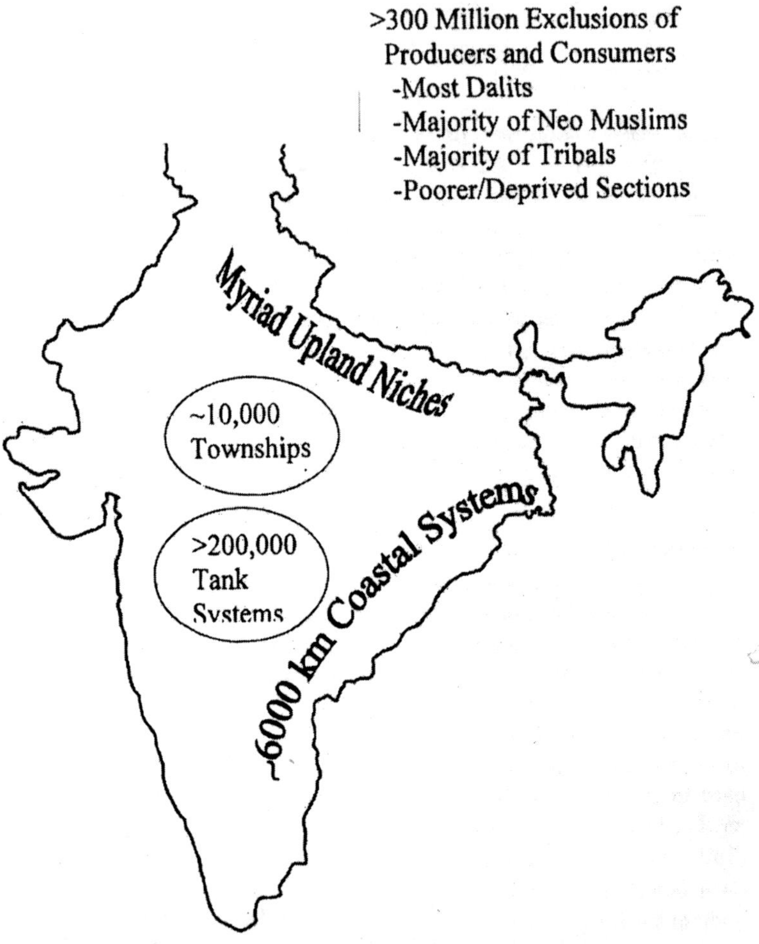

Fig 6. 5. Swadeshi Approach for Ecological Renaissance and Development of India - The Focusses

Chapter - 7

Biodiversity: Perceptions and Proactions

A. Introduction

From times immemorial, plants and animals were inhabiting and evolving on the planet Earth. During evolution, many of these became extinct. When biological death occurs to all the individuals of a species, the species is said to undergo genetic death, extinction. By the time our knowledge systems came to stand on sound scientific bases, only a fraction of the original endowment of plants and animals remained for man to study and exploit. During the last few centuries, colonialism, especially during the last decades, and selfish science have further made them dwindle in numbers. In other words, biodiversity has declined rapidly, and to many experts, this is ominous. Therefore, all-round concerted efforts are warranted to sustain the existing variety, prevent losses, and increase it wherever possible. These efforts have to be in the habitats, at microlevel, for obvious reasons. The promise of biodiversity hinges therefore on two foundations: The proactive steps- such inputs that will ultimately increase or at least preserve biodiversity. The other is preventive steps-such actions that will reduce its depletion.

B. Spectrum of Information

As a background, one has to know the total estimated number of plant and animal species - (a) originally must have been present on Earth, (b) extincted, (c) endangered presently, i.e., at the verge of extinction, (d) economically significant to humankind, and (e) related varieties in the wild for each valuable species. The International Union for Conservation of Nature (IUCN) has strategies for maintaining biodiversity. The International Boards of Plant and Animal) Genetic Resources (IBPGR and IBAGR) have the task of enumerating and preserving germplasm resources of species of economic utility.

From the above sources, conservation needs and conservation per se, variations in the issues on ecosystem and nationality premises can be understood. The story becomes complete with the superposition of international issues like GATT, Dunkel Draft, and Biotechnology, all weighed in favor of the Eurocentric world view.

Let us meanwhile run through the fundamental aspects of biodiversity- what is biodiversity? Why biodiversity? What can we do and should be done in our own local niches and regional units of political power and administrative jurisdiction.

C. Two Functions Of Biota

All organisms have two roles or functions in the biosphere - personal or individual and social or ecological. Individually, each organism lives and ends its life as per the dictates of its genetic endowment, during which it produces and reproduces its kind. This is productivity- qualitative and quantitative. Ecologically, each organism has its specific place, the niche, with specific role in the various chains and webs of an ecosystem. That is how all biota are interconnected in an ecosystem and all processes are cycled through non-biotic environmental factors. These roles are settled during the long march of evolution, through trial and error. This is the niche role in the community.

Thus a plant may have qualitative features of utility (medicines, dyes, resistance, growth responses, etc.) and quantitative uses of biomass and seed yield, both useful to humankind. The same plant may serve as a foliage or fruit food to some animal, repellent to an insect, habitat to a bird, and a fertilizer to the soil from where many animals and plants obtain sustenance. In this contest, one can understand why considerable ecological endangerment takes place whenever even a single tree is felled in a forest. Likewise, an animal may be a pollinator in some plant, seed dispersant in another, a predator on some animal, and prey to yet another.

There two properties- individual productivity and community role of plants and animals, become richer with the greatness of numbers and poorer with reduction in numbers. Both these have consequences.

D. Benefits from Diversity

Individual productivity and community roles of biota have three positive consequences on the ecosystem, which are vital for human welfare.

1. **Stability of the System**: More are the links and interdependences, more intricately bound are the components of the system. Thus greater becomes the strength of the system.

2. **Buffering Ability**: When the system is strong, it can withstand both expected and unexpected pressures. There pressures can also be offset by greater carrying capacity, which is also a result of stability.

3. **Carrying Capacity**: When the system is diverse, the productivity is wider in variety and larger in quantity. Then it can support larger needs of others, i.e., carry a larger burden of dependent human society.

The gross national product (GNP) should be essentially gross natural/ ecological product (GEP). When this is not so, it means there may be exploitation and oppression through long terms of unnatural power- power stolen by brute power of man. All the ecological crises of the present world can be traced to this ultimate source. Therefore, it is the outcry for ecological sanity, sustainability, ecological rights, feminism disarmament, and peace. This is the reason why all anti-establishment activities in the North, ultimately streamed into the ecological movement of 1970s, the Greens.

E. Consequences Of Extinction

In nutshell, biodiversity means having as great a variety of organisms as possible, alive and kicking. Whenever we loose one, any one present, it leads to two consequences.

1. Loss of a gene pool (or loss of a genome), which amounts to loosing a pool of characteristics before science could understand and document them. It means loosing a source of economic utility without humankind ever knowing it. What an irony?

2. Loss of an ecological link, which would mean a loss of an intermediate in the ecosystem without science ever understanding it. Therefore, the chain reactions or the consequent dislocations of chain activities will be unpredictable. The cumulative effect will only be negative. The future humankind that suffers it will never be able to know the causes to manage what has befallen them, Ecological repercussions of species loss is far greater than loss of its economic use.

F. Agricultural Perspectives

Of the few hundred widely recognized economically important plants, about a score are food-giving plants. The modern world is mostly dependent on four to six of these crops- rice, wheat, maize, barley, etc. All these cultivated plants (and domesticated animals) have geographic centers of origin taxonomic relatives and varieties cultivated for centuries and millennia living in the wild. Yet the only selected few and their genetically improved varieties are widely cultivated and as monocultures with an intensity of water, fertilizer, and pesticide combines. This modern agriculture, while boosting yields for consumption, threatened the continuous existence of traditional varieties of crop, which have their own virtues. Consequently, the genetic endowment of crop plants received by humankind through evolution is getting endangered through disuse; thus, the base of modern agriculture is being eroded. The pesticide practices are endangering the predator-prey relationships among the animal species of the system and render crops prone to newer pests and pathogens.

The answer to these seeds of decline in modern agriculture lie in maintaining traditional germplasm and resorting to ecological agriculture, at least in specific pockets. Such agriculture, obviously had to be subsidized philosophically, scientifically, and financially for survival, for every crop plant germplasm "banks" exist. They are the heritage to humankind. The modified varieties of these are not available equally free. Since these could not have arisen without the former, which are freely available, how can the latter be not free? A moral question. This is one case to illustrate that ecology is morality and equity, in this case, on the issue of biodiversity.

G. Target Systems For Conservation

In either case, our targets of action are natural systems- forests, wetlands, wastelands, hills, etc., as well as man-made systems- parks, agroecosystems, gardens, irrigation channels, tanks, etc. A third and more important system pertains to development processes: educational, housing, health, industrial and various engineering works. All these destroy flora and fauna- but at the same time, it has potential powers to reassure and conserve them.

H. Agenda for Action

The agenda for an action, therefore, will cover all target systems. The actions may be proactive that will increase or preserve biodiversity positively or preventive that may reduce depletion of biodiversity. These are as follows:

1. Inventories of plants and animals, their ecological and economic importance, agricultural lands, and their varieties.
2. Repositories of plants, animals, and valuable plant varieties through herbaria, zoos, museums, gardens, parks, germplasm gardens, etc.
3. For the above two, mechanisms are needed:
4. Policy inputs at provincial level encompassing all relevant spheres of governmental control and responsibility.
5. Voluntary altruistic effort that covers as many relevant groups as possible.

I. Mechanics Of Action

1. Organization

(a) Each district may have an environmental society affiliating many Ecoclubs.

(b) All these must have access to panchayats, municipalities, and district-level authorities, such as collector and chief of forest, agriculture, education authorities, and local bodies.

2. Associates

The following interest/activity groups, all with considerable clouts in society, have to be associated with the voluntary activity.

(a) Educational institutions.

(b) Allopathy, Siddha, Ayurveda, Homeo, and Yunani practitioners.

(c) Temples, mosques, and churches.

(d) Industrialist, land owners, and house owners

3. Activities at ground level

(a) All gardens and parks existing now must be improved by expanding them, increasing the varieties of plants, and including local plants for maintenance.

(b) All medical practitioners together must be given land and facilities for raising and maintaining a medical plant garden.

(c) Temples, churches, and mosques shall have to raise sacred groves of wilderness where diverse plants and animals are allowed to develop in security of religion and god-mindedness.

(d) All educational institutions, housing colonies, and industrial units are expected to initiate vegetational programmes around their units

(e) All water bodies have to be protected, and maintained with relevant biota.

(f) The farmers or big land owners shall be asked to maintain traditional varieties of crop plants and they shall be subsidized for the such ecologically sound agricultural practices.

(g) Inventories and repositories of all species with ecological and economic notes shall be maintained in the educational institutions.

(h) All the above have to be coordinated by the District Environ- mental Society- an altruistic and Voluntary voluntary organization in close association with district leadership - political, administrative, and knowledge

J. Legal Aspects

The above works cannot be realized unless some legal changes take place. To implement the agenda, the local bodies and local courts have to be delegated with more powers.

1. Preventive powers against antiactions: deforestation and pollution;
2. Subsidies for ecoagriculture and growing traditional varieties;
3. Land leasing for ecological purposes, i.e., reforestation, medicinal or germplasm gardens, etc.

4. Vegetational prerequisites for licensing housing, industries, engineering works and educational institutions;
5. Intervening with religious institutions for the growth of sacred groves, gardens, etc.;
6. Treating ecological societies as ombudsmen;
7. Bringing jurisdiction to the district level - the public interest litigations, involving ecocrimes.

K. Unified Coactions

Without co-ordinated and idealistic approach, no ecology question can be answered, therefore:

1. Ecoclubs may be established with grassroot-level activities. These may federate into a district society. Or the district society may organize the sub - units. The district units may eventually confederate at provincial level. The right hand must know what the left hand is doing, for mutual help and for societal happiness.
2. The society may interact with leaders of all political parties, chiefs of all departments, religious leadership, professionals, and media.
3. Lobbying, education and awareness campaigns will be the major activities of societies. Actual formation and maintenance of and improvements to biodiversity can be taken up only through governmental inputs.

L. Concluding Remarks

Biodiversity existed and still exists. Here is an irony, when the humankind needs more varieties of productivity at population level and more stability, carrying capacity, and buffering ability at ecosystem level, necessitated by an increasing population and their expectations, the diversity is declining. Meeting this paradox is a challenge that beleaguered all concerned citizens for the last few decades.

Backgrounds, problems, methodology, and solutions concerning various aspects of biodiversity are well established. Yet the issue of biodiversity is beset with not only ecological degradation but also with legal and political undercurrents. What has been plaguing is the inactivity at the microlevel compared to the richness of knowledge systems. Therefore, charities must necessarily begin at home grounds. In India, it is the district where leaderships are most wanted for conservation and development activities. In this connection, the following may be considered.

Role models in conservation activities have to be studied and then only followed - Auroville, Pondicherry, Kishore Bharati, Hoshangabad, Vanrai,

Poona, Bhagavatula Trust, Visakapatnam, etc., so that mistakes are not repeated and mechanisms of proven worth are followed.

The leadership has to be truly natural and idealistic. Leaders have only altruistic mottos and nothing else. A small nucleus can then be built within a wide cytoplasm. When violating viable systems have become a national pastime in India building anything ecologically sane, as it is a common good, it will invite great adversity. Unity among all fraternal units with ecological motivation is deepen important need.

DO NOT GO WHERE THE PATH MAY LEAD;
GO INSTEAD WHERE THERE IS NO PATH AND LEAVE A TRAIL

EMERSON

Chapter - 8

Chemicals Worse than Radiations

The importance of energy is such that the rate of social development is often correlated with energy consumption. Now *that* the word "development" has become *a* guiding spirit in every sphere *of* human activity, more so after the liberalization of economy, the power sector is slated for major leaps. Of these, nuclear energy has been under special favor. But *the* Koodangulam project, proposed to be *set* up *in* southern Tamil Nadu, is under a cloud.

Most types *of* energy involve technologies *to* detect, convert, *and* transport before actual utilization. During these stages, radiation and/*or* chemicals are released affecting the environment. Among these, *the* genetic impacts *on* biota, including human beings, set in motion a chain of events. They produce no warning signals and are irrevocable, some of them compounding with the passage of each generation. These *ex*ceedingly sinister impacts are, unfortunately, underestimated.

Chemicals and radiation catch human beings in the ambient and occupational environments, through food chain, usage, consumption, and finally accidents, and modify the genetic material This results in the initiation of cancers in the same generation and birth defects and genetic diseases in the succeeding generations. It is said in the coming decades that 25 percent of the population will develop cancers and over 10 percent the other two afflictions. The energy sector is also a great contributor. Harmful releases may occur during the production, conversion, or utilization of energy. Data suggest that chemicals actually cause greater and wider harm to human genetic health.

Nuclear energy has been feared from the times of Hiroshima and Nagasaki, though *the* lasting biological damage from radiation *was* known 20 *years earlier,* thanks *to the* pioneering work *of* Mullen. Rather than the trail *of* death *and* destruction that radiations left, what became significant *was* the fate of *the* survivors. Today, *we are faced with a* situation *of* slow *ex*posure *to* low *doses that is* suspected to *be* the result of all nuclear energy establishments, within *and* outside. Consequently, the genetic material of the exposed people is altered. This exposure may be direct, through *occupation,* or indirect, through water and herbivorous and/or carnivorous food chain. The Chernobyl disaster could be cited as an example here.

Data on cancer victims among people associated with any aspect of nuclear energy from mining to pollution are considerable. There is hardly any nuclear establishment in the world without this scourge making a preferential visitation.

Sterility among the exposed and birth defects among the children of such exposed parent(s) are well-known. Diseases of genetic origin appearing at different points of age are also documented. Much of our knowledge on gene-based ailments has been a direct consequence of studying radiation effects on biota from the mid-Fifties. Thus, it is understood that mining wastes, coolant waters, fallout of testing, occupational exposure, ambience because of leaks, and food contamination, all carry a genetic hazard to the exposed populations. And what is even sadder and more sinister is that most of the victims are yet to be born.

Though nuclear energy is a comparatively *re*cent source of energy in India, contributing about 2.5 percent of the national requirement, it is slated for an increase from the present 1,250 MW capacity to 10,000 MW in the next 15 years, The present number of units, 10, is therefore expected to triple. What then would be the extent of genetic hazards?

Now, coal is different. Here, the noxious chemicals mimic radiation in their effects on human populations. Evidence is building up to point out that several processes of conversion and utilization of coal release substances are genotoxic. Thus, during the liquefaction and gasification of coal, pyridines, quinolines, anthracenes, and fluorenes are released contaminating air and water. Therefore, a regular occupational exposure and a slow ecological exposure to these genotoxins has become unavoidable. The burning of coal releases, besides the oxides of nitrogen and sulphur, polycyclic aromatic hydrocarbons, and nitropyrenes, are also highly genotoxic in the experimental systems.

Coal has been the chief source of energy in India, as it is in most parts of the world. The present consumption of 150 million tonnes a year, which is three times more than three decades ago, is likely to double in the next 15 years. Not less than 70 thermal stations are fed with this resource. With about 120 billion tonnes of estimated reserves, coal will continue to be the thurst of our energy efforts for many decades.

Petroleum: The processing of petroleum releases wastes, many of which are similar to those from coal and contaminate air and water. Combustion of oil produces, besides the oxides of nitrogen and sulphur, polycyclic aromatic hydrocarbons and nitropyrenes. With its unavoidable use in all forms of transport, urban systems are a veritable sink of genotoxins. These toxins being ambient in the air, and epidemiological studies cannot be taken up. Most hazardous could be the chemicals released during experiments involving fossil

In India, petroleum and natural gas will continue to be explored, extracted, refined, transported, and used. The former is estimated to be around 2.1 Bbl and the latter 2.0 Btu. Most of the basins earmarked for exploration and extraction are offshore. In all its stages of production, petroleum damages the environment-water and air.

Combustion of biomass releases, as in the case of fossil fuels, nitropyrenes, especially benzapyrene, a highly dangerous genotoxin. Women doing regular kitchen work are said to inhale daily benzapyrene equivalent to 20 packs of cigarettes. Studies conducted in Hawaii have suggested that gene-based diseases could occur in women this way. It has been confirmed in the studies conducted in some Gujarat villages.

Fuel wood will continue to be used by the poorer sections. Whether the new technologies of wood conversion or efficient direct combustion will alter the compulsions of genotoxin release and exposure can only be surmised at this point of time.

The problem with biogas is entirely different. The slurry may pose problems through the food chain when reused in agriculture and fisheries. Unless biogas picks up as a source of energy and the slurry is extensively used, the genotoxin problem may not attend this source or energy.

Thus the workers in nuclear, coal, and petroleum industries and cooks in oriental hearths carry enormous genetic risks leading to cancers during their lifetimes and defects and diseases among their offspring for several generations to come. Further, anilines, anthracenes, fluorenes, pyrenes, pyridines, polycyclic aromatic hydrocarbons, quinolines oxides of nitrogen, and sulphur are also released into the ambient urban air, subjecting the population to spells of episodic exposure and to a continuum of low-level ambient exposures. These same and other chemical components may be chemically and metabolically modified in the contaminated water and food -chain. The attending threats to the human genetic system from this source are rather inestimable. It may be appropriate to point out that a 0.5 percent mutational load 20 years ago reached a high 10.5 percent now, cancers excluding.

At this point of time, generation of data on the habitats, other ecological situations, and human populations under stress seem to be the immediate necessity. A survey of the human populations under stress and an epidemiological assay for genetic ailments are needed. The nonhuman biological systems may have to be subjected to in situ monitoring wherever energy prospecting is a current activity. Chemicals released by different energy technologies can be tested for genotoxicity as is done in the West. Investigations of the food- chain, wherever effluents from energy establishments are reused for aqua cultural and agricultural purposes, are needed. Hardly any work is carried out in India on the genetic hazards from energy resources and technologies. An entire field is thus open for research.

Researches on the monazite belt of Kerala, which is contiguous with Koodangulam, started in the early seventies, and yielded some profound initial results. But in the next two decades, nothing really worthwhile emerged and so the Department of Atomic Energy closed its project.

Towards the fag end of these unsuccessful efforts, however, came the discovery of a micromutation in the hemoglobin of human blood, the significance of which is not known. Yet this is not enough to damn radiation. On the contrary, the absence of this significant effect can be cited in favor of nuclear energy.

Some of these and other in situ situations were subjects of an Indo-German Binational Conference organized by the Department of Atomic Energy in 1984. With more data, the whole issue was conceptualized by this author in 1990 (Environment and Genome, ed: Mendelsohn and Albertini Alan Liss & Co.). It was revealed that chemicals were omnipresent genotoxins while radiations were locale-specific occasional/rare possibilities. Chemicals not only remain stable, concentrate, and biomagnify, but also transform to worse species. And all these not merely mimic radiations but outdo them in inflicting a lasting damage.

Given the far wider variety, spread, quantum, and exposure of human populations to chemicals, which could be as genotoxic as radiation, are we justified in branding nuclear energy the villain? While the hazardous chemicals are inseparable from the technologies, radiation occurs only during safety lapses.

Can one have a technology-intensive, development-oriented liberal economy, on the one hand, and oppose nuclear energy, on the other hand, **when it is the only way to meet new** India's energy needs?

The Hindu 05 August 1997

ONE CANNOT BUT WONDER HOW AN ENVIRONMENT CAN MAKE PEOPLE DESPAIR AND SIT IDLE AND THEN, BY CHANGING THE CONDITIONS, ONE CAN TRANSFORM THE SAME PEOPLE INTO MATCHLESS PERFORMERS.

MUHAMMAD YUNUS

Chapter - 9

Chemical Warfare

Introduction

We owe our existence to chemicals of sorts like water, whether or not we are conscious and aware of it. We have always been intolerant of others since we began to be. It is also a fact that we are on a constant search of newer techniques of mutual extermination, i.e., war, the most recent being the use of toxic chemicals. The issue with all its inherent obnoxious elements has become all the more burning at the wake of revelations of the vietnam war.

Use of chemicals is called chemical warfare, cw, and has been defined as intentional use in war of toxic chemicals to produce casualties either by incapacitation or by death. Incapacitation may be reversible or not. Recovery, if any, may be rapid or slow. Effects may be delayed or immediate. The chemicals may persist over the target areas for long or vanish very soon.

Historical Aspects

It was the germans who first used them during the first world war; chlorine alone was used then, and caused 15,000 causalities and 5,000 fatalities. A total of 125,000 tons of chemicals were used causing 1,300,100 causalities. About 30% of the american causalities were due to cw. During second world war, ethiopians chinese and mongols suffered in the hands of italy, japan, and russia. The most criminal chapter in human history was the story of gas chambers during nazi regime. Recently, egypt intervening in yemen employed a few types of Gases Against The Royalists. The CW Currently Resorted to by U.S.A. Reads as a Ghastly Tale, Almost Destroying 10-15% Surface of South Vietnam Alone, Since 1961. Ironically Enough, Most of our Knowledge about CW is Derived from the above Experiments on the Vietnamese Land and People. So Far, CW was Resorted to by Different Countries, Twenty-three Times Since 1914. Most Gases Used, Except in Vietnam, are Smokescreens, and Tear, Irritant, and Blistering Gases.

Agents

Let us now come to the devils themselves, for most of the doom triggers are chemical. Explosion of tnt is a chemical reaction, so is the combustion of napalm. Most of the available information on the agents of this silent death comes from U.K. and U.S.A., their most premier institutes being stationed at porton down and port detrick. In U.S.A., edgewood arsenal published a four volume report on properties of war gases, in 1956, similar to the 'red book' of the chemical warfare station at porton in u.k. These two are classified and expected to contain all information on chemical warfare. Depending on the nature of effects, the chemical agents have been classified.

Vomiting and Tear gases: These cause irritation of different regions of human body and are lethal only in high concentrations. That is why they are often referred to as riot control or harassing agents. They may irritate eyes (lachrymators), skin (orticants), inner surface of nose, upper breathing tubes (sternutators), or the upper reaches of the alimentary canal (vomiting agents). The effects are immediate and the agents do not persist in the environment. Some such gases, in the coded terminology, are: (1) CS (orthochlorobenzalme-lanonitrile); (2) CN (chloroacetophenone); (3) CNS (chloroaceto-phenone + chloropicrin), and (4) DM (adamsite). The effects were so disgusting and painful that some victims even tried to kill themselves to avoid suffering.

Chocking gases: These irritate the lower reaches of the breathing tubes, and even destroy the membranes of lungs and respiratory passages. The effects are more far-reaching than are apparent. Initially, they breach the body's defenses against invading microbes and block oxygen uptake causing slow death by pneumonia or bronchitis or asphyxia. Some such agents are chlorine, phosgene, trichloromethylchloroformate, and disulphur decafluoride.

Blood gases: Agents like hydrocyanic acid and cyanogen chloride, when inhaled, are absorbed into blood circulation, very fast. This causes the blockage of oxygen circulation and ensues death with dramatic rapidity.

Nettle gases: Chemicals like dichloroformoxime are a sort of skin irritants. Exposure to them has been compared to being thrown naked onto a bed of stinging nettles, and hence is the name. Dichloroformoxime is a skin and eye irritant in low concentrations, but blisters the skin and even enters the blood circulation, at high concentrations. Its combination of highly painful skin attack with the possibility of a lingering death by asphyxiation makes it one of the nastiest agents available.

Vesicants: As against the above four types of agents, some substances persist long after their dissemination causing burning and blistering of eye and skin. Arsenicals like lewisite give warning of their presence by their characteristic odour while mustards have neither smell nor initial pain, and are absorbed through skin as well as respiratory tract-(both sulphur and nitrogen types) Some vesicants like the infamous 'Q' agent (a,w, bis-alkenes) can burn blind, and even affect lungs.

Nerve gases: Some chemicals affect the nervous system followed by the loss of control over the affected part Further, the chemical gets transmitted within the body In other words, body is first incapaciated and they forced to poison itself. Tabun, Sarin, and Soman are some examples. These are called G-agents by Germans who first developed them during World War. Tabun is persistent and Sarin is nonpersistent Soman is the most toxic. Tabun went to large-scale production in April 1942, a date more important than July 1917 when mustard shells were first fired. Americans call nerve gases as V-agents, their most potent being F-gas (phosphonoshiolates). This is fatal even in invisible doses on skin. Inhalation toxicity is five times that of Soman. Originally this was a British discovery, by the German Scientist, Gerhard Schrader, who was responsible for the development of nerve gases in the first place, in Germany All nerve gases are effective through the lungs and skin in very small dosages and prolonged exposure is unavoidable. They act quickly and create dangerous dosages upto 100 km down wind and kill civilians largely. All these gases are organophosphorous compounds discovered accidentally by Schrader of Germany during investigations on insecticides. Later on, DFP and fluorotabun were discovered by Saunders of U.K. All these affect dastardly without giving any warning. They are volatile, rapid and have neither smell nor irritation while penetrating the skin. In due course of time, compounds 10 times as lethal as Sarin were developed-derivatives of Systox similar to acetyl choline.

Incapacitating agents: These put the enemy out of action without harming him. Incapacitation can be of mind or body. Mental ones are the hallucinogens or psychodelics. Body incapacitation could be by denying movement (paralysants), causing fainting (hypotensives), etc. Psychogases appear to be more a weapon of public relations than of war. L.S.D. renders a cat cowering before a rat. Now we can imagine the state of manpower in war?

Because of their nonlethality, the psychedelics are ideal in certain circumstances, especially for major strategic use. They are expected to enrich the mind and enlarge the vision. It would be very interesting if they are applied on two warring sides by a disinterested third party. So far, 690 psychodelics are known, of which 90 are hallucinogens. Rest may be tranquilizers or energizers. Some hallucinogens like TMA (Trimethoxy-amphetamine) produce antisocial responses and precipitate homicidal violence. Some others like indoles cause delayed effects on chromosomes on the offsprings of the exposed. Of course, there cannot be something like a safe chemical. ***Even water can kill.***

It seems that one bomber or rocket with a ten ton load can deliver thirty-three effective doses of L.S.D. (Lysergic acid diethylamide) or one dose of atropine for every present inhabitant of Earth (of course the armaments stockpiled not by big powers can kill every human about 10 times each.) Other delivery systems could include contamination of a major water supply. This is too crude and unpredictable in its effects because of the variable affinity for water of different human beings, especially the decision makers who are the major targets, Air conditioning systems prepared cigarettes are selective and suitable in small groups. A microphone prepared to deliver an aerosol could help to discredit a politician in the eyes of his own community without any of his adversaries being present.

Detection of the psychodelics is difficult, as they are tasteless, odourless, colourless, and active in minute quantities. Sensitive animal indicators, such as the Siamese fighting fish, which responds characteristically to LSD, do exist. Dogs are already trained to smell Cannabis. But a general detector based on physical techniques for the presence of vesicants and psychodelics is more useful.

Recently, Americans standardized an agent called BZ. It has multiple effects, such as slowing of mental and physical functions, dizziness, disorientation, hallucinations, coupled with skinflushing, speeding heartbeat, urinary retention, constipation, etc. Chemical nature of BZ is not known but indications are that it may be a mixture of phenyl glycollates.

Defoliants: These are called herbicides or defoliants, removing the leaves from the plants and thus killing them gradually by denying the food-making machinery. All our knowledge about these substances is derived from their large-scale deployment in Vietnam. They were expected to be selective although they do not comply with our expectations. Examples are 2, 4-Dichlorophenoxy acetic acid (2, 4-D), 2,4,5-Tri-chlo-rophenoxy acetic acid (2,4,5-T), Cacodylic acid, etc. The range and type of effects is very wide. Employing them is a sort of anticrop warfare designed to deny cover and food to enemy and then intimidate them to starve or surrender. In fact, wartime scientists, during 1940s, dreamt of these as future weed killers.

Herbicides may be accumulated by plants that may be eaten by man and grazed by animals like monkeys, deers, birds, and insects. Some herbicides may persist in the soil and affect succeeding generations of crops. Sometimes, they may be transformed in the soil into other substances, which could be more deleterious. Quite often, they affect a susceptible link in an important food chain. For example, microbes are killed on which fish and crabs feed. The injury to the leaf bases that precedes defoliation produces volatile ethylene and its' effects on the neighbouring crops can be catastrophic. So far, about 30 types of plants are known to be destroyed like rice, beans, cabbage, tomato, and rubber. Thus by employing defoliants, we not only destroy the crops for a few generations, but also bring about permanent damage to its wildlife, soils, and streams.

Napalm: Napalm, in fact, is a physical weapon, producing its effect by blast and heat, although the reaction involved in its development is a chemical one. But circumstances in Vietnam have led to its association with chemical weapons. Napalm is a gelled petrol, i.e., gasoline in American terminology. The word is a combination of the first parts of Naphthanate and Palmitate-two fatty acids originally thought to be the active principles of the thickner. But it is a misnomer and the thickner actually contains all the fatty acid of the cocoanut oil, especially lauric acid. In contrast to many modern weapons, it can be produced from cheap common ingredients and can be delivered by simple devices.

The blast is self-limiting and incendiary agents are self-propagating, having a prolonged duration of action. Napalm's main action is by heat, carbon, monoxide poisoning, and kidney failure. Sometimes, white phosphorous, the igniting agent in napalm bombs, may become embedded in the tissues and continue smouldering long after the initial trauma. Thus, panic was found to be more in napalm victims than from other agents, in Korea. During World War II, in Japan, more deaths occurred due to napalm than by the atomic attacks.

Screening smokes: In fact, chemicals used to produce smoke need not be considered as a weapon but for their use in some strategic purposes in war. Chlorosulphonic acid mixtures and titanium tetrachloride are some substances sprayed from air for production of black, white, or grey smokes. Even phosphorus and coal are used. All these are employed generally either for signaling purposes or concealed in friendly forces and positions, like towns and installations.

Munitions and Dispersal

Munitions to disseminate the toxic chemicals are developed so as to create toxic environment, converting the bulk load into even distribution of particles of optimal sizes and velocities within a certain time. Sometimes the agent may be delivered as a cloud of vapour. Volatile nonpersistent agents like phosgene are the simplest to disperse and the heat-sensitive solid agents intended for lethal effect by inhalation are the toughest. Within 5 minutes of the fall of mustard gas, second exposure to it produces week-long temporary blisters, and 2-minute exposure causes death by skin absorption.

Disposal of munitions is ensured by a great variety of means like hand grenades (20-30 yards), rifle grenades (100-200 yards), and artillery and mortar shells (for longer distances). Aircrafts and missiles too are used. Warheads or missiles of .a thousand mile range scattering hundreds of bomblets have been developed, of recent.

Role of Environment

Effects of high explosives follow within a fraction of second of a detonation, while those of a chemical may be delayed for minutes, hours, or even days. In this lies both the strength and weakness of chemical warfare. On one hand, a toxic atmosphere enveloping the whole target area may be set-up or the entire load may be blown off uselessly by a sudden wind. Therefore, knowledge of target conditions is essential. Unexpected wind profile or temperature gradient may render the whole attack useless.

Defence

Defenses can be physical, chemical, or medical, although targets cannot be adequately defended, even by a combination of the above.

Physical barriers against the entry of the toxins like gas mask and impermeable clothing are the oldest known devices. Ointments containing reaction chemicals for smearing over the skin after an attack are also known. In this case, the toxin is chemically decomposed before it enters the body. Chemical purification of the contaminated air is very difficult. Iodine monochloride with polyvinyl pirrolidone is one known to work against mustard-gassed air. Medical defense (prophylaxis) consists of the destruction of the toxin within the body before it exerts its effort. Except the British antilewisite, this is almost impossible to achieve satisfactorily.

Thus, it appears that occupants must either wear all enveloping, gas-impermeable clothing and wear them all the time or they must be warned by sensitive and automatic chemical alarms to so protect themselves. Such alarms, however, are extremely meagre. Dogs have been taught to bark at Sarin, but BZ defeats them. Of course, any man not so protected would be beYond Help Before Medical Help Could Reach Him.

Useful by-products

Researches on chemical warfare yielded some useful information which could be utilized during peace time, if the policy makers choose so. Of course, the uses are negligible compared to the accumulated negative results. It is now possible to detect any foreign matter in the lower atmospheres of the earth. Expert knowledge on the toxicity, methods of analysis, dispersal of effluents, and protective clothing had accrued gradually. So too is the application in biology and medicine of the knowledge of the toxicities of organophosphorus compounds, gold mercury, arsenic, etc. And their cure by atropine and oxime. Some used were detected in the problems of cell division and cancer too.

Perspectives

Currently, Researches are mainly oriented to develop chemicals or mixtures of them so as to incapacitate the targets for long periods followed by complete recovery. Of course, there are other developments too psychic poisons like mescaline, methidine, and lysergic acid derivatives may be improved.

Some complex carbamates, 10-15 times as lethal as Sarin, are developed. Snake venoms, poisons from plants, fish, etc. and bacterial products are finding their way into the warheads.

Some of the current trends are centered around developing a mixture. Toxicity of the main component is increased so that the lethal dose of the mixture is lower than the sum of the lethal doses of the components -a principle followed in developing insecticides. Sometimes one of the components in the mixture hampers the medical treatment of the poisoning produced by another component. For example, Russians found that thiourea slows down the reactivation of acetylcholinesterase inhibited by nerve gases. Canadians developed adjuvants to promote the penetration of the skin by chemical agents. They found that the lethal dose of Soman in dimethyl sulfoxide is six times lower than that of pure Soman.

Vietnam War

Use of chemicals in Vietnam by Americans is unprecedented in the history. Not only the variety, but the scale and pattern of employment too. Chemical weapon designers developed CS weapons for every delivery system in use in Vietnam They sent batches for operational evaluation and then decided which to develop further. The result was that an extraordinary number of weapons have been designed around a single range or weapons have been designed around a single irritant agent to serve the functions of harassment, casuality production, and area denial. By 1969, more than half the procured CS was used for area denial. The powdered agent has been dusted on to the ground in high concentrations, so that any one passing over it stirs it up into an intolerably irritant cloud. However, CS was mostly used as support for other types of weapons. The gas forced the civilians as well as Vietcong to come out to the open-rest is mass murder by artillery and napalm. They sprayed CS, CN, and DM gases even in tunnels and shelters, through trucks, helicopters, and aircraft, and by every conceivable agency. Women and children were killed even by the above tear and irritant gases. The quantity used escalated sharply with time. In one operation, near the Cambodian border, three helicopters, each spread more than a tone of CS onto a small area thought to contain Vietcong! So far, Americans used $ 82 million worth of CS, CS2, and CSI gases.

Maximum damage is caused in Vietnam by the use of defoliants. Ever since, it has started in 1961, about 15 million gallons of antiplant agent has been sprayed across 20,000 sq. km, i.e., 10% of the surface of the country. They accumulate in water, on ground, in plants, and animals. As the time goes by, people become exposed to ever greater doses of these chemicals. What will be the ultimate outcome of such an en masse contamination?

Dr Bridge (a Nixon adviser) admitted that 2, 4, 5-T defoliant is the cause of a high rate of fetus deformity in mice and rats, which could happen to humans too. Dr Rosenhead of U.K. reported a dramatic increase in the rate of abnormal children being born in Vietnam- so much so that the Minister of Health in Saigon classified the statistics as secret!

During the Operation Ranch Head, during 1961-68, 2 million acres of land was sprayed mostly by 2, 4,-D, and 2, 4, 5-T, half of which is rice crop. On one hand, fighting starvation, and on the other, creating it! The affected areas include DMZ (demilitarized zone), borders of Laos, Cambodia, and Canals rivers of Mekong Delta besides the surroundings of Saigon. Herbicides employed costed about $ 40 million and imports worth Rs. 5 million from U K Plants were being affected within a range of about 30 miles from the target.

Defoliants are bringing permanent damage to Vietnamese wildlife, life soils, and streams. Four species of animals are in danger of extinction because of the spray program. Knowledge of herbicides is from studies in other environments- far different from that of South East Asia. So unpredictable are the results that when rubber trees were unexpectedly damaged in Vietnam, USA paid the French Owners Rs. 100 per tree!

Danger of laterization of the soil, which has already started, and poisoning of water resources is enormous, repairs from which is almost impossible at least in the near future.

The extent and pattern of herbicide treatment of Vietnam has no precedent. Mechanism of defoliant action is not clear. Therefore, Americans are setting in motion on a large scale a biological process whose consequences no one can foresee or control. This irreversible chain of events would continue to affect both the agriculture and the wildlife of the area-and therefore, the people-long after the war is over. USA thus is meddling with things that support people in peace or war, independence or subjugation, and whatever their form of government. This is something very revolting to the human conscience.

Public Opinion

Public opinion generally has always been against CW owing to its obnoxious effects-short- and long-term. Of course, it was not so much against Italian use in Ethiopia (1940s) as it was against French use in Morocco (1920s). Reaction to current American use in Vietnam is also very adverse. Although Governments stockpiled them, during Second World War they were not used on a military scale for fear of retaliation on other fronts and cities and adverse effect on neutral opinion besides many other such reasons. But it should never be forgotten that Germans employed gas against the civilian populations of occupied countries and produced millions of victims.

Role of Scientists and Intellectuals

It may be, noted in general that distinguished scientists who can be forthright in denouncing unclear weapons avoid the subject of chemical warfare. Probably they do not want to believe that their fundamental work can be perverted. Scientists enlisted for war were not overburdened with scruples. Germans lend themselves for putting people into gas chambers, or using concentration camp victims for trying out nerve gases.

Napalm was developed by Prof Fieser of Harvard University, a distinguished Organic Chemist. He first developed a formula for a petrol gel using raw rubber as a thickner. When the rubber supplies were cut off due to Japanese invasion of East Indies, he developed the synthetic one. And he never expressed his concern for his part in its development! So too is Dr Gerhard Schrader who developed nerve gases under the auspices of two Governments-Germany and England. Clear that conscience is the casuality.

Let us now survey the current scene. In 1988, twelve famous plant physiologists of America led by Prof Galston presented a memorandum to President Johnson urging the ban on the use of defoliants in Vietnam after visiting the sites of bombing. Pentagon responded to the challenge by releasing a report by Dr Minarik (Director of Plant Science Research at Port Detrick), which minimized the adverse effects. So too did the DOD (Department of Defense). However, AAAS (American Association for Advancement in Science) set up a committee (to investigate the impacts of defoliants on Vietnamese environment) whose report is awaited.

Thus there are two coexisting sides among the elite. It is said that the ethical scientist or physician should have no truck with them. Dr Howard Levy's case reads ominous, besides being interesting. He was court-marshaled and imprisoned for refusing to train combat soldiers in Vietnam, for the training would result in a net increase in disease, disability, and death. He pleaded that it is against his medical ethical responsibilities pronounced by Geneva conventions. The plea was not allowed and he was convicted for disobedience. But what happened in Germany? Doctors were convicted at Nuremberg trials in spite of the plea of having done under instructions from the State. Well, the bench had certainly judges nominated by USA. A case of hypocrisy suiting convenience.

What else are the intellectuals doing? British scientists formed a Society for Social Responsibility in Science. Late Bertrand Russell's International Tribunal of American War Grimes in Vietnam was well known. Activities of Pughwash Conferences are attracting increasing attention of public as well as of governments. However, the most notable are the activities of SIPRI (Stockholm International Peace Research Institute), very imaginatively established by the then Swedish Premier Earlander, in 1964. It was campaigning for disarmament in CW and also conducting verification through budgetary inspection, literature research, photographic reconnaissance, etc. Teams were sent to inspect at production stage and during field testing. So far only eight research and six production establishments of nine European countries were visited.

SIPRI endeavors to enliven the old and forgotten treaties that banned all types of chemical warfare, treaty of Trianon in 1920 banned all types of CW. Geneva Protocol of 1925 prohibited chemical and biological warfare although interestingly enough did not deal with Napalm. And USA was not a signatory to the Protocol and stockpiled warheads in Okinawa (Japan), West Berlin besides on its own territory. But what about Britain, Canada, and. Australia, who signed?

How does the American Government react to the adverse criticism against its flouting of Geneva Protocol. It is one of fooling on a mass scale. One cannot help feeling so. On one hand, people like Dr Macarthur of Pentagon hope to develop a synthetic biological agent against which there is no natural or acquired immunity, within 5-10 years, at a cost of $ 10 million. This sounds quite alarming. On the other hand, public is being attuned to terms like "humane" and ***"antibellicose***" gases-called "***off the rocker"*** and "***on the floor***"- a homely way of describing psychochemicals and paralyzers. And if they would not stay on the floor, there are the proper lethal gases too.

American Government often laments that chemical warfare ought to have better public relations. It means that people should be taught to approve of Napalm, defoliation, 'Police-gas,' and so on. In the words of Ritchie Calder, a noted British intellectual and a Kalinga Prize Winner, "this aims to produce the acquiescence of ignorance, which leads to escalation and the legitimization of the next stage the point where the military can do what they like. It is an easy progression from justifying the defoliation of jungle so as to deprive the guerrillas of cover to treating hundreds of thousands of acres of rice crops to deprive guerrillas and the people of food."

Behind these ominous clouds hone lurks enormous repugnance that the CW arouses in most people, but it is presently limited, though future enormous-potential would seem to make both the possibility and need for some action imperative. Alerting as many as possible to the issues involved might help in the aim.

Chemical Era, April, 1975, 11 [3] : 14 -19

CONSCIENTIOUS RESPONSES

Chapter - 10

Environmental Movements

A. Introduction

1. Earth and its components are heritage resources of humankind. Due to owning up of these by some sections of society, severe negative consequences fell on others. When governments are unable to prevent these avoidable problems, before they become irreversible backlashes, rearguard action was initiated by intellectuals, concerned citizens, and locally affected people. This substitution for the responsibilities of a welfare government, by Gandhians, Marxists, scientists, conservationists, and victims, in fact, constitutes a new dimension to political theory. These actions may be political, social, educational campaigns, public opinion builders, and policy changers, but, all for better environment to more people and for longer duration. In fact, a million mutinies must have occurred in the world against any damage to the local environments, but are unrecorded and unheralded.

2. Globalization reduced voluntary altruist movements to mere non-governmental activities, and reduced education to a trading activity. Even the great USA never played with quality and ethics in education even during its worst crises! If ecological resourcefulness is protected and best education is ensured, all others will take care of themselves. The finest humans improve even bad systems and situations. Bad Humans will destroy anything good!

3. There is mountainous evidence to show that huge projects provide only a very short-lived relief. In the long run, each project disturbs the ecology of the area it serves, so much that, clean air, water, and food cannot be taken for granted. So much so, human progress at what cost has became the central theme of the debate raging through the whole informed world. A farmer of Andhra Pradesh said to Nehru at the time of inauguration of the mighty Nagarjuna Sagar Project: ***"You lighted a lamp in our homes,"*** little knowing that his children were soon to suffer from a disease called "*knock-knee.*" The disease was traced to an increase of fluoride in the water from the project.

4. If nature's endowments, such as habitats, diversity, migrations, resources, resourcefulness, sustainability, carrying capacity, buffering ability, and balances among them, are tampered affecting the people, it is an ecological

injustice. A protest or an uprising, in defense of their natural rights, is called an ecological movement, even if the tampering is done for the development of society As a consequence of development, if the resultant environment is turned detrimental to the people, such as pollutions, toxicities, human, and animal ill-health and climate, it is an injustice for not preventing the backlash. Resisting these is called an environmental movement. In many cases, there will be a mix of concerns and both have implications to local societies and ramifications beyond the geographical and political boundaries-social, economic, and political.

5. History is full of protests against injustices, from the times of Moses and Spartacus dealing with religion, women, peasants, and workers, ranging from social liberation to political freedoms. But the movements for ecology and environment are very recent and encompass all people. They are neither left nor right, unlike the others, but are moving towards the front, confronting the challenges. This is the strength as well as the weakness of this subject! The violences experienced in Punjab, continuing in Assam and in several states in the name of Naxalism, Srilanka problem, and the Palestinian violence, can be traced basically to ecological reasons, such as land, forests, water, fossil oil, and human migrations. In all cases, some natural resources are lost or are under threat. These lead to loss of balances in the niche and harmony in society. And strife begins here. Civic protests lead to political actions with sporadic violence by the lunatic fringes. If not tackled in time, they mature into organized violence and even war. Terrorism is only a type of Guerilla war with roots in ecology.

6. Ecology does not recognize political boundaries - let us learn this from Europe; we are doing the exact opposite. If we cannot be ecologically correct, we will not be politically right ultimately. Such leaders will lead their people to doom. It is endeavored to acquaint the uninitiated of these aspects following a pen-portrait approach.

B. Backgrounds

The improvement of living conditions of humankind, through the exploitation of natural resources, economic growth, and social development, has been happening since the industrial revolution. However, this accelerated progress since the late 1940s is considered very limited and expend soon for FIVE major reasons by the Club of Rome and supported by the Global 2000 Report- overpopulation,- insufficient energy,- depletion of resources,- hygiene and sanitation,- and pollution. Therefore, a brief knowledge of ecology, environment, resources, their systems, development, and consequences is desirable to understand why social movements occur in defense of environment

1. Ecology: Ecology means a continuous symbiosis-mutual dependence of all the constituents of the planet Earth- on water and land - forming a common pool utilizing each other yet replenishing the common pool of resourcefulness. Therefore, all constituents ultimately balance each other - soil, water, plants, animals, minerals, atmosphere, energy, and humans. All these are distributed on Earth in different combinations and permutations. Each of such units is called ecosystems with its diagnostic and distinguishing characteristics.

2. Environment: Environment suggests that human species considers himself as external to the symbiotic milieu of all other components and treats them merely as his resources for his satisfaction. Therefore, SWAMPEAH became the environment of human society

3. Resources: Water, soil and land, plants, animals, microbes, minerals, and atmosphere constitute the resource base of humankind besides themselves. These have a variety of distributional patterns and productivity profiles both in quality and quantity.

4. Ecosystems: The above resources are distributed on the planet in about more than forty ecosystems accommodating different types of the above and developing distinguishing features of themselves in due course of evolution, such as forests, deserts, wetlands, seas, islands, rivers, grasslands, Savannahs in equatorial, tropical, subtropical, temperate, and other climatic and geographic zones and influenced by longitudes and altitudes.

5. Development: Resources present in various systems are exploited for human consumption and development by the use of knowledge, experiences, and experimental knowledge through science and technology. This usage is done by a planning by the political authorities following a political process using political mechanisms taking into consideration the social and economic well being of people. Consequently, the distribution of resources, including the products of development, constitute a basic ingredient of all these processes for equity and justice to be ensured. Industry, trade, commerce, and markets, therefore, are inseparable. During these processes, several consequences have resulted.

6. Consequences: Developmental changes though designed to be good to all concerned have bad consequences in time and space on [a] other components of Earth, Swampeah, as a whole, i.e., ecosystems and their ecology, [b]on other peoples people of other ecosystems, [c] on people of their own ecosystems, and [d] on themselves.

(a) The first consequence is a change in the existing scenarios- depletion of resources, degradation of the systems, and ecological imbalances where the ownerships of resources and/or development change hands sowing the seeds of injustice, inequalities, discontent, and ferment.

(b) The second consequence is economic- reorientation in the internal production and distribution, external trade deficits, internal and external loans and investments, debt burdens, and the economic backlashes on sections of people who have nothing to fall back.

(c) The third consequence is social-inequalities, hope and despair, divisions, conflicts, hatreds, and violence. As the traditional values disappear along with the traditional resources and usage patterns, the world gets triaged into rich nations, poor nations, rich people, and poor people, with buffers everywhere becoming fixed stars, while hopes and despairs alternate periodically as per the whims of people in power. Goodness and other human values continuously vane inexorably making most lives worthless. This is the inherent meaning of Dandekar's poverty line.

(d) The fourth consequences are the backlashes in the environmental and ecological processes, which often defy even a comprehension leave alone alleviations. Once these are set in, it is difficult to reverse, e.g., acid rain, ozone depletion, etc.

7. Responses: (a) Consequently, people react and respond to the challenges to their ecological security. This happened in the past and continues to happen in different parts of the planet Earth. These are outlined in this chapter.(b) The responses were first expressions of concern, advises, and cautions. They gradually grew into protests and resistances- local, regional, and even global in character and magnitude. These resistances assumed legal form, social movements, and finally crossing the normally accepted political mechanisms of change. These are the social movements in defense of human environment or environmental movements for social harmony.

C. Historical Aspects

(1)There have been many agrarian resistance movements and tribal revolts through out the world against perceived or real natural injustice. A few tangible examples, however, may be in order.

(a) There were at least 110 documented revolts by the natives in India between 1800 AD and 1900 AD by the peasantry and tribals against looting of or perceived injustices pertaining to their natural resources. And most of these are ecological in nature. They were all suppressed by the British by what Trotsky and later Mao lamented, as Cretinism.

(b) In 1847, Red Indians in North America who were supported by the Irish Protestants resisted settlers. The devastation of natural habitats was so great that moved conscientious people to nature's defense. The methodology adopted by Europe in capturing the entire North America, victimizing every protester of injustice in their relentless pursuit of plunder, is unparalleled in history

2. There were also protests in USA from 1950 against the point sources of pollution, and disturbances of niches especially by native Indians.

There were many disasters, accidents, pollution incidents, and resource crunches between 1950 and 1960. Therefore, came a Club of Rome, Limits to Growth, Global 2000, Futuristic Models, etc. At the same time, many dissatisfactions in the accepted norms arose. Well-settled groups faced dissents-Marxists, feminists, theologists, anarchists, etc. giving birth to **"Die Grunen."** Ultimately political parties adopted green agenda in Europe.

3. The state of the planet Earth can be summarized as a patchwork quilt of nine worlds brought about by four World Wars, the fifth one ongoing. This scenario will be characterized by a ratio of 2:3:5 between the upper, - middle, and - lower classes among the global populations as the final settlement of all wars in history. The major victim is the Planet Earth, the mother of all resources and the major villain, Consumerism, the mother of all problems. The only solution for all these is ecological sanity characterized by continuous resistance.

4. The above scenario has many authors. A few of these are Huxley, Vogt, Ehrlich, Commoner, Leopold, Boulding, Mead, Club of Rome, Greens, Gandhi, WWF, Auduborn Society, Sierra Club, Geographical International, etc. All these, however, were preceded by Vedic wisdoms, Buddhism and Jainism rooted in India which was taken to heart by Beat poets like Dylan Thomas and Alan Ginsberg, ISKON (Prabhupada Bhakti Vedanta Swami) and other similar followers of Indian thoughts. Modern Indian philosophers without much roots in Western materialism voiced similarly- Radha Kamal Mukherjee, Sisir Kumar Ghose, Krishna Chaitanya, Pyarelal and other Gandhians.

The most significant of all is the Chief Seattle who warned the Europeans usurping their lands by introducing the term RAINBOW WARRIORS for ecological fighters. And this is the name rightly chosen by the Green Peace Movement for their Monitoring and Scouting Flagship!

5. The environmental movements have no left or right. They have only front and back- Thinking thinking globally but acting locally- with heritages in mind...Therefore, the altruists have organized nongovernmental organizations (NGOs) committed to social justice and ecological security and recovery first as adjuncts to governmental activity later as better Agents of Change now as weak political forces. Following the experiences in the developed North, an environment movement has developed in India rooted in its heritage and the democratic system. It is characteristically pluralistic involving social activists, media, courts, and some times masses and academics.

D. Movements: Initial Stirrings

The first wake-up call, a warning, came from a Red Indian Chief Seattle during mid-18th Century to the practitioners of Providential Ecology, owning nature for exploitation, as a God's decree. However, the excesses on nature and the consequent protests, feeble, continued and resulted in the conservation of left-overs by SIERRA CLUB, AUDUBORN SOCIETY, NATIONAL GEOGRAPHIC SOCIETY, etc. More serious efforts came from Europe, in many forms, both emotive and scholastic- PUGWASH in Sweden, DIE GRUNEN in Germany, and CLUB OF ROME in Italy during the late 1950s to late 1960s. There is a long list of protests, mass movements and armed resistances, but the present descriptions restrict to some representative examples such as Sarawak Tribal Movement (Harrison Ngao), Forest Survival in Brazil (Chico Mendis), Movements in China (Mao) and Malaya (Martin Thong), Chico Movement of Philippines, Southern Nigeria's Resistance to Shell (Ken Sarowivo) Greens of Germany (Petra Kelly), Green Peace and Friends of the Earth Movements, etc.

E. Other Countries

1. Sarawak Tribal Movement: Sarawak is an enclave of Malaysia in the island of Kalimantan (Borneo), which mostly belong to Indonesia. The government has been felling trees on a large scale for export to Japan and Europe. The trade profits went to some contractors of Malay origin by executive orders. The mainland people were debarred from access to Sarawak by law. There was a veil of secrecy about the whole operations. The natives of Sarawak, victimized both by a depletion of resources, as well as deprivation of the accrued profits, were outraged by the censorship of the information flow also-adding insult to injury. There was a popular upsurge against timber mining led by an educated young tribal, Harrison Ngao until some concessions were extracted. And the outrage on resources and profits were mitigated considerably and the whole world came to know of this macabre episode climaxing with the alternate Nobel Prize to Ngao. The role of SAM-APPEN, an NGO based at Penang was a model for drilling ecological wisdoms without offending any one!

2. The Rain forest Protection in Brazil: Rain forests in Brazil were under constant stress of cattle ranches, mineral prospecting, rubber cultivation besides timber mining. In some parts rubber plantations were replacing pristine forests forcing the local tribals into cheap labourhood in the plantations. This injustice was resisted by the local tribals under the leadership of Chico Mendes. While the governments and media were alerted and some semblance of justice was forthcoming Chico was killed by the Rubber Mafia. But the movement was not in vain.

3. Movement of Tree Planting in China: During the cultural revolution in China Mao Tse Tung realized the value of trees in revitalizing the ecology and provide for a welcome environment to humans as no marxist leader has ever realized. Consequently, 500 million trees were planted all over China by the young revolutionaries and were mandated to maintain them until developing permanency in the habitat. This is a telling example of the impacts of cadres of a political party if their leadership realizes that ecological rules are more valuable to follow than political exigencies for obtaining economic justices.

4. Resistance in Mexico: As a sequel to NAFTA among Mexico, USA, and Canada, the agricultural pattern was sought to be altered in the South-East Aztec Region of Mexico. The local tribals resisted the implementation of any changes in their traditional mores with an armed rebellion. NAFTA itself was questioned forcing the government to intervene in their favor Resorting to violence at the very first provocation is the special feature of this resistance. It only means considerable pent- up emotions have been in wait for expression.

5. Chico Movement of Philippines: Chico river in Northern Philippines was sought to be damned for electricity generation and irrigation of the plains. The construction was objected because this region was uncared for despite being undeveloped for decades. It has been therefore agreed by the government to develop the region simultaneously on a priority basis. But nothing of this has happened. The workers on the dam site struck work inviting repression from the government. The response was a counter action by the local people. They realised that the irrigation and electricity benefits other people only. The counter violence gradually became a Guerrilla warfare. The Government was not able to quell this armed rebellion in defense of their natural resources from getting hijacked for other people. And Church had to mediate a reconciliation!

6. Southern Nigeria's Resistance Movement: Southern Nigeria is a contrast to the rest of the country. It is richer in resources; better educated and cultured; preponderously christian and politically uninfluential. The local people resented the dominance of the muslim majority and siphoning of their oil resources. A movement for human rights was organized by them under the leadership of Ken Sarowivo which grew in popularity. The movement demanded a share of the revenues from petroleum from their region. The companies as well as the government became restive as the field operations came to a stand still and a little violence erupted which is natural in all confrontations. The leadership was rounded up and the supreme leader Ken Sarowivo hanged very cruelly ignoring the protests from global leaderships, by preceding his public execution by four mock hangings when the rope gave away each time ! The SHELL COMPANY was considered to be behind all this macabre drama.

7. Greens of Germany: By far the most important of the environmental movements in the world is the 'Die Grunen" of Germany. This group was formed in 1970s when many sectorial groups with inherent internal contradictions discovered a common cause in ecological consciousness. Thus leftists, feminists, anarchists, Marxists, liberation theologians, hippies, atheists, nonconformists, and the like joined together with a single motto of Ecological

Sanity, which resembles Marxism and Gandhism together. This evoked such an opposition that they were ostracized clubbing with terrorists and anarchists and debarred from holding public offices! But they enjoyed a popularity among the commoners. Finding this of no avail to change the suicidal policies of governments, they entered civic elections and captured some city corporations. Yet they could only be partially successful. Fearing that they may enter mainstream politics, the establishment reacted by imposing a minimum 5% national vote for any party if its representatives were to sit in the Bundestag. Yet they obtained over 7% vote and the un-fancied Greens entered the parliament. This a watershed in the evolution of ecological movements in the world.

Green movements spread all over Europe. The traditional political parties included ecological agenda in their manifestoes. Consequently The Greens fared badly in the next elections but societies in Europe became sensitized to ecological issues. The recent elections in Germany saw the Greens back into reckoning. They now share power along with other parties. Entire Europe has similar outfits now. They brought Earth consciousness, developed an unopposable and universally acceptable agenda, introduced electoralism and civic environmentalism, green manifestoes. In short, ecology was politicized and politics ecolized. One of the foremost leaders of this movement was Petra Kelly. She was murdered along with her friend and mentor and the criminals were never found!

8. Green Peace Movement: Green Peace Movement which has origins in Europe is more down to earth than others. They are more practical and direct. They have championed many causes in defense of human environment following a variety of methods. Some of the movements they carried out with varied success are :against whale hunting by Japan, nuclear testing by France, copper mining in Brazilian rain forests and generally in favor of disarmament and against radiation hazards anywhere. This is a great role model for other movements elsewhere.

9. Sahabat Alam (Friends Of The Earth) Malaysia: Friends of the Earth Malaysia (Sahabat Alam Malaysia) was one of the earliest wake-up caller and whistle blower in the world and probably the first in the third world, to realise the dangers of loosing resources and ecology to the so called development. They subsequently established a network through out Asia and the Pacific. What has started as the upholder of rights of native people to resources matured into the area of consumer rights through various experiences. The famous Sarawak Tribal Movement against the International Timber Politics was lead by this organization.Very interestingly a muslim of Indian origin Idris and a Chinese Martin Thong founded and lead SAM. Martin Thong happens to be one of the leaders of the famous White Flag Armed Resurrection in the Northern Malaya for political rights for native people. He was one of the first people in the world who realised that Marxism shall care for the nature for caring the people better. SAM also happens to be an inspiration for our very own CSE of late Anil Agarwal.

10. Nobel Laureate From Kenya: A few years ago, Nobel prize for peace was awarded to a Kenyan woman Professor Wangeri Maathai for her contribution for a better environment. Despite endowed with great riches of nature, Kenya gradually became a pathetic remnant of the past needing inputs in many areas. She has just done that. Educating local people on the importance of forests, it has resulted in the considerable reduction in felling accompanied by large-scale tree planting. Environmental health through hygiene and sanitation are the other aspects of her work. Her efforts are multiplying and there is a hope in Kenya that some restoration of ecology will take place helping people find greater peace.

11. Indonesian Fires: The forest fires in Indonesia constitute the other side of the timber politics. While several laws were made in the neighbouring Malaysia to discourage the trade following agitations, no such efforts are deemed necessary in Indonesia. Governments there thought for years that felling and clearing are necessary to create spaces for the migrations from the Javanese Islands. In fact the biggest transmigration of people in history under governmental patronage happened here from Java to Kalimantan. There was no popular dislike what to talk of opposition by ecology conscious groups! But this has changed when greater felling took place illegally resorting to burning and the resultant haze engulfed Malaccas, Andaman Sea, and affected navigations. The international concerns were heeded to some extent, especially from Malaysia. Yet it is very strange that there are no indigenous peoples' movements against this practice unlike the other parts of the world. Perhaps moving space for masses is considered more important than any other consideration.

12. Regions of Unconcern: Unfortunately, there are not even concerns at developments in two regions. There could be some more. The Ethiopian region is one of the major centers of origin of cultivated plants. The countries here were beset with crisis after crisis (Ethiopia, Somalia, Eritrea) endangering the ancient germplasm of many useful plants, even their remnants. Another country is Afghanistan. This has last even the soil Seed Banks, whose implications are not enough appreciated.

13. Other Contributors: World Watch Institute, World Resources Institute, Friends of the Earth, and a score of others in Europe and America have been either conducting information campaigns to help initiate policy changes or environmental movements. Most of the people who were felicitated with Right Livelihood Awards, popularly known as Alternate Nobel Prizes, were activists in defense of human environment-Rosalie Bettrell, Vandana Shiva, and Harrison Ngao et al.

F. Movements in India

(a) In India, it all started when Indira Gandhi presided over the UNCHE at Stockholm on 5 June,1972. After the formation of UNEP in 1974, a Department of Environment in 1976 was initiated and UGC opened these subjects at JNU and AU. But the real awakening of the conscience began only with the formation of Centre for Science and Environment by Anil Agarwal, a friend of Martin Thong, Ahmed Idris and Mohan of Malaysia's SAM, the Friends of the Earth! The long continuing Chipko Movement in the Kumaon Hills articulated by Bhat and Bahuguna, the Bishnoi Ethos, Swami Jhambeswar of yore came to light. This consciousness of ecology was captured by INTACH and CEE. Many more cropped up gradually. And the CSE continues to lead India! (b) Concerns for Ecology and Environment and justice have found wider and deeper expression because these were our heritages. The argumentative Indian has protested too little. With the excavation of Pyarelal's New Horizons and the rediscovery of Gandhi as an Ecological Mahatma, protests became a necessity and a reality, now degrading into a fad and fraud in the garb of NGOs under the influence of globalization. Yet efforts protesting the undiminishing onslaughts on nature and people, continue. They are educational, judicial, civic, and political, substituting for the duties of the elected representatives and political parties (CSE, CEE, Appellate Courts, Exnora, NBA, etc) Even the past incidents demanded a relook. Thus it appears now that among the 110 + peasant resistances against SSZs during colonial times, many are patently ecological just as the current protests against GM crops, Bt cotton, seed monopolies, pesticides, dams and water, cropping patterns, aquaculture, export-import practices, etc. (c) Ecology and environment are considered every one's business but not any one's constituency. Therefore, a brief notes on about thirty movements, for acquaintance and contemplation. However, these movements are too meagre to meet the massive problems India faces.

1. Chipko Movement: Spearheaded by the Dasholi Gram Swarajya Mandal, the movement began as a protest in 1973 against the policy of the government of Uttar Pradesh to auction forests. Chipko activists stand for people's rights in forests and have since gone on to organize women's groups for afforestation. Chipko essentially meaning- HUG the TREES to prevent felling, remains the most famous environmental movement in India launched by Sunderlal Bahuguna and Chandi Prasad Bhat.

As a sequel to the above movement, several wild life sanctuaries have been carved out and poaching in these areas was made a cognizable offense. Indeed, laws that prohibited killing endangered species of animals and birds were passed through mid-sixty's. The Gir lion, the Bengal tiger, the great Indian Bustard, and a few other species benefited a great deal through such laws. For some species, they came too late: for example, the Indian Cheetah and the musk deer. But the questions raised by environmentalists are far more basic and it is in question so raised that the quintessence environmental movement lies.

After several years this experience had inspired another movement called "Appiko" in Karnataka again to hug trees to prevent felling around the Western Ghats.

2. Save The Silent Valley Campaign: This is the first major campaign against a dam in India, which started in the early 1980s. It successfully saved a genetically rich and one of the last remaining rainforests in Kerala from being submerged. The campaign was spearheaded by the Kerala Sastra Sahitya Parishad, and supported by all specialists in India.

3. Save the Taj Campaign: Environmentalists feared the pollution from the Mathura refinery, located 40kms away, could damage the Taj Mahal. The heat they generated forced the authorities to take precautionary measures and monitor the monument from any sign of deterioration from air pollution.

4. Save The Soil Campaign: Known as the Mitti Bachao Abhiyan, the movement was launched in 1977 against the waterlogging and salinity caused by the Tawa dam in Madhya Pradesh. The campaign mobilized local farmers to demand compensation for the lands affected.

5. Thai Vaishet Campaign: The setting of the world's biggest urea plant just 21km from Bombay at Thai Vaishet evoked enormous opposition from city groups, notably the Bombay Environmental Action Group, which feared that the plant will increase Bombay's pollution and over congestion. Their concerted efforts could only delay the project for two years.

6. Bedthi Campaign: This hydroelectric project located in Karanataka was the second in India- Silent Valley - be abandoned after environmental protests. The project would have submerged tracts of forests and prosperous areca nut, cardamom and pepper gardens. Local farmers and eminent scientists from Bangalore campaigned against the project.

7. Stop Bhopalpatnam-inchampal Dams: These two dams on the Indravati river in Maharashtra were stopped due to local protests from tribals led by the jungle Bachao Manav Bachao Andolan -a local coalition of activists and social workers.

8. Doon Mining: Limestone mining in the Doon valley and Mussorie hills has left permanent scars on the famous hill, destroying forests and permanent water sources. The Rural Litigation and Entitlement kendra in Dehra Dun filed a public interest case in the Supreme Court and the Court in a historic judgment ordered the closure of the mines on grounds of environmental destruction.

9. Rayon factory pollution in kerala: The case of "mayur" in kerala illustrates what law can do. The birlas built a rayon factory there taking advantage of abundant availability of bamboo in that area. Soon the local population began to enjoy a better level of prosperity than before. The factory let out the effluents into the nearby river. The water consequently became non potable. An agitation forced it to close up. The township went back to its earlier poverty. A new agitation to get the factory reopened succeeded ensuring steps to prevent air and water pollution.

10. Kaiga campaign: Opposition to the nuclear power plant at kaiga in karnataka started in 1984. In spite of the government's decision to go ahead with the project, local groups comprising farmers, betel nut growers, fisher folk, journalists, and writers wanted the project to fold up. The project continued with considerable improvements to ally the fears of local people.

11. Gandhamardhan bauxite mining: The proposal to mine bauxite in the gandhamardhan reserve forest in orissa even through formally cleared by the government, has been stalled because of the intense agitation of the local tribals who do not want to see their forests destroyed.

12. Narmada Bachao Andolan: This campaign against the massive river valley projects on the Narmada river, one in Madhya Pradesh and another in Gujarat, evinced enormous public interest. In a now famous meeting in Harsud in Madhya Pradesh, several thousands of campaigners across the country, gathered to express their solidarity to the cause.

Sardar Sarovar and Narmada Sagar Projects are designed to take water to Kutch in Gujarat. Four states are involved in the execution of this Project-Madhya Pradesh, Maharashtra, Gujarat, and Rajasthan, Gujarat being the major beneficiery. The mandatory report on the environmental status dealt with eight relevant issues: catchment area treatment, compensatory afforestation, command area development, rehabilitation, flora/fauna, archaeology, seismicity and health aspects.

All these are self explanatory and do not need elaboration. However, the work in respect of these obligatory duties had been so slow that all the four states tell that the completion of these studies and subsequent formulation of plans will take some years. This report pointed out that while the environmental and rehabilitation aspects should be implemented simultaneously with the main construction work, the construction work progressed faster by Four years while all others are far behind schedule.

Of the 1982 families from Maharastra which are to be resettled in Gujarat. So far only sixty-three families are taken care of and for the rest, the land where such resettlement can be affected is yet to be identified. Studies on flora/fauna, identification of the site where the archaeologically significant temples are to be rebuilt, and the question of health hazards, such as appearance of malaria and spread of water borne diseases, are not just given due consideration yet.

The worst was resettlement of the oustees. Therefore Medha Patkar has launched an agitation. Meanwhile the dam height was sought to be raised. Every trick conceived was faced squarely until World Bank revised its stand and the government of Madhya Pradesh softened its stand. It is against this background that pleas of environmentalists like Baba Amte, Sunderlal Bahuguna, and Medha Patkar have to be heeded. And the Andolan continues!

13. Save The Western Ghats March: This padyatra, jointly organized by a number of environmental groups in 1988, covered over 1,300 km across the states of Maharashtra, Goa, Karnataka, Tamil Nadu, and Kerala. The March focused attention on the environmental problems of the Western Ghats - local people, watersheds, wildlife, etc.

14. Save Water—Save Life: Organized by the National Fishermen's Forum, this padyatra spanned the entire coastline of India from West Bengal in the East and Gujarat in the west to Kanyakumari in the south. It drew attention to the problems of water pollution and of over exploitation of fisheries by commercial trawlers - the two major problems faced by the country's traditional fisherfolk.

15. Tehri Dam Campaign: The construction of Tehri dam in seismic Himalayas has been challenged by many environmental groups. Protests by the Tehri Bandh Virodhi Sangharsh Samiti, a local organization, have continued for over 20 years. Setting of industries and power plants has increasingly become a cause of protest because of fears of possible pollution and destruction of the surrounding environment.

16. konkan controversy: konkan railway is a 760 kilometer long broad gauge line constructed on india's west coast from roha in maharastra to mangalore in karnataka. at one time work came to a halt on a 50 km heavily populated section in north central goa after spending 700 crores of rupees. certain interests wanted the track to have a more eastern alignment through less populated areas as the present alignment could be danger to the way of life and livelihood of the people.

However, experts felt that the present alignment will cause far less damage to ecology and environment than one suggested. The reasons for the opposition are more sentimental than scientific.

Dr desai, director of the national institute of oceanography, panaji, observed, "from the environmental and ecological point of view, the present alignment will cause far less damage than the others." the reasons for the opposition are more sentimental than scientific.

India's assessment of environmental implication of the konkan railway through coastal goa reinforces the present state of decay of khatan lands. for years these lands were useless for agriculture. it has estimated that 2500 sq. metre mangroves loss. twenty percent of carambolin lake is lost.

17. Chilka Bachao Andolan: Chilka, Asia's largest brackish water lake lies on Bay of Bengal in Orissa state and is 60 km long and 30 km broad at farthest points with an area of 1,200 sq.km. during monsoon and 900 at the height of summer. Winter brings millions of migratory birds from the far corners of the world. almost 60 km from tip to tip and half as much across at the widest point, it lies sandwiched between the railway line and the Bay of Bengal -132 species according to one estimate -from the far corners of the world, including Siberia.

Therefore, one part of the lake, along with Nalaban Island, has been turned into a sanctuary. The lake's mouth into sea is rich in dolphins. With all this, Chilka has been a major attraction for tourists and nature lovers for a long time. But over the last few decades things turned murky converting it into an area of conflicts and violence. Rival mafia groups have clashed with bombs and fire arms over the fishing grounds where prawn breed in profusion.

It all started by the Tata Project with Orissa Government - a large shrimp culture complex in lake with a Rs.30 crore turnover. In response a former Orissa Revenue Minister Banka Behari Das launched a Chilka Bachao Andolan. The Andolan contended that the project [a] would block local fisherman's access to the lake [b] take away a vast grazing ground of the local cattle [c] pollute the lake with organic nutrients and fermented feed at regular intervals - killing the marine life and drive the fish catch already falling due to siltation and declining salinity, and [d] drive away the migratory birds because of the large number of diesel pumps of high horse power to be used for maintaining the flow of water.

India is a signatory to the Ramsar international treaty on wetland preservation where chilka was identified as one of the world's most important water bodies on account of its unique ecosystem. Therefore, in October 1992 central government has decided not to grant environmental clearance to the Tata Project. And in November 1993, came a judgment from Division Bench of the Orissa High Court in defense of Chilka Lake Ecosystem. Meanwhile, economic pressures made many non-fisherman turn to the lake for survival. The Chilka Bachao Andolan, took up the cause of the local fishermen with renewed vigour involving then local villagers on a large scale. This movement was supported by environmental specialists by and large.

18. East Coast Road: The necessity of a road along the coast connecting Madras with Kanyakumari was fel for long. This exercise involved connecting the existing stretches of road ways and modernizing them all together into one unit of about 700 Km. The road is at last in place for about 150Km and very popular. It is true that many otherwise unemployable locals are busy with some livelihoods along the stretch that has completed, over 25% of the slated. The negatives are also in full force - While the money came from the ADB, locals gave the land and the nature gave away its resources, the beneficiaries are the money bags -the rich, neo-rich, and the consumerists with all comforts, pleasures, as well as crimes that they breed.

There was an agitation by some NGOs against this as it affects the ecology necessary for protection against cyclones and the other coastal calamities. Another opposition was based on potable water dynamics. Unfortunately for these Protectors, water is not a problem in this stretch. The vested interests incited the locals. Even after the intervention of the judiciary, opposition could not stand on itself. The governments came very heavily on them indirectly and directly. The prejudice was so great, that even the alternate road alignment proposed by the ecologically concerned was just ignored as if it were a waste paper ! Because no one disputed the necessity of a coastal communicating system and connectivity with the hinterland for the benefit of all. Consumerism and sectarian profits won over the age-old wisdom of ecology and continuing healthy environment

19. Mullaiperiyar: Movement and Counter Movement: This is a century old issue engendered by the then British Power and the Travancore Princely State. All has been going on well until a few years ago. There was also a similar legacy of the remote past on the use of kaveri waters. In this the upper riparian state Karnataka brought more areas under cultivation, that too, of more water consuming crops correspondingly reducing the water availability to the lower riparian state, TN. The crux of the issue is' the expansion of cultivation by water-intensive crops

In Mullaiperiyar -Parties are different -consequences are different, but the issue is the same. TN brought more areas under water-intensive cultivation as Karnataka did with Kaveri waters, necessitating greater storage of water in the dam's reservoir, which is possible only by raising the height of the dam. Kerala fears, as the lower riparian region, the safety of the structure itself due to its age. This is an issue of fear because if anything untoward happens, the sufferer is Kerala, while the beneficiary has been and will continue to be TN.

There is a farmers' movement for greater water which can be done only by raising the height of the dam. For partisan reasons many political parties are adding fuel to this competitively -even threatening Kerala state of some thing or the other. On the other side, fear psychosis of common people is further fanned. No one seems to bother how to satisfy the needs of the farmers on TN side and how to assuage fears of people down the river.

The movements are not even really political but plain verbal warfare for sectarian benefits on one side and real fear of a disaster on the other, on a truly ecological and environmental issue. The holistic people are silent for fear of repercussions!

20. Sethusamudram Shipping Canal: A grandiose project in sea waters, a parallel of which occurs no where in the world, but dear to the 'Tamil Heart' as the political label, while most do not know what it is all about! Unfortunately, ecology knows no language, sect or culture and development projects are not conceived on parochial grounds but on real value to the people.

(a) The money spent on it could modernize over 100000 non-system and system tanks of TN, Karnataka and Andhra Pradesh and generate enormous developmental activity everywhere, affecting people directly. The money the project is supposed to earn in future, no one knows when, will have to first be used for the pay back and then shared by several parties. This may ultimately serve the jingoistic appetites of egoistic politics of fooling all people for all times. No one dares question this project nor debate it seriously, with the government's opposition alone in attendance for fear of repercussions!

(b) In fact, it is not a badly needed project for people. The biodiversity of the entire region and their ecologies will be affected and how no one knows, for, there were no precedences nor parallels. No one knows how the coastal fisheries and the livelihoods of people there will be affected. Our experience with EIAs and EMPs during all these years showed that they are dishonest.

(c) There is a continuing clamour occasionally against the project by the regional fisher folk. The Intelligence Bureau made a perception survey whose outcome was not announced for obvious reasons! The challenge was met politically invoking the spirit of Tamil glory. First the Die was cast. Next it became a fait accompli. Now the finishing lines and dates are being announced in a fashion reminding us the kings and emperors of yore. The maritime glory of Tamil civilization is only TWO years away from staging a come back by becoming the first step uniting its two sections separated by waters which will not be insurmountable any longer.

The voices of dissent continue expression in the vernacular press occasionally. The electronic medium maintained a safe distance. The more informed, the influential, and the traditional champions of workers and the poor maintain golden silence - all want to be politically right, lest.

21. Tamilnadu Green Movement: TNGM has come to some prominence recently for destroying the Genetically Modified Crops. Whether this move is well thought out or not, it is noteworthy that it found approval by the farmers as well as the general public. I can only comment that there is a continuing hallow of suspicion and confidentiality about most products of genetic engineering.

TNGM has spear headed many movements in the past. Bhavani and Noyyal river pollutions and the problems of Ooty Hills and Indira Gandhi Sanctuary around Top Slip are among them. The members are dedicated to the cause as marxists are ! Their activities are always in soft key and very cultured. Educating and spreading awareness of the issues is also among the movement's major interests. All major actors in this activity are professionals or employed and has other major activities

22. Bhopal Gas Tragedy: The delivery of MIC gas from a storage tank and effects on thousands of people along the wind direction in Bhopal is well known and documented. What I wish to bring to your notice are two facts:

1. HOW difficult to get Justice - redressal of grievances and compensation for the losses - for common people even after getting the backing of all good souls of India - from a combine of money and political power.
2. In and Around every factory there must be some pollutions - quality and quantity varying. occupational, ambient and ecological effects are therefore attendant. Are we attaching any value to this subject of toxicology in Environmental Sciences ? How much of our detriment load is due to these causes rather than from metabolic and infectional sources for which we seek medication ?

This subject was fully studied and documented by the western countries as a cooperative activity and by the UN. How much India use these and apply to tropical situations?

23. Chattisgarh Movement of Sankar Guha Neogi: Sankar Guha Neogi organized tribals of Chattisgarh against exploitation of every kind the most prominent being against the profiteering from forest produce. While he mixed environment with trade unionism and representative politics, his thrust has always been ecological security for the local people and tribals. No wonder he has been killed, unable to bear his successes, by the oppressive elements.

24. Jharkhand Movement: The tribals of South Bihar and the adjoining regions have always been resisting exploitation of resources from their region destroying their forests in the process. Since this cannot happen unless the political power rests with themselves, they have been agitating for a separate state.

25. Auroville Movement: Auroville is a small settlement of concerned individuals of many nationalities bound by the philosophy of Aurabindo and ecological conservation and security. This has been functioning for the last 30 years. Just by example, they have been able to resurrect both concerns as well as efforts for better environment. Restoration of degenerated soil, harvesting solar and wind energies, recycling wastes for better agroproductivity, alternatives to chemical pesticides, social forestry, organic farming, tank regeneration, watershed management, ecologically sound housing are some among their activities. In fact, this Auroville movement can be mother of a myriad peoples' movements in the rest of India.

26. bishnoi tradition: The chipko movement has derived inspiration from the tradition of bishnois in rajasthan to protect the trees and the wild life associated with them for which their women folk laid their lives long ago. even recently, the media reported how some film stars of bombay were hauled up for violating this sanctity under bishnois' initiative. this tradition is by far the most sustained environmental movement in india

27. sacred grove movement: This tradition, prevalent in many parts of india, protects the environment around sacred places like temples as properties if the divine. consequently, there are huge oases of verdant vegetation and wild life around many places of worship in tamil nadu, kerala, karnataka, maharashtra, etc. incidentally, these are known to harbour rare species of plants.

28.karnataka's joint sector projects on degraded forests: The karnataka government's decision to grant nearly 80,000 acres of degraded forest land and revenue land to a joint sector company, for afforestation was opposed by environmentalists. samaj parivartana samudaya, a local voluntary agency has filed a petition in the supreme court contending that people's access to government forest land is crucial to their survival and so the governments decision to undertake afforestation through commercial interests, affects their fundamental right to life.

29. karnataka raita sangha: The farmers of karnataka formed an association under the leadership of dr nanjunda swamy for protection of their interests some years ago. they have been working for the interests of common people vis-a-vis the natural resources with varying degrees of success. some of the causes dearer to them have been kaveri waters, cargil seed monopolies, foreign food industry (kentucky fried chicken), and piracy of indian wisdoms and heritages. these movements have alerted people at the executing as well as receiving ends.

30. Water Movements In Maharashtra, Palamau and Sukhmojori: These are comparatively smaller movements not aimed against any oppression. These are concerned about sharing the most sacred of resources, water, among all the needy. The Paani Panchayats in some parts of Maharashtra were successful. The experiment at Sukhmojori on the regenEration Of A Whole Village Based On Water Use Has Been Repeated At Palamau And Is Considered A Role Model, Shortcomings Not Withstanding.

In recent years, there have also been protests against nuclear power plants at narora, kakrapar, kaiga, koodankulam, and nagarjunasagar and against their possible setting in kerala.

31. center for science and environment: This center has been doing immense service to the cause of environment for the last two decades under the leadership of anil agarwal. though they have not directly organized any environmental resistance movements, they have championed their causes, provided information supports background materials, advices for policy changes, lobbying for paradigm shifts and attitudinal changes from presidents and prime ministers to commoners with exemplary positive efforts. without cse and its periodical down to earth reports many movements in india would not have been known to the concerned citizenry.

G. Singur and Marxism

1. If the peasantry evicted are compensated to their satisfaction much before the project is initiated, justice indeed is done for the first time in India to oustees ! Compare this with the free-bees in Tamilnadu and Pondicherry using the largesse from the Central Government - tax collections from whole India

2. Development does exhaust some natural resources, whether under Capitalism or Communism. The niche owners are compensated. The oustees are taken care of first and project begins only later. This was not done for any niche-consuming developmental effort in India so far. This is ecological justice for which Medha Patkar has been crusading for long. This can also be interpreted as the justice consciousness of Marxism

3. Therefore, if Marxism and ecological consciousness combine, there could be a future for not only these loosing roots but also to the bottom rungs. The precedence for this was none other than Mao, followed by Guevara.

4. Trotsky's Cretinism, which troubled Mao, but helped the British to suppress all mass revolts in India, now evolved from insularity to insecurity which is ominous for the continuity of our polity. Therefore, championing ecological issues and natural justice, Marxism becomes a hope at least by providing a human face to the Market March which has already begun turning faces into a mixture of cruel and cunning animals.

5. Marxists have to rethink in their own interests that revising orthodox communism and abandoning political expediencies in favor of ecological wisdoms is beneficial for people, ecology, environment and long term security of the country. Marxism can redeem itself as the God that has not failed, if only goes ecological, which they are in many respects.

H. Hot Spots and Hot Topics

1. All the 600 districts of India need attention. 200,000 tanks need resurrection. Coasts, estuaries, SEZs, industrial estates, MSW, and MLW need re-looks and assessments. Data from all governments need verification. Similarly, University Departments of Environment, Government Departments of Environment, and Pollution Control Boards warrant continuous vigilance.

These are the springs of pseudoscience, corruptions, frauds and crimes. Civic environmentalism for about 10 000 towns and cities. Empowerment, income generation and promotion of livelihoods using BCT and MSSRF are needed in all ecological niches. Judicial environmentalism is cleverly defied by indulging in a contempt of court by not introducing environment as a subject in all educational institutions.

2. We need Rain Bow Warriors - Amte, Bahuguna, Medha, Hazare, etc. on a large scale. NGOs became - GONGOS, frauds, feeble, tired, and/or livelihood earners. No one is joining a movement, even a mere private meeting, a small distance March, A public meeting or a demonstration WITHOUT RECEIVING PAYMENTS as in the case of political activities including voting. The state of India is extremely bad -we came to a stage, within 16 years of globalized marketing, when every one is maneuvering for a profit from every one else - mothers from children, wives from husbands, and children. Media Exposures and whistle blowing are urgently needed.

I. Activists and Activism

1. Local challenges must necessarily promote local response, normally. and these responses have to be articulated for achieving a betterment. always these don't happen. here comes the need for role models. fortunately, we have any number of them with a history of struggles and achievements - 40 such sources are given above. all these are different from one another in many respects but same as far as consciousness of ecological security of humankind is concerned. considerable inferences or lessons can be drawn from them for future use depending upon the exigencies and scenarios,

2. Inferences Although No Clear Strategic Development Is Discernable, There Seem To Be About Seven Or Eight Types Of Organizations. Many Of These Vos Do Not Co-ordinate Their Activities with one another or sometimes with the governmental agencies.

(a) Individual volunteer: Community and National or Global citizen
(b) Government Dependent: Supported fully or Partly, GONGOs
(c) Trust Dependent / supported: Own funds (activities) Aid others
(d) Autonomous Bodies: National / international, own funds and Agenda
(e) Scale and Scope: Small/big; Regional/National, International
(f) Activity: Educational, Social, Political, Relief, etc.
(g) NGOs moving towards an attack on fundamental causes, from being an assistance agency and from club outlook, profession, or hobby to radical humanism.

Very few NGOs provide immediate relief as humanitarian assistance. Others concentrate on developing the capacities of the people through self-reliant local action and assists government in implementing their policies more successfully. Some try to obtain changes in some policies and institutions at the local, national, and International level. Social and judicial channels are used to obtain ecological and environmental justices. In some nations they became political. When suppressed brutally, they become violent leading to guerilla warfare

3. Movement can be killed by bureaucratizing it through large funds And most of the mushrooming outfits are after these. Private prosperity on public property is often achieved that resulted in most NGOs becoming frauds in India. Therefore, are warranted Drainage Reports of Catherine Mayo & Beverly Nichols- as well as the New Horizons of Pyarelal.

4. While the Knowledge Bases must be continuously strengthened, fair knowledge of history of the country and region operating is necessary, which is not found in most of the cases.

J. India Today

1. Resources have been depleting very fast. The existing ones are inadequate. Common properties exist in tragedy. Alternatives are not developed. The products of development do not spread equitably. Yet their waste products and improper exploitation of resources left the ecosystems, degraded. Yet consumption is high necessitating imports!

2. The people deprived of subsistences, will soon constitute a half of India. These developments, have triggered internal migrations, social tensions and provoked fears of insecurity. Consequently, violences have erupted in our polity in the name of subnationalism or plain lawless selfishness of vested interests. ***All Violences Are Thus Ecological In Origins.***

3. Globalization became neocolonialism, liberalization, a license to exploit, privatization of property and poverty and prosperity from public properties. Our Human Resources ? Best migrate, the remainder labour here - facing quotaisms, insularity, insecurity, alienation, and selfishness.

4. Our cultural unity, which survived scores of political disunities, has at last broken down for the first time in Indian history. Can we then remain politically united? And How?

5. Immigrants into India behave like natives. Strong Indians behave lime future non-Residents. And many more Indians behave like missed migrants, therefore rootless : 100 million, > 10 %

6. Every where we find the Gang of Four, Dushta Chatushtayam, practicing dance macabre - nimbyism, nodmytism, nimyltism, and nerlicofisms, the three forms of escapism (not in my-backyardism, tenurism, lifetimism) and the only form of activity most dominant in india (favoritism based on regionalism, linguism and communalism). A***Nd Above All - Tourism Of Sorts!***

7. Government is responsible for development and progress. It has all the -knowledge, - decision making, - finances, - bureaucracy, party functionaries - external inputs, and expertise. Yet voluntarism is advocated, encouraged, and depended upon? ***Has The Polity Degenerated?***

8. How long can a voluntary Action sustain itself? A federation of NGOs is essential for enhancing the power of lobbying for and achieving the objectives. If ecolizing the political forces and mechanisms does not happen, environmental activists themselves have to become political forces. Otherwise ecological anarchy breeds social and political anarchies.

9. In the meantime judicial activism has come as a welcome innovation. Public Interest Litigation and Green Benches ensure to prevent assaults on environment especially polluting industries and provide necessary Oxygen to environmental movements to sustain themselves.

10. Not surprisingly EE are not election issues because they affect all and not the Vote Banks alone. No political party or candidate includes them in their manifestoes because ecology means some regression in development and advance preparations against its consequences through vision and knowledge - not withstanding the ritualistic EIAs and EMPs both of which are mostly jokes.

11. No wonder that no one really has time for these when the educators and the educated sell anything for personal profits in these globalized market times.

K. Imperatives

Markets and Consumerism deteriorate human environment. The challenges of depletions, degradations, and pollutions can be met only by the resourcefulness of knowledge, efficiency and ethics. Solutions consequently lie in the alerts and whistles, knowledge and education at all levels and Resistance Movements and NGOs. Promoting all these if not contributing is the pivotal duty of all concerned.

(a) Researches: 1. Peoples' movements is a subject for an elaborate and in-depth study - from not only ecological and enviromental angles, but also from social, economic and political perspectives. We then realise that India is racing down towards a visionless abyss ! (This is what is revealed both by Chipko as well as Narmada experiences)

2. In some respects, India is passing through the same experiences that USA underwent during the Third World War - 1946-68! Solutions for us therefore lie in the US history of recovering from those mad decades.

(b) Rainbow Warriors: 1. Dwindling natural resources, irreversible ecological damages, and environmental detriments to people on one side and consumerism without values on the other, do not square up; the apparent does not tally with the real. Ecological and environmental Movements should address these contradictions in addition to their microlevel objectives -and with empathy - Can this happen? Yes. It happened! We need RAINBOW WARRIORS - Bishnois, Bahuguna and Bhat, Amte, Hazare, Agarwal, Patkar, at least ONE for each of our 600 districts.

2. The Hot spots are waiting for attention: Industrial estates and SEZs, coasts and estuaries, MSW and MLW, system and nonsystem tanks, temples and their lands and tanks, urban jungles and 10000 Towns, PCBs of governments, Environment Departments of Universities and Governments - all warrant investigations, exposure, monitoring, and corrective actions in defense of people in general and their futures in particular.

(c) Fifth Estate: 1. The *Tamashas* in the name of education, ecology and environment need exposures, Wake -up calls, whistle blowers like the Tahelka, 24×7, TV-9, Aaj tak, etc. in a truly Rainbow Warriors fashion. Unless HRD grows ethically, degradations and poverty cannot be faced and alleviated.

2. How long Voluntarism? How far? A Need for Private Enterprise in Altruism - Altruists Private Ltd!

3. The first estate has petered into god-men, The second estate is failing dominated by crime and criminals. The third estate is more farcical, being the mother of the second degenerating with quotas and frauds. The fourth estate is loosing credibility as barons develop sectorial agendas and politicians owning media. The fifth estate has to emerge the Activists for Natural Justice and the honest among the Fourth Estate should reincarnate the first one, the GOD.

(d) Education: Education should cater to the requirements of globalization and meet its challenges. OR It should be like the European or American Pattern whose products are accepted all over the world. We cannot have a few cosmetic changes here and there - a way of self deception. The junk that is recruited into the Universities as faculty from early 1970s, (by compromising Mehrotra Committee Recommendations) and continues even now (by corrupting every reformist input), in worse form. It is bad people here that destroy India as Nehru always feared!

Many of the movements suffer from a basic weakness - lack of reliable data and knowledge bases due to apathetic governments and improper Education in Ecology and Environment, the ultimate generator of both data and knowledge. It is different things for different people when it is inescapably holistic. Consequently, everything concerning HE in E&E is beset with corruptions, frauds, sensationalism, and even blackmails/crimes.

L. Perspectives

The question of what to do emerges even after a casual perusal of the realities. See then what are the alternatives before us.

1. The first is to do nothing and wait for the mythical or the proverbial Kalki in the form of Ecospasms and triages - and both are happening. The second is to invite National Socialisms-Fascism or Religious Traditionalism or Subjugations. The third is value based voluntarism involving the fifth estate as well as the private enterprises.

2. The dilemmas facing India are really pathetic. No Government can adopt a "***go to hell***" attitude. It has to press the resources to the utmost. India uses only a tenth of the rainfall it receives annually and even with all the reservoir - building plans it will be able to use only upto a quarter of that resource in future . They save water and generate power. All this means generation of more employment opportunities. But it also means huge displacement, unaffordable financing submerging vast tracts of land etc. How does one balance these apparently conflicting issues. And this is the case with all the Natural Resources.

3. The polity is disintegrating despite all the holy dreams and pious visions. One step forward by a few is accompanied by two steps backward by many. The seeds of failure of reforms, LPG, are now deeply embedded in the body politic. The culprit is Higher Education, with its top-down mechanisms, which should normally be a national repository of hopes as human resources to face challenges, now pitifully hovers between the irresponsible and the criminal. This is because of a perennial Tsunami of unqualified backdoor entries into the university systems at all levels, which unfortunately has- multiplier, cascading and cumulative impacts on the entire society for decades- a Vaitarini of the epics !

4. The future of india will not be influenced by nris, forex, fdi, mittals, tatas, birlas, ambanis, infosys, wipro, sensex, and the like, but will be determined by the **thoughts, words and actions of the products of universities, the ecological victims and the backlashes** of development. ecological and environmental movements are expected to complement and not compensate for the failures of the lawfully accountable and shall endeavor to become the fifth estate along with the media which is the need of our times!

EARTH PROVIDES ENOUGH
TO SATISFY EVERY MAN'S NEEDS,
BUT NOT EVERY MAN'S GREED

MAHATMA GANDHI

Chapter - 11

NGOs: Altruistic Organizations

Voluntary Agencies are the only hope for ecological security and political stability

Introduction

India and its 900 million people are in a quagmire. They cannot be pulled out in one piece in a united action, for; there appear neither unity nor action. A thousand local actions alone can salvage the whole. The quagmire is the loss of our secure nationhood and credibility as a polity for a variety of reasons, mostly ecological. The rescue can be expected only from the forces of altruism. These forces should organize into voluntary activists to expose and prevent BADS and initiate and promote ''GOODs'. Fortunately we have several role models globally and one or two in India. The present levels of voluntary actions in India are highly inadequate. At least a doubling of efforts both in preparation, spread and quality of effort are needed to start with in view of the current scenario in our country. The anticipated consequences and imperatives for responsive action ask for more gradually.

The situation is challenging. If the altruists - Gandhians, Marxists, ecologists, and patriots of other hues donot take up the cause of ecological securities of our people, it may be impossible to preserve the Indian polity save by dictatorships - Fascist, Military, and such other strangleholds.

The Scenario: India Today

Resources have been depleting very fast. Water, soil, land, plants, animals, air, energy, and forests are all inadequate even to meet basic needs. Common properties exist in tragedy or owned up. The products of development do not spread equitably. Yet their byproducts and improper exploitation of resources left the common heritages, the ecosystems, degraded. As a natural corollary of all these, more and more people are deprived of subsistenses, who go down the poverty line (a dollar/day income). They will soon constitute a half of India. These developments, have triggered internal migrations, social tensions and provoked fears of insecurity and ill-being. Consequently, violences have erupted in various parts of our polity in the name sub-nationalism or revolutions or plain lawless selfishness of vested interests. All violences are thus ecological in origins. Our cultural unity, which survived scores of political disunities, has at last broken down for the first time in Indian history. Can we then remain politically united? And How?

Society and Politics in India

The politics, supposed to bring and keep people together, is in the stranglehold of Neclicofism (N4epotism, castesm, linguism, communalism and favoritism) fiddling Mandals, Mandirs, Masjids, and Moneys. The society and its leaderships are wallowing in consumerism with no iota of concern to fellow beings. Homo sapiens has truly become H. cannibalis. It is no wonder that the global forces - an euphemism to the most powerful and dominating societies of the North - are pushing India from a Third World to a 7th world country, true to the triage philosophy. Yet corruption and bribery have become omnipresent and omnipotent in India. More one is corrupt, more one needs power. - It is total corruption yearning for absolute power ! Unless one has money, or power one cannot live with reasonable human rights. This is our country now.

Decadence and Renaissance

Such declines befell several societies in history but they returned to Sanity sooner or later due to the leaderships of knowledge systems. These responsible people rarely buckled under the pressures of antivalues from political, economic and social forces and became rallying points for recovery. But in India these became forces of decadence themselves. Youthful leaderships of disvalues are released from our University campuses in thousands every year to convert all agencies of change into levers of power and profit. Our higher education and its leaderships not merely failed India as Nehru lamented, but accelerated the society's destruction, as never happened anywhere in the world. History of sorts! Our Professors, Readers, Lecturers, Deans, Directors, Scientists, Vice Chancellors, and others have been churning out only weaklings and traitors, so as to partake in the direct profits or fringe benefits from such patriotic functioning. Some of the conscientious people opt for serving in heaven (Europe and America) or rule in hell (as IPS, IAS, etc). All practice Nimbyism (not in my backyard), Nimyltism (not in my life time) and Nodmytism (not during my tenure) three species of one genus - escapism from responsibility towards the Indian masses.

The Grand Finale: Rao-Singh Package of economic reforms has come in the context of the inevitable marginalization of India through Triage. It is here for every one to see GATT and its Dunkel. For avoiding this as if, an attempt to totally liberalize economy letting all motivated and competent people do their best without hindrances has been initiated. It is a do-or-die situation, a last chance to revive responsible activity, fairplay of forces, hardwork etc. We responded to this by more bribery, more exploitation and a SCAM. Regulations have gone, but regulators remained. Industrial expansion, consumerism, neo-richness everywhere. Where is Ethics? Who protects People's subsitances, their ecology and environments?

Consequences: A natural corollary to all inequalities which will increase due to our improper responses to Singh-Rao model (Last resort towards a Patriotic Society) and GATT (a Global Inevitable) will be conflicts, violence and even Bosniation - because injustice will not be accepted by all people and for all times. The expansionism of development syndrome will ruin our resources, ecology and environment still further as everyone will be after the unsatiable goddess of consumerism.

Imperatives: Should the Homo sapiens allow these happen without resistance? Should not the helpless calls of the muzzled voices be answered? Therefore the Altruists have to come to the open in the form of voluntary organizations (NGOs) committed to social Justice and Ecological Security 1and recovery. Infact these first came up as adjuncts to Governmental activity or as better Agents of Change, than the political forces. In other words it could be argued that NGOs are an evidence for governmental failure towards its people - A failure of all political parties and their mechanisms. If these were good enough why NGOs at all? But the activities of such NGOs also are. highly inadequate compared to the necessities. With this backdrop of Indian scenario shall now dwell upon my perceptions on Voluntary Activity vis-à-vis ecological security.

Enough is Not Enough: There are many voluntary organizations in India perhaps a few hundreds. Some have done visible works, others not. Some are recognized for their works while many others, not. There is a whole range of very idealistic (realistic) Activities to inactivities. Over all these are inadequate in the face of the tasks. Most of them suffer from a lack of theoretical foundations, hazy perceptions and absence of tactical lines of efforts. As all other sectors- Academic, Administrative, Economic and Political leaderships failed India NGOs are expected to be a bit of all these but mostly and preponderously, social. The leaderships, motivation, agendas of action, areas of operations, cooperations, Resource persons, social strength, communication etc. of the NGOs need to be reoriented in the altered Circumstances, in India - Economic, Ecological and Political. Mere satisfaction at some work done has to be replaced by an objective of total success on any chosen issue. Taking issues to logical ends must be the rule.

Role Models: Local challenges must necessarily promote local response, normally. And these responses have to be articulated for achieving a betterment.

Always these don't happen. Here comes the need for role models. Fortunately we have any number of them with a history of struggles and achievements - Green Peace, The Greens of Europe, Friends of the Earth, World watch, World Resources Institute World Wide Fund for Nature, Asia-Pacific Peoples Environment Network - just to mention a few. All these are different from one another in many respects but same as far as consciousness of ecological security of humankind is concerned.

In India we have a Centre for Environment Education, Centre for Science and Environment, Bhagavatula Charitable Trust, Chipko Andolan, Bombay Natural History Society, Kerala Sastra Sahitya Parishat and Narmada Bachao Andolan. These also are very different from one another.- Yet, they have a common goal of ecological welfare of the society. Despite these models, the activities of a voluntary organization should always be guided by the local situations and strong individual foundations. Therefore a blue print/checklist of guidelines to NGOs are suggested here with for reflection.

It was suggested that the Altruistic action of the NGOs constitute the last hope for Indian people now that the leaderships of the knowledge systems too are vying with other sectors in antivalues. Perforce the NGOs have to constantly steer clear of the dark forces in society, if necessary confront them. Unfortunately most dark forces have pious faces, sweet languages and suave manners. John Bunyan's Pilgrims Progress alone can be the guide. Even one has to select the Resource persons very carefully, only the honest and the truthful. Consumer protection can become very handy for NGOs to accelerate their activities. Championing the cause of consumers wins their faith and establishes credibility. Enlisting people's support for other causes will be easier. The running ethos being a simple family norm good today, better tomorrow- The ultimate NGO ideology of Vande Mataram - saving the Mother Earth - will be easier to achieve locally. A thousand local successes constitute a national triumph.

How long can a voluntary Action sustain itself? A federation of NGOs is essential for enhancing the power of lobbying for and achieving the cherished objectives. Ecolizing the political forces, parties and mechanisms is a sure way of saving to the people their common natural properties and assuring permanent ecological security. If this does not happen, Altruists and their Associations themselves have to turn into political forces. Otherwise ecological anarchy breeds social and political anarchies.

If the NGOs do not increase in number and upgrade their activities in all respects - leadership, concerns, agendas for action somewhat in the lines of a Chipko or Narmada Activism the future of India will be drowned in violence and bosniation. History indeed beckons the Altruists to rally round any environmental cause dear to them. And there are myriad of them.

TABLE. 11.1. Guidelines to NGOs

A. THEORETICAL BACKGROUND: Regional and global

1. History of exploitation of natural Resources
2. Origins of Ecospasms - Their Progression, Consequences.
3. Altruistic activities elsewhere
4. Ideologies: Gandhi's, Marxism, Ancient Wisdoms, and theories
5. Basic Ecological principles

B. IDEOLOGICAL OBJECTIVES: local and regional

1. Selection of niche and Agenda for action
2. Social and Political objectives
3. Educational, Journalistic and Populist approaches
4. Service on the ground

C. CHARACTERISTIS OF THE NICHE: Local areas of operation

1. Ecosystem status and the resources
2. Political, Economic and Social history,
3. Ethnology: folklore, tradition, Religion, legends, myths,
4. Human resources
   (a) Artists, doctors, writers, Intellectuals
   (b) Bureaucracy
   (c) Peoples' Representatives
   (d) Educational Institutions
   (e) Hospitals
5. Industries: Raw materials, Products, byproducts, Technology
6. Agriculture: Diversity, Productivity
7. Any Problem(s) on any or all the above

D. ACTIVITY PROFILES AND AGENDA FOR ACTION Ideology dependent

1. Agenda: Resource depletions, human settlements, occupational exposures, pollution, ecosystem degradation, ecorestoration, etc.
2. Activities: Education, social service, field studies, lobbying and legal actions, agitational approaches, and political action.

E. MECHANISMS OF ACTION

1. Advertisements and esthetic appeals
2. Educational Techniques - audio visual
3. Agitational and political processes

F. ORGANIZATIONAL ASPECTS

1. Infrastructure
2. Financial strength
3. Personnel - Leadership and associates

Table 11.2. Ideological Preferences of Various Environmental Moments in India

Crusading	Ecological	Appropriate	Scientific	Wilderness
Gandhism	Marxists	Technologist	Conservation	Enthusiasts
Polity Republics	Village Protectment	Dictatorshipwith women, low castes	Decentralism	—
Decision	Highly centralized Dispersed	Decentralized	Centralized	Centralized
Society	—	Equitable	Equility	—
Economy	Misused	State	Mixed	Mixed
Enterprises Small (mostly)	Large (Mostly)	Small + Large	—	
Consumption (Moral)	Limited equity	Limited ecology	Limited Unlimited	
Tech Change	Low	High	unlimited	—

Chapter - 12

Cataclysmic Kalki in India

Ecological Degeneration is Harakiric to Indian Polity

A. Introduction

There have been concerns on the fate of planet Earth and its constituting nations due to the improper exploitation of natural resources ranging from extreme fears to nonchalance and optimism. Some of these are sectorial and others holistic either in causes and consequences or in imperatives and mechanisms, sometimes in both. These perspectives, however, have been voiced by various religions, philosophies, myths, heritages and literatures. Discussions on these were accelerated after the Earth Summit in Brazil in 1992. The present attempt is one such pertaining to India and the ancient wisdoms. A few words on these wisdoms -religions, are therefore appropriate here.

B. Ancient Wisdoms

All religions were rooted in nature and its forces. They had their own concepts of creation and role of man in it. Christianity was the most anthropocentric considering all nature's resources as materials provided by providence for man to expend. On the other extreme is the Vedic-Hindu Ecocentrism that considered man as essentially a part of nature and must perforce live by its diktats. Even the Lokayata, the Ancient Hindu materialism is limited by it. As long as this Dharma is followed as a tradition, peace shall prevail.

The Santana Dharma extols the mother Earth and all its components, for reverence inhibit excesses. Pridhvi Suktam (hymn to Mother Earth), Panchabhutams (five major elements of nature), Vanaspati (Flora), Pashu-Pakshi (Fauna), Ritu (Cycles), and Autariksha (The atmosphere and above). All elements of nature were raised to divinity and ordained that they be not hurt by Pradushana (Pollution) for fears of consequences. If it happened, they must be propitiated, by making amends. Such were the dicta of Vedic heritage in favor of ecology. As long as Dharma (Ethics) is followed Shanti (Peace) is assured. Inherently, excessive exploitation of resources is always Dharma, untraditional, irreligious unbecoming and therefore suicidal.

C. Kalki the Exasperated

Thus the Vedic scholarships of Ancient India (Table 1 and 2), voiced the most serious fears on the future of man vis-a-vis his environment. Of these, the all embracing is the one associated with KALKI, the Tenth incarnation of God dealing with the end of humankind. These incarnations underscore not only the concept of evolution by animal stages and human phases, in time and space, but also suggest an increasing severity of problems from the actions of ruling classes. Consequently the righteous had to make progressively more serious efforts to ensure social welfare. However the tenth incarnation, KALKI is supposed to act in total exasperation at the close of the Kaliyuga, when evil totally eclipses society. The Bhagavatham has given about 10-15 characteristics of such a society (Table 2). And these tally well with the current eclipses to resources and ecological relationships. Therefore KALKI is synonymous with, eco-catastrophies that societies invite on themselves. The current global ecological morass that the civilizations perpetuated proves the accuracy of this line of thinking (Tables 3 & 4).

D. Ecological Quicksands

Population explosion, Technology, Chemical spread, Urban Krakatopia, Pollutions, Resource depletions, Consumerism, and warfares were blamed for various abnormalities, menaces, crises and scourges by Ehrlish, Commoner, Carson, Mumford, Polunin, Nicholson, Leopold, Russell, Huxley and others. However, the appropriative psyches of the ruling classes behind all these led to recognize the limits to growth (Club of Rome) and predict a holistic doom for humankind (Hardin).Yet a rare promise of boom (Kahn and Simon) and a steady state of the current statuses of development and ecology were also hoped (Brown). But these are intercalated by 'If' in the form of the concept of Spaceship Earth (Boulding) an advocacy of partial doom by triage philosophy (Paddock) and Gaia hypothesis (Lovelock) which bestows infinite enduring and recovering capacities on Earth. All these released a galaxy of terminology such as - Lifeboat ethics, Tragedy of the Commons, Triage, Spaceship Earth, Population Bomb, Ultimate Resource, Closing Circle, Silent Spring, Green Paradise, Sinking Ark, Green Politics, Future Shock, Common Future, Futurology etc. Prefixes of 'green', 'eco-'and 'environmental-'are employed for anything within our fancy confounding the confused. The biggest bandwagon in history that of a consortium of these thus was given birth!

E. Hinduism and the Kalki

Reverting to the main theme, it is generally agreed that the Vedic literature and the heritage that is bequeathed through it, 'Hinduism' in popular parlance, have all the essentials of conservation and selective usage of nature for human welfare. More importantly it has cautioned of a 'final solution' in the form of a suicidal destruction of humankind, unlike other religions. This is supposed to be brought about, as per one opinion, by the 'Dancing Shiva' - an allusion to the angry retributions from the celestial God of Destruction. But Shiva is placatable by his powerful consort, AdiParashakthi and the propitiating devotees, leaving a hope for retrieval from Pralayam. But the KALKI does not leave hopes - consequently more relevant to the current crises.

The KALKI of 'Kaliyuga' galloping on a horse with many sword arms is considered an unpreventable marauder - the horse suggesting speed and strength, the sword, destruction and the arms, different routes. In other words, the humankind faces attrition on many fronts as consequences of depletion and destruction of resources and ecology. Already the world experienced scores of spasmodic ecological boomerangs - varying only in extent, speed and toll (Table 4). Yet the greed of Homo continues unabated. Without either reducing their avarice, or acting at the sources, the ruling classes work hard on the covers-up. Indeed humankind is ingeniously self destructive.

F. Pessimism: A Well-informed Optimism

Meanwhile, the three worlds have gone through five global wars (1914-18; 39-45; 46-68; 72-90; 90-) to split further into about 8-10 worlds (two first-worlds, two second-worlds, one gulf and four third-worlds, at least). Consequently some countries will never suffer. These will assist a few others to recover partly or fully. But others will be helped to perish leisurely, allowing some segments to survive and flourish at the expense of the remainder. These nations, thus selected out through a triage operation, will face KALKI in their own lands, among their own people. Such a self destruction - slow, gradual, and inevitable - has been expected, forewarned and experienced. Yet the self indulgence continues as if under the spell of a death-wish, through a 'Hara-kiri'- preparing for it and exhibiting it - to the utter glee of the creators of triage.

Thus KALKI acts only in some societies like India where the predictions of Kaliyuga have come true. And KALKI acts through variety of ecospasms. Therefore, India has only an unecological future of imbalances, disharmony, and disintegrations (Tables 1, 2, 3, and 4).

G. Nimbyism, Nimyltism and Nodmytism

A question arises - Why so much pessimism on the ecological future of India and its nationalities? The answer is not far to seek. As a sequel to development, problems crop up. These are preventable from others' experiences and soluble by knowledge systems, but for the excessive consumerism (without fulfilling basic needs) and corruption to achieve this (ignoring natural justice).

Even these can be inhibited from injuring all the nooks of society. For this, the Education, Educators, Scholars, Scientists - in short all Intelligentia - the role-models for progress, justice and peace are ethical, vigilant, committed and crusading. Even if all else in society and nature are destroyed, the ethical higher education can redeem and retrieve all losses. India's problems exactly lie here, hopelessly.

H. Nimbyism, Nimyltism and Nodmytism

Plain criminality or acquiescing with it, blatant irresponsibility or promoting it, are patented by Indian Universities, Institutions and their leaderships. These Jekylls & Hydes indulge in unnecessary, repetitive, fraudulent, careeristic or fascistic research activities. Genuinely humanistic pursuits are extremely rare. The right and the just are persecuted and prevented from helping the society. Crime wears the apparel of innocence and the victims are branded guilty and punished brutally. While dealing with problems, Nimbyism (not in my backyard), Nimyltism (not in my lifetime) and Nodmytism (not during my tenure) the great escape-routes of the selfish in India, are the ruling ethoses in all fields in India. Law is often used to strengthen injustice and administration as a tool to humiliate the honest and the cultured - the hopes of tomorrow. Most professors, scientists and vice-chancellors vie with each other in leading this ethos and profiteer. And their trainees, work than them, spread out in the entire country, like cancerous metastasis.

While all crime - social, economic and political- can be prevented or punished, academic crime cannot be. ***(It is like the acts of a mighty blackman going for crime in dark with a well-oiled body, a mask on his face, a dagger at his loins, but armed with a sweet smile and a pleasant speech- who can catch this species?).*** There are no mechanisms even to protect or prevent from victimization in this field. This sector, the one that feeds all others, is deeply strangled by six BUGs - Power, Patronage, Profit, Pleasure, Popularity, and Prominence. One of the worst intelligentia ever to be found - leader-ships, institutions, trainers and trainees in history- self centered and unaccountable- dominate India. No wonder that truth and common sense became casualties when most needed for an ecological security-a security that leads to all other securities. With ecological insanity, destruction of all values, opportunism and cunning both among the young and the ruling old of today, where are the reasons to hope for social reconstruction and a recovery of ecological resourcefulness?

I. Kalki and Bosniation

Thus the Kaliyuga is here at its final stages - notwithstanding the years 'allotted' to it by the Ancient literature. This is already closing in, through divisions, conflicts, hatreds and violences. Villainy is the hero and heroism, a blasphemy. KALKI is continuously destroying the sustainability of resources, carrying capacity of the habitat and the buffering ability of all these for a lasting brotherhood. Bosniation of the country appears inevitable sooner or later, accompanied by violence. If any part of India survives, it has to break off from the federal statehood, lest metastasis overtake it. Like the existing and preceded violences, the future one also will be ecological due to dishonouring and disinheriting the ecoistic forces euphemistically called 'God' and ignoring the lessons from Bunyan's Pilgrim's Progress.

Wherever time-honoured ethical, cultural and social values are vanishing, resourcefulness of habitats and web-chain forces of ecosystems breakdown. If knowledge systems, leaderships and participants do not make efforts to retrieve society from such breakdowns, ecologists and 'God-like' humans also cannot rekindle the loosing hopes for ecological security. It is not the Krishna of Bhagavadgita who will be the final arbiter but the Kalki of Bhagavatham. Politics deals with power and ecology with balances. When the former is not used for the latter, the latter avenges it. This is unstoppable. This is the sum and substance of the concept of KALKI already in force inexorably in India, as per the prophecies of our traditional mythology (Table 2). Or was it a futuristic abstraction of concentrated wisdom ?

J. The Imponderables of Our Society

Coming to the current times again, the cries of Shivaji Rao regarding Madhura Refinery and Tajmahal went on deaf ears decade ago. A Bahuguna or a Patkar had to threaten a protest-suicide to force the ruling classes see reason. Extremism in development, i.e., consumerism by the 'haves' breed extremism from the 'have-nots'. Floods, droughts and famines, plague the country all the year round in one or the other State where children, women, native people continuously suffer. And begging, doling and profiteering in between has become a pastime. Non existent interstate problems are stirred up periodically to divert attention and charge people emotionally against reason and sanity. Watching the specter of Indian leaders begging for foreign money for their own states from their own linguistic sections among the NRIs, India seems to exist only in Andamans or imprisoned in its cellular jail.

K. Imperatives

Despite the writings on the wall, wisdom demands - what is to be done? - A consciousness of imperatives. These imperatives are many, urgent and demand great courage, hardwork and sacrifice. Should there be takers, I believe that the violence in Kashmir, Punjab, Bihar-UP, East India, Andhra Pradesh, TN-Ceylon are all ecologically justifiable because the ecological securities of people were played with. If these were to be prevented India needs to be reorganized and developed on ecosystem basis. A marriage between Marx and Gandhi on one hand and ecology and Indian heritage on the other has to be consummated for which CSE, Bahuguna, Patkar, Shiva, Alvares, BCT, Kunrakudi, etc. play the cupids and midwives. Ecology-consciousness has to be fully adopted by all political parties and administrations as they did with population control.

Regarding the all important knowledge systems, a Tarkunde or Krishna Iyer or a Gajendragadkar commission must expose them without waiting for some Shahs to commit hara- kiris. The great truth robbery indulged by Universities, Institutions and their leaderships for half a century must be revealed to the public. This can be done only by investigative journalism where journalists are not purchased by Academia - Industrial complexes, a new phenomenon. The lockers of the Auditor or Accountant Generals, University Grants Commission, Department of Science and Technology, Council for Scientific and Industrial Research, Central Bureau of Investigation, etc. must be broken open for all the skeletons to fall out and reveal themselves. There cannot be a micron of space for anti-politics in knowledge systems, as they are the ultimate arbiters of the fates of nations. Meanwhile leadership monopolies by professional Vice Chancellors, Deans, Directors, etc.should are stopped under popular pressure. Some of these sit in dozens of Committees, as if they are indispensable. Yet they do not implement in their own empires what they recommend as Committee members and Chair persons. Needless to say that the leadership is often a continuous hobby for occasional presence and power-wielding! And they are mostly busy building empires, strengthening mafia and enjoying themselves at India's expense! All these had to be urgently stopped.

The Publishing racketeering in books and papers that is ongoing in the field of Environment is so shameful that one wonders if Indian intelligentia can ever be redeemed without punitive violence! If possible Universities should be closed for some years for better purposes of illiteracy removal and spreading ecological messages. Restoring ecological sanity in all walks of life by following the Vedic values is the only remedy if a tomorrow is to be lived for (Table 1).If knowledge systems are purged of all dirt by all ethical forces and popular movements, Indian society will fully recover to face KALKI and stand up again with self-respect. Otherwise not. This calls for courage and sacrifice of the highest order in Indian history. Or, shall be accept the inevitability of disintegration and seek our destinies as nation-states by hastening the inevitable?

Table 12.1
Resources of the World and India

	Global status	*Indian scene*
1.	6.5 billion population	An unbearable 1 Billion
2.	Arable land increases by 4%	Only statistically possible
3.	Increased yields only source of food	Fertilizers, Pesticides & Power for irrigation and machinery -?
4.	25% increase of Bio-energy	Not possible- needs outstrip possible supplies
5.	Fossil energy finite, National disparities	Coal satisfactory, oil decreases
6.	Increased need of Atomic energy	Increased need
7.	50% world needs 100% increase in water	Nearly same
8.	40% Forests will be lost	Similar scenario
9.	Agricultural Soils deteriorate	Similar scenario
10.	More pollution of air	Greater pollution & more serious health effects
11.	Greater quantity of wastes	Similar scenario
12.	20% Germplasm will be lost	Losses inestimable

Based on Global 2000 Report

Appropriation of resources and their development gradually increases population under poverty line to 50% or more The above problems cannot be eased because of migration or oppression of the good and talented, leadership by the incompetent and selfish. These lead to social tensions and violences, besides failures. Destruction of values, ethics and justice Condemn India into Bosniation.

Table 12.2
Vedic Wisdoms to guide Leaderships towards a lasting Ecological Security and Human welfare

Sources	*Defects to shun*	*Pitfalls to beware*	*Virtues to cultivate*
Vedams(4)	Lust	Sensual enjoyment	Wisdom
Vedangams(6)	Anger	Success at others	Truth
Meemasams	Greed	Regret after charity	Self control
Nyayam	Benefits	Bearing indignities	Scholarship
Puranas	Sensuality	Lack of will power	Equanimity
Sastras	Jealousy	Misogyny	Objectivity
Upangams(4)	Conceit	Sacrifice	
Ayurvedam	Sorrow	Charity	
Dhanurvedam	Avarice	Righteousness	
Gandharvavedam	Arrogance	Sense of shame	
Arthasastram	Envy	Control Sense of	
		Self-pity	

Defects and pitfalls of humankind's leaderships in all walks of life brought about the present series of crises. If allowed to continue, KALKI destroys humankind for ever. To avoid this, virtues are to be employed in all walks of life replacing leaderships. This is the essence of Vedic Wisdom.

Table 12.3
Ecospasms in the world and India

	Name	*Consequence*
1.	Depletion of Ozone layer	UV ray exposure of people in South India.
2.	Green house climate	Many types of problems from a raised temperature, Coasts
3.	Germplasm depletion	Reduced resources of plants and animals
4.	Acid precipitation	Adverse impacts on all ecosystems
5.	Soil erosion	Productivity of the habitats of origin, siltation of reservoirs
6.	Water scarcity	Most basic need unsatisfied, ill-health.
7.	Sea water intrusion	Permanent loss of resourcefulness
8.	Pollutions and Mutations	Chemical stresses on all life processes, cancers, birth defects,
		Genetic diseases
9.	Hygiene and Sanitation	Compounds poverty and ill- health
10.	Population Explosion	More to share less leading to violence
11.	Floods and droughts	Poor people suffer - homeless resource-less.

(a) Illiteracy, poverty and ignorance compound the problems.

(b) Immorality, corruption and irresponsibility, makes life hopeless and helpless and justify violence ecologically.

(c) Children, old people, women & the poor are most marginalized

Table 12.4
Prophecies of Kaliyuga

1.	Truthfulness, mental and physical purity, forgiveness and compassion will decline.
2.	Wealth alone determines the status in Society, morality and merit. Lack of wealth will be seen as impiety. Hypocrisy replaces goodness.
3.	Physical cleanliness and good health will be replaced by perfumes, deodorants & make-Up
4.	Infatuation chooses partners in life.
5.	Homes will seem desolate for want of hospitality.
6.	Gluttony and sensuality will be goals of human life. Skill will be judged by the ability
	to support one's family. Virtuous deeds will be done for gaining fame.
7.	Might will be the right and fair. Trickery alone will be the motive force in Business
	Dealings, politics and social interactions.
8.	When the world is overrun by wicked people, the most corrupt will be the rulers.
9.	External appearance and dress identify people's spiritual development. People lose sensitivity to purity.
10.	Justice will be thwarted because of people's inability to gratify those in power.
11.	Ability to speak convincingly will be the only criterion of diplomacy. Audacity of Speech will be the criterion of veracity.
12.	People will perish through drought, famine, excessive cold, storms, scorching Sunshine, rains, snowfall and mutual conflicts.

From Bhagavatam

Chapter - 13

Feminisim in an Ecological Age

A. Introduction

It can now be said that the age of ecology has arrived. The United Nations Conference on Environment and Development concluded a few months ago. Regional Meetings on the follow up are now taking place. Almost everything in the several worlds of this single planet is being looked from an ecological angle. The problems of society at large are being reassessed from environmental angles of origins, progression and consequences. The ceaselessness of our society appears a manifestation of the life of the human female just as the story of Mahabharata in essence is the story of the Great Karma!

The strifes of society are age old. The sufferings of the women are age-old. As the Great poet Sri Sri wonted to remark often, God seem to have created the Poor, the Old and the Women to bear all sufferings of society so that the remainder remains happy! But if the women are poor and old too! The Great Uterine Injustice becomes poignant indeed! In view of the fresh look that many problems are receiving in the current ecological age, a reappRaisal Of The Women's Issues Is Attempted By A Teacher Of Ecology Hoping For New Imperatives To Emerge In The Process.

B. World Views and the Ages

The ecological age must have its forerunners a historical background which may be considered as follows. Until the religions and literature emerged there was no question of a world_view. Living with nature following its laws was the creed, if we can call it so. Religions Hinduism, Judaism, Buddhism, Jainism, Christianity and Islam may be considered constituting the religious world_view spanning about 3-5000 years. Faiths, beliefs, Gospels and their spread was the ethos of life during these times constituting the Age of Religions. This gave in to the age of Science where Darwinism - biological and social was the running chord of life preceded by Laconism. This age of reason and determinism, hardly 2-500 years old, has almost brought the world to the brink of Hara-kiri through the ultimate pyrotechnology, but for Soviet Union's abnegation. Still we are not sure what biotechnology has in store for us.

C. Ecogical Age: Perspectives

This brinkmanship of Science and Technology resulted in the emergence of an ecological world view. Though as yet no such definite world view of this kind could be formulated, shades and versions were available for the last two decades articulated by the Club of Rome and the United Nations Conference on Human Environment in 1972.The running theme was that man can no longer consider to be the owner of the planet and that he has to observe the laws of ecology to survive and benefit from science and technology. Just as religions and all that go with them continued through scientific age, distorting it continuously and influencing it occasionally, science and technology are an integral part of ecological age for, they alone can redeem what they endangered! Obviously, religions and the like have little to be around still.

In our ecological literature today we have thus- limits to growth Spaceship Earth, Life Boat Ethics, Sinking Ark, Global Commons, Triage Principle, Gaia, Deep Ecology, Green prefixed- terminologies, Biosphere, Ecosystems and the like. While perceptions vary, perspectives differ, imperatives range between a backwalk to pre-religious nature living to forward march through Science and Technology. At the same time they all aver that ecological relationships and balances have got to be maintained at a sustainable level without exceeding the carrying capacity of the biosphere.

D. Human Female and Ecology

Perceptions on the Gender Divide: The Characteristics of ecological age can be summed up as three commons - common property, common justice and common sense .The properties bequeathed to humankind are evolutionary, historical and therefore ecological. From understanding these, intellectual properties of recent times developed. These are global commons, the rights of humanity without any distinction of creeds, classes and genders. Thus Justice is common. A realization of these, the greatest common sense, must ideally pervade all leaders and sections of society. But it does not always happen, often in the name of realism because of psyche of appropriation towards these properties, the resources. This has to be squarely faced. Just as there was a call for rights for freedom two centuries ago by Thomas Paine the author of "Common Sense" now is the time to fight for rights for Resources and their development. This is the common sense - TWO! Ecological Common Sense. This tells us first the state of the human female.

1. Scenario of Labors as Fixed Stars: The female child is trained in traditional chore from childhood through the birth itself is resented mostly.

Education is little or not given, but used as labor at home or elsewhere. Married off quickly and if possible with a value in cash on her, as if disposing off a handicap - passing a 'scourge' on to some one else's responsibility. She gathers fuelwood, water tends cattle, collects dung, makes the cake, works in the field, maintains the house, washes clothes, cooks food and clean up, and she bears, rears the child and cares for all others around. Suffers pollution in the kitchen, male chauvinism at home and work place, second rate treatment wherever cohabitation with male is the norm, widows woes are all the more worse. As if these are not enough they endure male maharajas and forced to bear violence at home from the male and absence of support quite often from fellow women!

The above vary depending upon the country, race, religion, economics and history. There are thus strong exceptions. There were Presidents, Prime Ministers of countries, Nobel Laureates. Many women became firsts in male domains. These achievers too were exceptions. Even these women, mostly came out of their privileged groves earn positions and fame equal or more than of men. They rarely come up from unprivileged bottoms - excepting a handful. From an ecological point of view, in this current ecological age, how do we conceptualize?

2. Ecological Duties: (a) Labors in the Niche: The Fuel wood collection, water gathering, household chores, animal husbandry, aquaculture agriculture traditional marketing had to be coasted by giving appropriate value to each of these. No wages are earned for these. These are merely wages saved for the man, the housing elder. If any income comes through marketing or, generated, yet they belong to man's Kitty. However, that some of these becoming maternal property exceptions are well known.

Involving in the labours of one's own niche is an ecological truth, but so much gender bias? Ecologically every work is productive quantified and has its utility the value. Here it is a different story.

(b) Wages for Productivity: Women working in other's households and agricultural fields seasonally to earn a livelihood for the family but does not constitute female property in reality and often is appropriated. A kind of parasitism. Very rampant. As parasitism is ecologically unsound it is rare in nature. But in human society, it is the norm.

(c) Living in disturbed habitats: Air pollution in the kitchen and insanitation within and outside the homestead are health hazards. Some of the workplaces of laboring women including the children are similar. While there is a certain ability to bear with, the costs are usually greater than the wages earned. While the woman gives what good she can, she gets, not what good she needs and deserves. She gets back a few undeserving and costly lashes this too does not happen in the natures ordain! Most animal in distributed habitats have their avoidance mechanisms!

(d) The herd instinct of defense against danger: When there are crimes, social or violent the women rarely get together to make a united defense - in sharp contrast with the animal behavior of Unity. Alcoholism gambling, smoking and other alphabets of the male, violence, rape etc to which the female is often subjected to needed resistance by the entire community- such straying are uncommon in nature.

3. Biological Duties: Child Bearing, Rearing and Family care: The convulsions the female body and mind undergo during and after pregnancy are well known - as if "you invite a disease, endure it and recoup from it" all by oneself. Of course this is naturalism all animals. In case of human the couples of birth perhaps are very hope! Bringing up the offspring is like tending and growing an agricultural field - a periodical recurrence. This is done for more duration than any other animal. Looking after a while lot of people - husband, children, in-laws of all sorts and in a way acceptable not only to them but the neighbourhood is the lot of most women in India and several societies of Asia.

Is there any species where the female is the worker and the male, the king? Despite a queen and workers in a bee society equality and division of labor is the norm between the male and female of a species, when they are separate. But the genders are the equal suggestions by the hermaphroditism, the concept of Arthanaareeswara of the Hindu legend. What is the value attached to these Biological and ecological duties? How can one cost these? Are women getting their Due? Unequal differentiation between male and female in their duties and rights is therefore ecologically untenable as it happens nowhere except in human societies. This may be the reason why we became hazardous to ourselves while no animal societies were self destructive and unequal in giving and taking.

4. Reality of the Gender Divide: One of the basic ecological commandments is that every organism takes what it needs from the biosphere and gives back what it can. The human female gives what it can but prevented from taking what it needs since times immemorial. Thus the most Anti-ecological single issue today is the state of the female of the Homo sapiens. How can there be genuine development, advancement towards beauty, peace and prosperity, without one half of the human species! Without the Female? This is the crux of the problem now with us.

5. The Uncommon "Commons": The Commons mentioned above had been in a state of distortion, divorce, decay or disuse. The common sense of common justice and common property is absent considerably across the world. Common development therefore was nonexistent unless and until forced. This has been the story of the female in most of the world for most part of the recorded history. This is not to ignore that in the North the condition of the female has obtained a great advancement. But this too was an achievement of centuries old struggles!

6. The Common "Nonsense": Man is the only animal, perhaps, to kill the spirit of species in his female. This gender discrimination is anti-biological - as man alone cannot do everything on his own, a truism in biology and therefore in Ecology, doling out in return a subjugation adding insult to injury is ecological imperialism. Ecological degradation further affects women. Then it is ecological fascism which, they endure. Now that an ecological age has heralded, it should be made the age of female equity which however cannot come on a platter, but through struggles. Therefore radical feminism is a necessity to the ecological age and is the most pivotal part of ecoism itself. All ecologists have got to be Feminists - their radicalism depending upon their degree of ecological consciousness.

Having stated that environmental consciousness incorporates female welfare as a central feature, we have to examine whether the feminism as understood or represented is ecologically sound. The answer is an affirmative. Yes! for feminism to me is a crusade for equality denied by force-soft or brute - by the male Maharaja!

7. Ecoism is Feminism and Vice-versa: Though environmentalism means several things to several people, commonalties in all these constitute Ecoism- the interactions, interdependences and successions resulting in a climax of harmony among living creatures, communities and their non-biological environment. If we limit this to human society, it would mean that peace in society is possible as a climax, not as a wane that veers, only if successful interdependences and interactions are practiced among all sections of society. Which sections is more vital and basic than the female? Thus ecolism at the basic level means Feminism. Since the problems of the female cover half the humankind, it is the only front ranking problem and the ecological philosophy deals with the problems in the front. It is neither left nor right and perhaps both!

Whether it is the Bishnoi women of yore, chipko movement on deforestation, or the Santal women's preference for fruit trees over their men's choice, or protests against state promoted alcoholism or women's protests against discrimination and movements against wars, pollutions, nuclear tests, armaments in the west - all these are manifestation of outraged souls overflowing the limits of endurance across years, decades, centuries, perhaps millennia! Yet very small protests, But significant, because a philosophical framework has emerged during sixties and seventies in the form of women's liberation movements and later in the green philosophies. Consequently most ecology movements across the world have women leaders and greater women participation.

Mead, Ward, Carson, Bertell, Kelly, Spretnak, etc., in the west Patkar, Shiva, Sarala Devi, Sugadha Kumari, etc., in India were all ecological activists basically trained in other disciplines, drawn by the stirrings from souls wounded by the never ending exploitation of nature, the human female and the seeming permanence of inequalities. Thus feminism and Ecoism are more or less synonymous. Ecological issues are best articulated by Females and Feminists. Symbolic indeed. If mother arth is wounded there may be downfall for humankind. If The Mothers - the women are outraged, yet, in a similar way, the same would befall. If it has not happened, it is only because of some occasional efforts, though feeble, yet effective for sometime, a constantly discontinuous effort ensuring pockets of sanity. This minority sustains the hope in others. As is often said. If this wretched world is still surviving with a hope, it is because of an exceptional minority of ethical people and women's endurance!

8. The Thesis: Being women, entrusted by evolution in nature, with the continuity of human life they ought to be bestowed with certain basic biological rights. This is natural justice because natural rights go with duties nature conferred on them. Being a pivotal link in all ecological activities constituting ecoism, the female is entitled to ecological rights since they perform their ecological duties, most sincerely. If the above are prevented, by the aggressive tendencies and malpractices of males they are entitled to seize justice. As the eco-degradation continues all-over the world, any philosophy and action to retrieve has to address women first - as the first sufferers. Environmentalism means, first, Justice to the female! Therefore green the female first.

E. Origins of Male Dominion

1. Evolution of human behavior: From a biological point of view the differences in the behavior of human male and female can be traced to the animals. Animal Kingdom is full of examples of sex related differences which gradually evolved to a pattern characterizing their habits, habitats, taxonomy and ecosystems. The closest to human society is thus the primate behavior. No wonder therefore so many accents are laid on this group preceding Hominids whenever human behavior was considered.

Some Tribal societies around the globe represent the post faces of Hominid social ancestry even to this day and help us to construct links. A peep into these of primitive societies tell us the evenness of gender status - a kind of justice, status, equity based, if not equality. Thus it appears that not withstanding serious sex-related differences, social status and interactions remained on an even keel, upholding natural justice perhaps as a biological heritage of living in similar heritage. However, it should be realized that the ancient societies differed from each other owing to their inherent racial geographical and ecological differences. Consequently the social or cultural evolution took different characteristic directions, and velocities.

2. Status of Women across the Globe: As a sequel to differential cultural evolution across the globe, the societal mores pertaining to human female varied considerably in different groups. Thus differences exist among Caucasoid, Australia, Mongoloid and Negroid races and among Nordic, Anglo-Saxon, Semitic, Latin, Central-Asian, Arabic, Moslem, Hindu, Buddhist, Christine, Chinese, African, Nomadic, tribal, societies both in the status of woman and her treatment by the male counterpart. Thus the 'niches' of women today is a blend of primate ancestry, hominid roots, ancient social organization, ethnicity, geography, history, ecosystems religion, knowledge systems and ultimately and above all, the male ant politics and female voluntary subjugation. The last two have become so dominant and prominent now that we generally loose sight of the other factors where these two were rooted in the post of biblical. Yet it is clearly not known when and how the domination of man over women began.

3. Male Dominion over Female: (a) Biological and Ethno-ecological Roots, That the male and female of any species are not the same or identical is known from biology but they are nevertheless equal with specific divisions of social and biological labors. This ecological principle of division of labor was followed in ancient societies to an extent more pronounced than in other animals due to cultural evolution. Thus it is the brain that made all the difference! Times changed adding more people and more understanding of nature's secrets. The female tended to be less dynamic by lagging behind in muscle and mobility due to maternity. Here it is the uterus that made all the difference. Man's dominion, thus came as an instinct - a territorial instinct arising out of handicapped female confinement and increasing knowledge nature's usefulness. This gender in equality must have started when the psychic pressures were insistent enough from an enormous female staticity and male self confidence. These could not have happened except at a time of great acceleration in population and man's knowledge, and exploitation of his ecosystems. Thus domination of nature and female appear to have started simultaneously. It is poignant irony that revolt to male chauvinism sprang up in parallel with the first signs of environmental degradation.

(b) Prehistoric Scenario: Ancient societies are characterized by passage from hunter-gatherer stage to agriculture, from signs, symbols and sounds to languages. Simultaneously with these, the human settlements, families have arisen as units of existence, subsistence and procreation. Interactions within these units with others in the settlement, with other settlements, with ecosystems and their facets - productivity, utility and cycles followed gradually. The last one became a source of wonder, enigma fear, hope, resource, beauty, pain, peace progress and harmony. The literatures came up from human experience and imagination encompassing all the above. Empirical knowledge started building up. Concurrently the division between male and female deepened. The female got burdened with uterus, its sequels and consequences. Man went vagabond in all his activities, female having become a captive house keeper, bearing, raring and caring the brood. All these must have taken several millennia.

(c) Ancient Literatures: Yet the Sanskrit, Greek, Latin, Hebrew and Tamil Literatures before the advent of Christianity, bear testimony to the equity of the female compared to the more recent times. Females were known to function even in knowledge systems and endowed with as much power as men savants. At least the female was a force to reckon with. The Mahabharata, of course has introduced to us the existence of polygámy, polyandry, premarital procreation, property rights, casteism and Adharma. Yet the Geetha extols the Dharma for the continuity of the human race. The Dharma with Vedic roots and routes was dominated by the law of Nature's pre-eminence over man. This pre-eminence of nature, Ecoism, is almost synonymous with Feminism today as we shall see later. Therefore the righteous conduct, ecoism and feminism have the common consequence of human welfare for ever. In such a land of ancient Dharma, subsequently a place of Hinduism and later Buddhism and Jainism, how did this gender inequality and injustice arise?

(d) Biblical Times: Biblical times tell us of feudalism, Injustice, prejudice, Vendetta, concept of sin, slavery, migrations, legality and legitimacy, etc. The scene was Western Asia and Eastern Europe, the place of Judaism, Christianity and then Islam. A region of many tribes, territorial urges, feudalism, military campaigns and persecutions. In such contexts protection of the physical weakling, woman, for increasing the population strength of the tribe became a necessity. Thus by the middle of the first millennium after Christ male dominion over female was established, in the Middle East, Mediterranean Europe and Central Asia - cradles for many civilizations.

(e) India's Bhoomata Concept and the Captive Female: Refreshingly the story was different in the Indian subcontinent. Mauryan times, Buddhist ascendancy and Vedic Dharma ensued female individuality and self respect. But the status did not lost long. The Muslim invasions through North West - Turks, Afghans, Moguls, etc., for pillage of natural resources, and other weathers of India through conquests and spread of the faith if necessary. Taking our women for creature comforts being far away from their homesteads was natural converting them to their faith to raise local populations became a necessity with the invaders in due course. This challenge was responded by female protectionism of the Hindu. Glorifying her virtues, the necessity of upholding dignity and honor even if it means sacrificing life became an ethos. The concept of Habitat as mother developed.

Unfortunately this protectionism deteriorated into captivity. Over a millennium approximately, the status, *locus standi* of woman in Indian Hindu Society became nearly slavish, menial inferior and the like. This continued, scripturized, ritualized- Adharma legitimized as response to Moslem challenge and ultimately legalized as an enjoyable imperial convenience at home through social sanction.

(f) British Times: With the reign of British establishment the scenario has changed with India coming under the rule of one disinterested in hegemony over the native females. The injustice to women in homesteads neighbor-hood and society at large began to be realized. Societal leaders turned inwards rather than preoccupied with territorial defenses. Millennium old, strange hold on the female partners andro-centric society was sought to be liberalized. The efforts to rationalize or reform began in Bengal with the help of the British rulers. The rest is history of modern India - A Summary perhaps will suffice.

F. Problems, Hurdles and Setbacks

1. The Black holes: Man Eaters, Escapists and Opportunists: Womankind has to guard themselves against their fourth columnists, nay, and the betrayers from their own ranks. We have seen many women, after high education, do not pursue Career, for a career gives not only an economic freedom but also a social self respect and a chance to lead and liberate others. A few others are happy to use the attainments to win a "Good husband" and be happy for ever. Many happily go to the Heaven on Earth (USA) and continue despite the undeveloping India. A few make a facade of social work, albeit for self propulsion. Many more virtually do nothing, waste their times at the desk and if necessary specialize in Man Baiting or Man Eating in the name of Women's Liberation. This man eating syndrome consumes in the name of female liberation or individually several stances and proportions. Acquires this with them amounts to their undeserving promotion and patronage. Disagreeing with them is tantamount to anti feminism male chauvinism. In the name of female rights, there lot exists as privileged parents what is to be done. Many of these vices are common with men too. But men can afford being the chauvinist successes for centuries. Should women emulate? No. All these does not meant that these are not women leaders (Ms. Durgabai, Ramabai, etc.) For women's causes. There are quite a number but not effective enough as per the demand due to the pitfalls from fellow women, mentioned above!

2. The Great Marxist Setback: A Challenge, an Opportunity: Marxist applications in communist countries did considerable justice to women. Even in countries where Marxists were not in power, they took up women's causes and women leaderships were encouraged. At least they did not yoke the women. Now that the monolith collapsed and Marxism is hastily retreating due to a kind of pessimism and fatalism from failure, women's movements receive a set back. This is a challenge to believers that Marxist persuasion alone is totally women-friendly in social progress, while others can be partial at best. Through this an opportunity to revive themselves is offered. This also adds a great responsibility on all privileged womanhood to articulate women's rights and lead them to some fulfillment. In fact they are subject to accountability on this score!

G. Modern India

1. Indian Panorama: Here One Step Forward & There Two Steps Backward It may be pertinent to make some general observations on the efforts of some Mahatmas for the social progress of women in recognition of their great value to the entire society, and as a protest against unnatural injustice to them. Ram Mohan Roy and Vidyasagar in Bengal, Karve and Phule in Maharashtra and Veeresalingam in Andhra took up women causes in writings and actions and achieved unimaginable successes though small in proportion to the tasks. In Kerala the task was made easier for the reformists owing to the matrilineal system. Both in Bengal and Kerala the successes of Marxism contributed to the comparative emancipation of women in those parts of India. Is not Marxism then ecologically sound as far as the half humankind is concerned. Or is it that the revolutionary spirit in Bengal - Social, Cultural and political - bred by Tagore, Bankim, Sarat, Vivekandanda, Subhas, Surya Sen, Aurobindo and the matrilineal system in Kerala contributed to women's liberation which then eased way for Marxism to take root? Perhaps they contributed to each other.

2. The Andhra Scene: the Perpetual Rebels: In Andhra Pradesh, however, the revolutionary spirit came from Veeresalingam and Gurajada as a revolt to heinous feudalism and its social and cultural ramifications. Ghandhian inspiration to men and women alike in freedom movement generated a large number of local revolutionaries of all sorts. On this fertile ground germinated the seeds of Marxism. Andhra developments are more like grass root developments - the real need of today. A few words on Veersalingam probably are relevant here. He was a thinker, teacher, writer, reformer, and Activist and therefore a leader. Where? in a small town, Rajahmundry not Calcutta, Delhi or Bombay! He was no rich man and had no patron props and no healthy man. His small town was a personification of orthodoxy with all vices-Devadasis, child marriages the attending intrigues, widows, untouchability, casteism etc. And he was a Brahmin. Yet he did his work. He happens to be the only leader who could brave a Sankaracharya camping in his small town to stifle, ostracize him thus fail him. But the Acharya himself has to leave the town before dawn! Yet when Veeresalingam tried to infuse reformation ethic at Madras, seeking a bigger Canvas he failed to move the city into sense miserably and had to beat a retreat back to the small town of Rajahmundry only to continue his work with more vigor.

No wonder that two violent Marxist revolutions came up in Andhra in 1947-50, 1963-70 and the third is on. Women took central role, equal or greater to men in these activities for social change. They were totally free and liberated as no where else in India. A Sri Sri and a large number of his like filled Andhra. Perhaps the greatest emancipation is seen in literary circles. The largest group of women writers in India is perhaps from Andhra. Almost every town, in the past, used to have a spectrum of activity from one to the other political extremes - all aiming at social progress.

3. GORA: the Ecological Marxist: Directional change in practices: Gora is a phenomenon, none of like traversed India, Gandhian, Marxist, Botanist, Feminist, Teacher and Ecologist down to earth! Ecological because, he practiced and preached thought and wrote Atheism. Atheism to me is the recognition of forces of nature - evolutionary and historical of materials and biologicals and therefore ecological. Due to ignorance or awe of natural forces God and Religions were created - as concepts first, dogmas next, superstitions later and Fads and fashions now. He was the only force recognized by Gandhi, Marxists, Social reformers and even the Dravidan Movement of the south for whom a brahmin is an anathema (Brahmin to me is the man of wisdom and an ascetic, Brahmin as per scriptures has to be a Marxist. Then why the prejudice? This is another question.) His work is there for every one to see - Science, Society, Values, Nature, Females, etc.

4. The Dravidian Movement: a Great Historical Tragedy: The movement that influenced about 60 million people in the South Indian State of then Madras and now Tamilnadu was popularly known as Dravidian Movement. The main plank of it was restoration of self respect and rights to Tamilians usurped by others especially Brahmins as they were Aryans in contrast to their own Dravidian origins. This is patently ethnic and racist ignoring the fact that races mixed up in India during the last 5 centuries even agreeing that races flourish and rule 25 centuries ago were patently Dravidian. Yet it has to be recognized that this movement by E.V.R. Naicker and later articulated by a score of others, has revitalized the Tamil nationhood in India. Supposing that the movement concerned itself with the social emancipation of women and economic emancipation of the poor - what could have been the consequences? Instead of Brahmin hatred, Hindi hatred and the like, if hatred of poverty, male chauvinism, injustice, inequality, etc., are the main planks, we would have got the most just social segment in our subcontinental in Tamil country. A great historical opportunity was thus lost, or wasted and ultimately corrupted as we discern today! The strength of EVR's movement and its pervading influence even today is perhaps greater than Gandhi's for EVR's followers can be seen today in almost all villages of Tamilnadu. For such a man, GORA was an object of friendship and reverence! This behavior is a testimony for the scientific and correct attitudes of GORA!

Dravidian culture and literature that existed even before Christ deserved international reckoning from historical and evolutionary standpoints and therefore from ecological angles. But it is steeped deep in sectorial morass! Leadership itself has done injustice to the theme they wished to propound and serve. Thus the so called Dravidan movement reduced itself into a neo-fascist variation when it deserved a place equal or near to Greek, Latin, Hebrew, Sanskrit, Arabic oriented cultures of the world. It has reduced itself to the level of a political expedient when it could have liberated the women and the poor of Tamilnadu. No one could have stopped its upward march. And Indian peninsula would have become an ecological paradise, with justice, instead of paradise lost. GORA has tried to combine the issues of poverty and gender for his social onslaught but the canvas was limited, expanse was limited and resources limited. Articulations were also limited on one side by Gandhism and the other side by Atheism - a curious mix that did not sell!

H. Ecology and The Human Female

1. Message of Ecology: In summary, ecology tell us that each organism gives what it can from the habitat and takes what it needs from it. There is no gender difference except in biological functions and the consequential duties. Over dominance or parasitism are very rare. There are niches for all. Division of labor for all by participating in the webs and chains. They live in the community maintaining their characteristics. A shared ecosystem attains a climax, highly balanced and continuous maintaining a constant harmony. Even in death, the organism adds to the fertility of the habitat and the health of the system. Is not Marxism i.e. Communism, then, basically an ecoism prevalent in nature?

2. Ecosystems and Evolution, Society and Progress: Ecosystems have, in time, evolved. All organisms had their niches, roles, needs from and contributions to the community. When ecosystems degraded, they could be restored only after understanding and following the mechanisms by which ecosystems were built. Compare this with human systems analogy. Human Societies have, in time, progressed. All societies had their innings in making the human history. These societies have degraded and are degrading due to Anti-female practices, among several others. They can be retrieved only by an understanding of ecological principles of gender equity that men violated (by initiating antifeminism or male imperialism). The apparent analogy actually underscores the inherent and evolutionary homology between human society and natural ecosystems (i.e. Natural communities) which man has divorced during his cultural evolution.

3. Historical Antifeminism turns ecological: The woman is loaded with excessive burden - a kind of subjugation. If endurance is wanting, if efficiency or utility is wanting, if disagreement and revolt crop up, they are subjected to oppression and violence. Sometimes they have to slog in a resource-less ecosystem or a degraded habitat or withstand toxic environment. Yet behind most successful and great men were women and vice-versa! Does gender equality manifest only by high intellectual attainment, supreme humanism or by their combination? Is this the psychological reason behind the non working ethos of modern women, outside home and woman bashing by man at home for entirely unrelated reasons! Both these are due to a general wearing down of ethics in society despite education and literacy. This is catalyzed by consumerism and excessive sense of security. Both these in turn are due to excessive exploitation of natural, resources and the consequent degradation of the environment- both threatening security consciousness. Historical anti-feminism becomes thus ecological now. Therefore ecology and woman breed each others welfare or ill-fare.

If environment is endangered, niches are lost, extinction may take place - both of organisms and the guiding principles, and Conservation is needed. So too of the human female who labours and suffers - duties and no rights. If pollution arises, organisms suffer from toxicity and after effects. The ecosystem is endangered. The human female too undergoes this experience. Both conservation and de-pollution have to be done by women themselves- from masculinity, consumerism, hedonism, chauvinism, escapism etc. Philosophically therefore, the Gaia hypothesis (ever resourceful mother Earth), motherland concept does not hold, if the human female is not emancipated and restored to her right place with the true spirit of Arthanreeswara through the Angry Kali mechanism.

I. Resurgence of Female Consciousness

1. Slavery and the Female Response: There has been slavery in the Ancient times of the Vanquished by the Victors. It lasted for some time. Colonialists used the negroids as slaves. This too ended. But the slavery of woman to man seem to be everlasting in most places. As I said before, women themselves must revolt, lead, must articulate using the humane males as complementary agents of change. But Escapism or female chauvinism will not serve any purpose. Man eating, is no answer for woman eating. They have to cease to be an eatable Commodity. Woman plus man becomes a human being. In this context we must remember Gandhi and Marx and recall Premchand

"When man tries to be woman,
he becomes a Mahatma.
When woman tries to be a man,
she becomes a Demon".

2. Female Consciousness: Tribals, India, World: Currently we see the primitive tribal societies more justice conscious vis-a-vis the human female - as a part of their culture and as if a heretage from natural systems such as the forest. The non tribal modern female is still steeped in inequality! Yet tremendous awareness of gender in equalities and efforts to remedy them has spread in India in recent times. Peoples' Movements, for Women liberation, Ecology, Revolutionary changes in society and other causes in the western countries have all conceptualized human female as equal to man and even actualized to different extents. A brief Summary may perhaps be needed.

3. Greening the Female First: Slogans serve a purpose. They represent an ethos, Green Industry, Green tourism, Green bride etc. But Green Feminism shall come first because if the Natural resources and the ecological balances among them are not fertile, the human female suffers most. Carrying with biological as well as ecological duties where does she go under social repression, if the resources of the Earth, human rights and policy privileges are not available to her? A case of multiple troubles. It is more. First a woman, then poor, followed by illiteracy, non-availability of resources from the retired ecosystems, living on marginal habitats. All these get inherited socially. Greening the female is most immediate. It is from here reckless retaliatory violence and protest emerges. I think this is the reason for the successful and widespread spear-heading of anti-liquor agitation by women of Andhra (In this context, I feel that at least all NGO's must be statutorily required to engage in female and ecological issues in addition to their other preferred goals).

J. Imperatives for Action: Humanism

Having summarized the sad state of the female gender the questions arise - So what? What next? As we have been developing an ecological world view ushering in an ecological age, the imperatives have to be ecologically sound. In other words, justice, equality, community etc., have to be ensured.

1. Uterine Handicap and the Great Opiates: The male brutality invited by uterine handicaps of women has to be compensated for in various ways. The Masculine imperialism leads to female protectionism. The female itself has to change radically.

Religion and God were once considered as opiates. Today Matrimony, Male necessity and Motherhood (forgive me for this) may have to be treated as the opiates, glorifying which through ages the male mind mesmerized the female. In other words the Feminist movement shall declare freedom from uterine compulsions and treat pregnancies as nine month diseases. Perhaps these are ecologically sound in our present society.

2. Ecological Rights: Social rights, economic rights, political rights and Human rights basically constitute the Ecological rights as they are natural as per evolution of life. In an ecological age, these rights must be given statutory provision at all levels - National to local - somewhat like what we did for the depressed classes! After all women have been performing their ecological duties since ages!

Enough safeguards are also necessary lest the socially better off sections may grab and monopolies the opportunities resulting in more inequalities. And the problems never get solved!

3. The Privileged Woman's Duty, one's own steam: It is the duty of the educated women to rally the fellow women and provide philosophy and guidance for action - A self-respect movement instead of enjoying their new privileges in the name of female liberation! The female must realize that they have to resist and fight on their own. Male cooperation can at best be treated as a bonus.

4. Top - Down Mechanism: Majority of the privileged upper class/upper caste/urban female combine enjoy their privileges and are happy to be so. A few try to do something to better their unfortunate cousins. But several of them are work-shirkers and pretenders. If only these sections, take up women's causes and fight for their rights, not only the man-eaters but also the Male chauvinists could be kept away. The future of women, as a vital link in the ecological web is assured. Women are glorified by all great men, not for nothing. She is the key for peace, progress and beauty in life.

5. Personal Law of All Religions and Regions: The state of women in India is different depending upon religion, regions, class and castes. A uniform law, if secularism has to come to people from the Government (!), is warranted, most urgently. Personal law must be the same for all. Religions and Gods, created by man outlived their life span. With the advent of ecological age, underlining the principles of nature to be all powerful, traditions based on superstitions are worth discarding. Though Marxism said as much, it is now fortified with ecological knowledge and realities. How long shall we discriminate women in the name of superstition-infested tradition?

6. Property Ownership: Self Respect: We have earlier mentioned the valuation of female labor and its costing. Even if all is done who pays them? None. They end as "Expenditure saved" or "Wages saved". Therefore certain economic rights are necessary for women - as it was in the matrilineal system in Kerala. This is nothing new and was debated for long - yet nothing concrete has emerged.

Now that a few lakhs of acres of wasteland is likely to be distributed to the village poor, why not give the entitlement to women instead of Men? With one stroke of policy thousands of families get land they have been tending and equal number of females, self-respect confidence and social responsibility.

7. Adult Literacy: A Spur for Action: This has been one of our thrust areas of government sponsored social activities. If these are made mandatory for all NGO's to literate women many a superstition will disappear and confidence to brave oppression and take initiates could be built. Perhaps this is the only way to build them against male imperialism, violence and looting of precious earnings for pursuing bad habitats e.g. alcoholism.

8. Female Constituencies: Opportunities to Serve Themselves: Just as there are reserved constituencies for depressed classes in all elected bodies there shall be some reserved for women at all levels - especially lower levels.

9. Female Infanticide in Ecological Euthanasia: I have tried to establish, as many others did before, that the female has been subjected to natural injustice for centuries-a kind of "murder"(historical and therefore ecological) which has become a cause for many physical suicides and uncountable mental suicides. In the light of these, what is wrong if a female child is killed before or at birth? Does it not amount to mercy killing, when nothing seems to make a dent into the hard injustices against the female? I could call this ecological euthanasia to avoid the ecologically unsound "murders and suicides" happening only in human ecosystems. The value of the female will dawn only when their populations have dwindled enough.

10. NGOs, Agenda and Funding: Every NGO should be required to incorporate a women's issue and an Environmental issue in their agenda of actions and service for registration. Such organizations alone should qualify for public funding. If the voluntary organizations are totally dedicated to Female liberation and welfare, the supports must be total, for environmental ill fare will be correspondingly reduced.

11. Bottom - up Mechanism: The un-privileged lower caste/lower class/ rural female groups have achieved more spectacular rights for themselves by sheer force of revolt from desperation, to my mind. But political articulation was wanting most of the time. Million mutinies therefore made no dent on the masculine monolith. A form of this is currently seen in Andhra Pradesh against alcoholism. Many such movements were successful mostly led by Marxist oriented political activity in the past in India. These have grown few and far between!

K. The Hopes

1. The God that Never Fails: The two great movements - Dravidan movement of EVR and Atheist Movement of GORA were Ecological nonstarters so far due to inadequacies in articulation to catch people's imagination. They are Gods that failed ecologically as their times were overshadowed by Gandhi, Nehru and Marxist thought. Now that ecological age has dawned itself with a myriad prophets, these two must take their prides of place, revitalize themselves with predominantly ecological content. Their time has come. Ecological content would mean that Feminism, Marxism, Gandhism, Ancient wisdoms are all taken care through science and Technology for development and stability of for peace, in an environment of interdependence which alone ensures equality, freedom, and justice. Ecology is our new God. Environmentalism our new religion. The poor, the women and the other un-privileged will be the focuses for greening - an euphemism for conservation, productivity and minimum consumption. If women are alright ecology will be alright, the society is alright. If ecology is alright, the society is alright the whole world will be alright. Therefore the Ecology God can never fail if only these two movements make a change of direction in their precepts and practices to accommodate ecological wisdom. Let us green the Feminism.

2. The Resurgence of Ecological Marxism: Often a question is asked. Is Marxism dead with the collapse of Soviet Union. To my mind, it will surge up if green feminism is taken up under Marxist precepts and practices. It will also surge up if the poverty and poor people got worse under the new dispensation, despite good noises. The hopes raised, if belied and ecology worsens - the double troubles, ecospasm, will result in the arousal and espousal of Marxism with stress on ecological resourcefulness and equitable interactions. In other words, ecology will merge totally with Feminism and Marxism in societies with the scenarios feared as above. In India with the record of dishonesty we witness, I anticipate resurgence of Ecological Marxism as a leading social thought in many parts of India, before the turn of Century. The female homestead labours have to be costed. Not that they will be paid prices. They may forever remain as wages saved. But at least they can be prevented from "paying prices for doing labours" a very strange socially legitimized ethic in India. Happy days indeed for women's causes thanks to Ecoism and Marxism and struggleful days for women, thanks to the un-ecological humankind and fourth columnists from within the female ranks.

L. The Perspective

Not withstanding the limited number of women who have made it to international positions of fame and power such those of as Prime Ministers, Presidents, Nobel laureates, the majority of our planet's women still struggle against unequal conditions and opportunities of life. From a feminist ecological viewpoint, human society in most parts of the world is unsound and seriously at odds with native's law which is overwhelmingly in favor of protecting and perpetuating the female species whether it belong to the plant or animal kingdom. Predominantly, in most of the natural world, ecological duties are distributed fairly between the sexes, although the well-documented dominance of the queen bee over her male workers revises this concept to some extent. It is difficult to think of any species where the female is the worker and the male, the overseer or king, although the freaks of nature sometimes occur. The subjugation of the female in human society can be seen as one example of a freak occurrence, which, over the long history of male dominance across cultures and faith, has been transmuted into a norm of surprising tenacity. It is possible to argue that the natural appropriation of nature's resource and deviation from the laws of ecological balance by human societies, most of which are male dominated has led to the present degraded condition of our planet's environment.

When a man wants to become woman,
he becomes a Mahatma
When a woman wants to become a man,
she becomes a Devil

Premchand

Chapter - 14

Ecology and Foreign Policy

A. Introduction

Foreign policy of a country is laid on the basis of many criteria bilateral and multilateral- and reflected in the policies pertaining to military, trade, finance, science and technology besides political activities at the global and regional levels. This article tries to argue out a case that the natural ecosystems and the environmental consequences of developing these systems are more important in the times to come as a criterion for framing and implementing a policy with other countries. Such a premise may be very useful in dealing with neighboring countries, as well as benefiting from the countries advanced in Science and Technology. However the present attempt is more of an introduction to the issues involved than an analysis for literature on the subject is little and far scattered if at all.

First, Second, Third Worlds and so on. We now have eight worlds and supposed to be in the 5th world war. Appropriation of resources and resourcefulness. Based on this acquired power, the world has been triaged into eight subworlds!

In other words, since most relations are economic and costs of development are ecological, ecological strength determines the durability of external relations. Only When this foundation is strong, as supposed to be 50 years go for India, rest gets well built. Now the basic structure is still in its place while the foundation is weak, subsided and degenerating. Yet super structures are being built in the name of globalization. Therefore a perspective is sought to be developed for reframing our foreign policy based on ecological realities and environmental consequences of development.

Resources are important for development and human welfare. These are in nature and are interrelated which we call ecology. Ecology is the science of home systems which have many constituents such as soil, water, air, plants, animals, minerals and humans. These are interlinked for food, habitat, reproduction and defenses and energy flows across all these systems. Thus these are all interlinked, resourceful to each other and exist together in harmony characterized by communities and ecosystems. The value balancing of these resources is the resource economics and their dynamics in development is development economics. Together they determine political economy of a country and all the internal policies. Any improper or imbalanced development of these disturbs the environment of man everywhere and therefore the political economy and policies. Based on the strengths and weaknesses of this economy a foreign policy is laid to take advantage of strengths and fill up or compensate for the weaknesses. Since every country does the same, interplay is ensued- bilateral and multilateral- using every ploy available since the beginnings of recorded history. Since the roots of human evolution and development are in ecology and environment and are much recognized and lauded by all traditional wisdoms including Vedic Cultures, why not give it a primacy of place in the thinking that makes a foreign policy and the debate that precedes its implementation. This may be called a Paradigm Shift- if one chooses!

B. Ecology and Environment

The Planetary Components Are Interconnected and resources for each other. They are soil, water, air, minerals, plants, energy, animals and humans occurring in various systems on Earth - Aquatic, Terrestrial, Aqua-Terrestrial, Forests, Mountains, Deserts, Underground etc. All these are interconnected needing and influencing each other. These are supposed to have reached a state of equilibrium in time through evolution. This is called ecology. Humans gradually became the center and began using all other components including fellow humans. Gradually the equalitarian ecosystem became the resource for the key stone component. Thus the term 'environment' is borne as an anthropocentric synonym of ecology. Due to the conversion of resources for societal purposes, development, the constituents of ecosystems have been depleting and their resourcefulness dwindling popularly described as environmental degradation.

Some of these systems occur across several countries. Events on the ground- natural or developmental therefore affect all of them. Even developments in far away places affect local scenarios. Therefore a policy to deal with these unforeseeable realities on ground is warranted. The systems supply resources for economic development, the basis for internal and external trade and backbone for internal security. This in turn provides strength in dealing with other countries because external policy essentially is an extension of the internal policy. If this sphere is taken care, all else will be secure.

Environment is the anthropocentric term for Ecology. Ecology is the balanced/unbalanced occurrence of biota, habitats, climate & geography spaced together and evolved in time. Resources occur together influence, depend on and affect each other- geography - climate -Habitat and biota (Ecosystems). State of Society depends of Ecology and development of the interlinked resources and their resourcefulness and the environment it creates or undermines. Politics has become a war without weapons. And war has always been politics with weapons. Both need resources and are in the Ecosystems. From a planetary point of view this is ecology. From an anthropocentric view these constitute the environment of man from times immemorial.

C. Internal and External Policies

1. In the interests of a country if it is decided to follow a particular foreign policy towards some countries, this needs support from internal policy. In Malaysia & Indonesia different internal policies are followed for an identical foreign trade policy on timber in Sarawak - Borneo Region! Which caused different types of consequences?

2. If World Trade Organization is an important part of External Policy of every country as it deals with foreign trade and local economy, the environment clauses of it are as much an integral part of this External as deals with ecology resources that influence local economy and external trade and development of societies. If this is not realized the country may be triaged in the comity of nations and people with in like wise as per the diagnosis and imperatives of Paddock envisioned during the First World War. I believe that this too contributed to the growth of people under the poverty line and may never cease. If the natural wealth of nations is not conserved poverty becomes ceaseless and welfare a mirage which are the duties of any polity and the forces governing it. For these reasons the international bodies such as UNEP- UNDP- WHO- ILO- FAO- UNCED- have internal and foreign policy contents as well as of environment.

3. Ecology and environment are important for both Internal as well as External policies. If you spoil within, you could also spoil the external policy. Rich unite despite barriers. Poor break up despite common suffering! If you are unjust and selfish in internal Policy, how can you be otherwise in external policy? External Policy is an extension of Internal Policy towards ecological resources and development of trade surpluses, balances, debt burdens, etc. Resource Economics and resource ecology are the two sides of the same coin. Assessing the role of ecology in its foreign affairs therefore assumes considerable importance. Only USA has done this exercise and very recently. But the document is not available.

D. Experiences: Other Countries

The Arab- Israeli War of 1967 became the mother of Club of Rome and Die Grunen of Germany which climaxed after two decades in the Earth Summit in Brazil via Stockholm and Nairobi. Accounting and valuing NR can be used as a pawn or master in politics of war- unsettle countries- induce positive thinking- realizing the limits to growth futility of war and inspire and people to limit negative actions and accelerate positive actions. Thus the now famous slogan- Think Globally and act locally has come into vogue accepting that truly realistic solutions are only idealistic.

1. U S A and U S S R, The Salt Talks And End of Cold War: Much has been said and written about the end of the cold war and the breakdown of USSR. The question arises has the USSR to break down if cold war were to end? Yes. If one goes to the origins of the beginning of ***detante***. If USA does not have the information of environmental degradation of soviet productivity systems for years on through remote sensing and ecological surveillance, they could not have provided for the needs of USSR and influenced them to the negotiating table for SALT. If these degradations have not caused enough strain on the social systems within and among the communist countries all those divisions would not have taken place. Economy and ecology are intimately connected and inseparably- both in nature and society, the nurture. If we ignore this political upheavals are difficult to prevent.

The degradations of environment and its mother- the ecological niches precipitated revolutionary changes in internal policies and consequently in their external policies in Eastern Europe. If Communism absorbed an iota of Gandhi and Darwin on nature and societies communism would have won over the world and everlastingly.

2. Timber - Malaysia and Indonesia - Japan - Europe: Timber from the tropical rain forests of Malaysia and Indonesia is purchased by Japan and send to European Community after processing and satisfying its needs. This external policy of EEC and Japan to save their forest resources, and of Malaysia and Indonesia to trade their age old resources for some national interest, resulted in the denudation of precious rain forests and other consequences symbolized by the recent haze and smog in Kalimantan and Sarawak regions of Borneo island- the Indonesian and Malaysian political units respectively. In between it is interesting to note the policies followed by these two countries to accomplish these trade benefits. Malaysia awarded contracts for forest cutting only to Malays among its nationals, restricted the movement of its citizens to Sarawak in order to give a free reign to the contractors. This resulted in local revolts by local tribals to such an intensity and concern that its leader Harrison Ngao was awarded the Alternate Nobel Prize, The Right Livelihood Award. On the other hand in Indonesia, the contracts were awarded to the nationals as well as foreigners. The areas denuded of forests by them were filled by the immigrants from Java Island to such an extent that it was often called the ***greatest government-sponsored migration*** of people in history.

E. Experiences: India

1. Colonial times: Colonialism essentially was meant to enjoy the resources of weaker countries through political and military hegemony rather than trading and sharing. At philosophical level a sanction was obtained by invoking Darwinism in the nature to human societies. In other wards the foreign policy of colonial powers was based on the resourcefulness of others to build a good environment for them. Every colonial outpost will tell us the materials they have taken from us and return what they sent us. Ports, rail roads, trunk roads served these purposes. East India Company's Biologists scouted for plant and animal resources such as Medicinal Plants and introduced them into homoclimes under their control or acclimatized for their use. India has saved the sugar industries of some countries and introduced it in some others for the welfare of the colonial powers. But it was the external / foreign policy of the then government of India by the British.

When the British left India in 1947 there should have been a thorough review of all the records of ecology, environment and resources and a study of British policies on these in their country. These were unfortunately carried out partly and leisurely!

2. Displacement of Ecological Niches: Migrations of people took place through out history from Exoduses to Trickles in several parts of the world for a variety of reasons. All of them have one common cause - the threat to their ecological security - security of habitat for food, shelter, reproduction and harmony. If these migrations are engendered under pressure we call them refugees. The biggest migration in history took place in India in and around 1947 when over 10 million people were uprooted from their traditional niches and socio - ecological balances. As a consequence of this the foreign policy of India has always been under constant stress. For India it has precipitated a schism and constant security threat internally because the niche displacement happened in Pakistan fully but not in India, the residue hovering as a threat always. Human Ecology thus threatens both our internal and external policies.

We had to constantly appease the so-called minority community internally and Muslim countries externally a drag on ecology of resources no other country has to face in the entire history.

3. Displacement from Ecological Niches: Another niche displacement was initiated with the abolition of Jamindari system without affecting land reforms. Consequently peasants were displaced from their rural ecosystems. This is further accentuated by the sucking and magnetic appeal of the parasitic urban ecosystems due to competitively consumptive pattern of development process during the Peruvian era. Consequently both systems suffered; we have a shot up poverty line which today consumes considerable productivity of others while they themselves rendered unproductive. If the central dogma of ecology was understood that every organism gives what it an and takes only what it needs which both Lenin and Gandhi understood the economy of the country, the materialistic other side of ecological coin, would not be under such stress that the country's gates had to be opened fully to the rejuvenated East India Companies of yore.

We had to have governmental cooperation with socialist countries while social, economic and political realities are out of step with those countries due to the niche displacement of rural producers into urban consumptive system when they have no purchasing power to consume. The army of beggars, rickshaw pullers and pavement dwellers became too great in this way.

4. The Great Irrigation Scam of India?: Experts without exception maintained that India will not run short of energy what with its many rivers and enormous rainfall. Even the confidential dossiers by the British to their India Office were said to maintain this. Where have this water and the hydroelectric potential gone? Were we nose-led by experts from abroad and the agreeable counterparts from India? In addition we have also lost great amounts of top soil which may be termed as the great terrain robbery! If we can't conserve utilize what we have how can we obtain from others? When we established nuclear programme we ought to have thought of the energy and water needs of people instead of industrialization and uprooting people for lessons in this direction are available from history. We just ignored Meghnath Saha's alternate constitution proposals and the visions envisaged there in.

5. Indus Waters and Others that did not Flow: The agreement with Pakistan on the sharing of Indus waters was a task well done. But it was not matched by proper internal policy inputs in Punjab. With excessive water and its use this agricultural land has every potential to become the central plains of USA a few decades ago- a dust bowl. The best way is to give its waters to the neighboring states in the south and also encourage rural migrations to these parts to render them into Ganganagars! Punjab's water, lands of Haryana, Rajasthan and U P with considerable Punjabi populations. The Indus agreement should have been followed up with a Thar Desert Agreement on desert research and development beneficial to both countries all along the borders of Gujarat and Rajasthan. And there was the Quetta Conference in the Fifties to initiate the foundation work. To make these contribute in the long run for peace in the region the internal policies ought to have aimed at overtly encouraging the Muslim intelligentia to catalyze an evolution of Islamic masses to the same level as Christians, Buddhists, joins, Zoroastrians and of course the Hindus. This multicultural plural society is what we could call an ecological society as is the norm in natural ecosystems.

Aksai Chin which we lost to China was described as land where "Not a blade of grass grows ". If ecology was known then this idiom, justification and the appeasement policy that preceded could not have happened!

6. Allies in International Affairs: India had allies in UAR, YUGOSLAVIA, INDONESIA and GHANA during the Nehruvian era who worked in tandem on many matters of the world, the third world in particular. Had they similarly studied each others' resources, traditional wisdoms and introduced technologies and initiated exchanging while conserving and restoring the lost ones, our situation would have been better for other countries to emulate. Many of the arms twisting implicit in the GATT would have been resisted more successfully than could be now accomplished. Look at the specter in Ethiopia, Somalia and Eritrea - a region which was once the center of origin for many cultivated plants, and therefore could still harbor Germplasm, presents a specter of tragedy of ecology with denudation due to ravages of war and famine. The existing co-operations among the G 15 countries are welcome but quite late.

7. Silence Issue: The issue of Tamils and Sinhalese in Srilanka is basically a contention for balanced human ecology based on resource sharing or appropriation. If the intelligentia understood this partly through Marxist paradigms and partly through traditional wisdoms, policies could have been engendered that would not have precipitated all that mayhem we have been witnessing and does not appear to abate. The attempt to solve it through Indian Peace Keeping Force was ill advised on the score that they had no knowledge of the ecosystems and the ecology of the region. Predictably it became a fiasco enhancing the miseries all around.

F. Inferences

Ever since the hunter- gatherers settled as a community near water and therefore a food source for self and domesticated organisms, all conflicts, battles, wars and campaigns in essence happened for greater resources and better ecological advantage. The territories and ecology changed hands with the outcome of wars- Campaigns deforestation, crop fields, and taking women, seeding women. Australia benefiting from New Guinea, SA from Namibia, Belgium from Congo. Now we recognize five global wars and fragmentation of the three worlds into 8 / 9 all resulting from a varieties of combinations of factors- internal and external. The necessity of reorientation of India's Foreign Policy on a more realistic that is ecoistic basis is obvious unless we have forgotten the role of Khyber and Bolan in the foreign policies of the invading countries towards India.

To suit WTO (GATT) following the fall of SU, internal financial policy in India's has been under change of course. This liberalization exposed scams, which would have remained buried for ever. 5 Billion dollars however, appeared to have been already siphoned out from India according to some Miami based Economists by under - and over invoicing imports and exports. All East India Companies are back in business - Neocolonization is back with a bang. This cannot be faced without competence and philosophy far greater than India ever had. i.e. when Gandhi was alive and all powerful.

G. Perspectives

Broadly we have to classify countries as neighbours, fellow developing countries and the developed North. There may be any number of criteria to further classify them, in addition. In order to deal with all these we have to see whether the internal scenarios exist and ready for interactions. Briefly our objectives should be as follows.

1. Developed Countries: We need them more, they too need us: The developed countries may have systems similar to ours and / or problems similar to us. In either case the relationships among resources, development and prosperity is a lesson to us. We have to obtain every information about these countries, analyze, select and settle for some relevant and useful to us. Coordination among the embassies and the ministries of forests and foreign affairs necessary. Where they need our experiences collaboration should be carried out to upgrade ourselves and learn related aspects. But we should not forget that they have better and more information about us than ourselves in many respects.

2. Developing Countries: we need each other: Of the few scores of developing countries, a further classification is possible based on the parameters in the Global 2000 report. While the natural resources of all countries are valid for us a few among them, the G 15 have many things in common. Here considerable lesson sharing and collaboration and transfers of information, knowledge and traditional wisdoms are possible. However an element of patriotic fervor is necessary both in environmental issues as well as obtaining ecological informations.

The link is obvious:

ecology-resources-development-economy-trade
-sequels-consequences- Quid pro quos- Internal Policies
- foreign policy and many such matters.

3. Neighboring Countries: we need each other; they need us more: The neighbors may be proximal or slightly distant either by land or water. We can give them more than we need from them therefore came the Gujral doctrine, a modification of Panchsheel of Nehru. With some countries the ecosystems are contiguous necessiating to work together. Not only we have to defend from others' activities and designs but pursue policies in tune with the long term interests of our country. We already erred in two respects. On the west is smuggling of arms and terrorism to undo our development efforts and on the east pushing out local people changing our demography, politics perhaps even polity. This would not have happened if we always followed the ecological principles strictly even with respect to human populations. Pakistan dug underground passages from their side to Indian Territories. Has India done.

No. Can it Do.? No. There are no local people there to back us. The appeasement and patronizing policy within our country towards Muslims during the last 50 years is ecologically unacceptable because their contribution to the productivity (development), stability and harmony of the Indian ecosystem is negligible if not parasitic and predatory. Here comes the validity of Madhok principle of reciprocity. When it comes vital and basic things of balances and social harmony the Gujral doctrine is ecologically unsound.

H. Some Imperatives

1. The defense department shall embark on ecological and environmental researches as a part of the foreign policy as USA has been doing for long besides surveillance on the neighbors, all these in support of the forgotten Madhok principle, i.e. Gujral doctrine extended to natural systems, but assert like the chief of a traditional joint family. A few examples are : Mountain health research for defense department, Bird Migrations, Behavioral ecology of select animals.

2. Turmeric and Basmati behind, and about a score in the pipeline it is necessary to have a cell in the external affairs ministry to document all our traditional wealth, ethno-botanical lore, germplasm of all those materials the colonial powers used to ship to their countries in the name of trade.

3. Central Asian Republics of the former USSR, Burma and Indonesia and modernized Islamic Societies should be studied and cultivated with a hope that Indian Muslims also secularize themselves and begin to contribute to a healthy internal policy which in turn help the neighbors see reason and live in peace without wasting precious resources in wars and war - like activities. To my mind Cuba and Castro should be models for the developing countries in maintaining a just internal policy (Human Ecology) without which one can never have harmony with outside.

4. Ecological surveillance on the neighboring countries and on their neighbors. This has two components. Firstly we must have a thorough knowledge of their terrestrial and aquatic ecosystems and their resourcefulness and stability- Thailand, Indonesia, Malaysia, Pakistan, Ceylon, Burma, Nepal, etc. Secondly Pollution in neighboring countries may affect our systems (e.g. Himalayan Glaciers, Nepal rivers, Coastal waters, Air pollution in the Indo - Gangetic plains from westerly winds Rhine pollution in Europe, US polluting of Canadian air, Russian pollution of Scandinavian forests]

5. A Possible Check list for developing strength so as to be able to build ecological security for viable and useful foreign policy may include the following among others : Space science, Agrosystems, Remote Sensing, Aquasystems, Forests, Electronics, Ecological principles, Industries, Environmental facts, Scenarios, Wild Life, Wind and cloud movements

6. Our government should endeavor to develop Multinational Asiatic Corporations on conserving, restoring, increasing natural resources through researching and developing. Indian industrialists should be encouraged to initiatives. These could lead gradually to Regional Trade Agreements, Regional Development Agreements, Regional Ecology Agreements and Regional Environment Agreements. Indian Ocean Rim States can cooperate with each other in germplasm conservation, sharing and owning. South African flora and fauna are very unique and unparalleled in many respects. As a part of our foreign policy all these must find considerable priority.

7. Through Embassies, Collaboration with Academics, Tourism and other traditional means we should endeavor to protect our systems, draw lessons from others experiences and be strong enough to teach lessons to the offenders. Media Watch, Surveillance through Academic Exchanges, Spying, Remote sensing, practicing Futurology and Projections etc. In fact a professional team of ecologists should be on our embassies to do all these with patriotic fervor.

8. The role of academicians is very crucial in almost all things considered so far. So far they seemed to have failed dismally. If the first 500 youngsters who were sent to Britain for training and research included some for ecological and environmental purposes we could have probably begun well for all of them returned to India and most worked in our IITs. We therefore lacked in the initial preparedness. Despite Dr Nagendra Singh's presence in the Club of Rome in late Sixties, and Gandhi, Pyarelal, Kumarappa, Radhakamal Mukherjee etc, before rooted in our very Vedic wisdoms, the message of environmental crisis came to India a decade late! Reports similar Global 2000, Environmental Quality, Annual reports on the state of the art do not exist even now but for the Reports of CSE, a very eminent NGO of India who has initially modeled on the Shahabat Alam Malaysia of Penang. Academicians - both of governments, privates, universities are all blissfully silent! Indeed a great disservice to the nation. Even of today when anything about environment is a fashion most studies, teachings, researches, publications, consultancies, projects and fundings are fraudulent and self propelling while country spins to fall.

I. Conclusion

I feel considerable biological materials are being taken out of country even now clandestinely as herbarium specimens, seeds, pollen. Many projects financed by foreign agencies are suspect for their content of eco-intelligence. This is in addition to the aerial surveillance that we cannot prevent! The entire subject of ornithology especially dealing with migrations is a potential biological warfare tool! If the speculation that the AIDS virus is from the African Green Monkey of the Congo, there is no assurance similar agents may not visit us in future. If the theory of Genepidemics proposed by me a decade ago is extended a bit which says that

> Pesticides mutate pests to other forms needing new pesticides so that we will have both pests and pesticides perennially. These two in addition will be functioning as vectors as well as sources of genetic change in man leading to cancers, birth defects and cancers.

Other countries can spread mutant forms of bacteria, fungi, viruses, insets, vectors, pathogens through birds! I personally believe this must be happening. Today's Malarial resurgence also is traceable to the WHO's project on Malaria two decades ago which was force-closed under clouds of controversy.

If military strength and preparedness is a part of foreign policy, Why should defense departments finance research projects on animal behavior, bird migrations, vectors, and other similar biological and ecological subjects ?

The above experiences are very important for our defense against such foreign policy onslaughts as well as tools for implementing our policies on others. Unfortunately Indian ornithology has not gone beyond Salim Ali's taxonomic thresholds. Ecological aspects of habitat, nutrition, behavior, breeding, genetics etc of birds, vectors and other animals of potential ought to have begun long ago. The mosquito problem is one of ecological imbalance more than pesticide lethality.

All such problems just keep the entire country and its resources bogged down in historical problems such that we really cannot make into a viable future!

We have to have a strong data base on the natural resources, ecological relations, environments of the past and present and likely changes in the future about all neighbors, allies among the developing countries, and developed countries. Also we should examine their relationships with other countries connected to ecology and environment. If we review the entire foreign policy on this premise, new insights to the old issues and proper outlooks for future might emerge. This exercise is no means easy. Firstly it needs a shift in the mind sets and secondly warrants great data collection and assessment first on our ountry and next on other countries one by one and systematically and patiently. A tall task indeed. But necessary in the next century when most wars will be environmental and ecological in nature.

Chapter - 15

Ecological Future of India

1. Origins and prognosis

Planet Earth was resourceful both biological and non-biological. They interacted, cycled and evolved. When Homo came on the scene, he continued the biological instincts but soon found that all others can be manipulated to his advantage. The quest for discovery, ownership, consumerism, invention, hedonism and the like grow side by side with the awe respect and wisdom of continuing a life of harmony with nature's laws. The material and the ideal contrasted, conflicted from pre-religious times, through long centuries to modern times and appear to continue into the past modern times and appear to continue into there of the next millennium. But this appearance is a semblance, a mirage, a shadow before sunset as the materialism has totally ransacked the idealistic forces- most often as Trojan horses. But ran will not die and decay. Some will do. Some will flourish. The remainders will be the buffers .This is the writing on the wall for all to check.

2. Settlements And Conflicts

Human settlements formed into tribes, assimilated into societies grew into civilizations. They lived together and fought divided for dominion out other societies, i.e. others resources. The ***story of man's travel was full of conflicts, hatreds.***

3. The Continuing Wars

Skirmishes, battles, wars and holocausts. The most major one was the First World War, followed by the second one. The third was unleashed by the colonial powers and their allies from 1945 to 1970s. Marshaling all sources for Marshal Plan of growth and consumerism left cracks on their own being. Therefore a Fourth World War for self regulation, repairs, introspection, co-operation, coactions with the help of science and technology if not reason and logic continued until lets sights. The 3rd and 4th were greater were because man's selfishness war for more explicit during these times will not war happens but villains. The intellectual properties of humankind were used against fellow human societies with no pause.

4. Fifth World War

The naturalism has finally won its last round on idealism which is however graceously allowed to mark its feeble existence on occasions in the form of social protests. Homo sapiens how evolved into ***Homo cannibalis or Homo rakshasanis*** both environmentally masking themselves as ***Homo hypocriteus***. Homo sapiens is truly endangered and getting extinct. Thus began a Fifth World War to make a final trichotomy a winners, a loosers and the buffers in between. This is marked by the disintegration of Marxist Politics in Europe reducing the world truly into several worlds.

5. The Several Worlds

We had first- an old and new world. We had orient and accident their came the First World and Second World. The scenario changed following the First World War with the emergence of Soviet Union as the Second World country and christen the colonized Nations as the third world following Second World War. But now each of these are subdivided. The First World has become G-7 and the Remainder. The Second one became the Slavic and Islamic. The third world became the ASEAN, G-15, LDC, UDC. In all these eight groups there are rich sections and poorer sections- the former decreasing and the letter increasing from first to eighth worlds. These divisions are on economic bases, realities of day to day life. In between comes the Western Asia, a world by itself. In each of these there are tribal memories recalled for security reasons and or fears which were little existent before. The resources, their interdependences and common utilities, vanished through appropriation, exploration. All resources, interactions, coactions, interdependencies, cycling, taking according to needs and living according to ability disappeared. In their place came divisions, conflicts, hatreds and violences. The system, ecosystem broke up divided. Thus all violence today, as always before are basically rooted in ecology.

6. End of the Journeys

But the planet will not collapse. Man will not became extinct. Earth will not end as it began, without Man (Strauss). It is being divided now, the process began, as an inaugural of the Fifth World War through Science, Technology, economics, trade, politics, communication Blitz. In short, by dividing Man himself into or Jekyll and Hyde but with some mutated characteristics- Idealistic Mask and Materialistic Sinews. This arrival has materialized through a few convulsive Journeys.

7. What is Ecology

The club of Rome phase, which exists even today served to caution, provoke and send. Ward, Stevenson, Boulding asked us to treat the planet as a spaceship and any sub-microspic mishap may track the system to everyone's chagrin. Thus came the Slogan "Think Globally act Locally". Other clichés of "***borrowing earth from children***" and having but only "one Earth", ecology or development "***Development without destruction***". "Green", "Envrionmental" prefixes and a suffix of- friendly - joined the jargon.

People have lost sight
that ecology means simply,
equity and an ethos of empathy.

8. Club of Rome and Tragedy of the Commons

The Club of Rome is realistic. Pessimism idealism of the Spaceship was confronted by Simon and Kahn through science and technology. But their optimism cannot be said to envelope the whole globe, but only sections. This was highlighted by padlocks philosophy of triage. The 175 countries and their people cannot be taken care of. Only some who could be helped either sustain themselves or redeemed through inputs, even if they are Crouches. That one need not worry about the entire world (having exploited enough for several centuries) was also advocated by Hardin. Hardin cautioned of Life Boat Ethics and the tragedy of the commons. The former was a Christian and Hindu religious myth and the latter a part of native wisdom in most societies and part of the linguistic idiom.

9. Gaia hypothesis

There is some popularity verging on the Craze out the Gaia hypotheses of Lovelock. That the concept of Mother Earth, Bhoomata, preceded in Hindu though centuries before the Greek is a different issue. The Earth is taken as an organism with the competence to correct itself whenever an injury is caused. The available evidence in favor of this may either justify the continuation of ecological series, may render the Samaritans slacken and complacent. Gradually, under its cover Triage will be implemented with military precision through Science and Technology and Trade and religions sanction of Hardinism taking advantage of the demise of Marx and burial of Gandhi and existence of Christ and Buddha only as status and Photos. The opponents of Darwin, implemented his credo through social Darwinism at home and colonialism elsewhere.

10. Kalki Concept

The above scenario appear to be also at predestined as per the concept of Kalki in the Hindu Mythology. He may not emerge from Tamraparni and ride a horse in quest of decimating humankind. But the Kalki comes, has already a mind, in the form of Ecospasms, scores of them across Globe disaster whose rests or routes are ecological. Both, suffering and the like continue on a large scale increasingly.

11. Why Kalki

Mendel is manipulated; Darwin maneuvered and disoriented; Religions compromised and Marx discredited all by their own followers Freud and the Greek Gods and Goddesses of glory, pleasure and passion rule through materialism. Human heritage stands subverted by the audacious Right of Might. This situation is destined to warrant a Kalki to end this age for the ethic of Yugam i.e. ecological living seems to gasp for breath.

12. The Ecology Question

Thus the ecology question today is not material Vs ideal, Spiritual Vs Mundane, Biological Vs other Sciences, Politics Vs Economics Hedonism Vs Altruism- It is psychological covering the whole lot of the above - that of morals, of ethics, of truth. When these are threatened, as the Geetha's Preceptor said, the destiny befalls in the form of checks- Ecospasms, the Kalki the destroyer- to create anew. When revolutions from Spartacus to Sitaramayya say the violence to destroy evil is necessary to create anew, who can fault them? Will Marx be ever dead- However fashionable it may be to discredit, to dishonor and jettison.

13. The Karma of Backlash and Feed Back

When we discuss a question like ecology we naturally gravitate to our own niche, India. From heritage point of view two things must be mentioned-level of Philosophy and knowledge. Hindus or Sanskrit or Vedic literature in the earliest literature extolling the virtues of nature - Panchabhootas, the Ritus the Gods of Resources and Earth as the Mother. Only when they are all in their places, there shall be shanti. Those who are not initiated into the ancient wisdoms can recall the signature slokam / hymn of the Serial on the Indian Television, Chanakya. The concepts of evolution, righteousness, right livelihood- equity, Dharma, the virtues and vices, behavior of plants, animals, (Panchatantra) Prithvi Suktam, Harmony and peace, work vs fruits were all part of Sanskrit literary ethos. Duties Vs Rights contribution Vs consumerism (hedonism) was the theme of Bhagavad-Gita. To implore on the correctness of all the above wisdom, drive it deep into the ethos of a nation, the concept of destiny and Karma and God as the arbiter had to be created. And they even now, 4 millennia later everything else, everywhere else collapsed and gave way. Our ethos still exists and lives.

14. Tamil Heritage

The Tamil literature, pre-biblical, similarly has many of the wisdoms of Sanskrit literature. While Sanskrit is dying following Greek and Latin, Tamil flourished. Unique vitality of the People heritage and culture.

15. Knowledge Systems

Together or separately these descend a prime place in the knowledge system of the world in-a-vis ecology environment and future of man. But it does not. It is full of European wisdoms and ending in Greek myths and roots. This might have happened in other fields of human activity- But allowing this continue on a matter so vital as ecology speaks of live Indianness of Today's India and failure of our recent knowledge systems.

16. Idealism Won Then

Yet why were/ are we behind the North. An answer can be found in Sankara, the Acharya. It appears that Sankara defeated all the then leaders of Knowledge systems and saved Hinduism. Who were they? Buddhists, Jains and a few other Hindu savants who were materialists or protomaterialists. This continued. On the contrary, to idealists in the west lost to Materialists several decades ago. The idealistic Philosophies and knowledge systems based on them promoted only advances or reforms or interpretations in these in India while the materialistic Philosophies promoted inquiries and probes into the secrets of nature, using the unraveled truths for the welfare of their society.

17. Materialism's Triumph

The Schism grew, materialism has won our idealism. We remained behind. This idealism should not be confused with ideals, theories, principles stand et al of our modern age. Let has more of metaphysical and spiritual content and less of mundane matters.

18. Neomaterialism

Unfortunately the neo-materialism in India devoid of spirit of enquiry but full of consumeristic appetites in India is corrupted by versions of this "idealism" of superstitions kind and the materialism of the west is not tempered nor harmonized by the wisdom of Sanskrit and Tamil Literatures. The first is an irresponsibility, a kind of Cannibalism and the record, an unfortunate state of history.

19. Unity in Diversity - Diversity In Unity

Indian Society is fragmented. During times when Indian sub-continent was divided into a myriad Kingdoms, the people were united culturally by the traditional Hindu ethos. Through colonial intervention, we became politically one. Despite being a political entity, we have become divided culturally. Can political vivisection be far away? Why this cultural divisions- Individual, social and Educational are the reasons.

20. How Many People Matter in India

Only 200 million people matter in India all with same power of sorts to mend and maintain matters. Of there about 10 million are decisively consequential i.e. about less than 2%. The most dominant or ruling ethos here can be summed you be revised version of Camus's lamentation, half a century ago. "when crimes wear the apparel of innocence, it is the victims who are condemned and decimated". The old adage that power corrupts and absolute power absolutely is reverse here. Corruption empowers and absolute corruption brings absolute power, Strangleholds, coteries, caucuses, rackets, Mafias and the like.

21. Beleaguered India

Educated man tries to become superman, i.e. Frankenstein and woman, a man i.e. a devil. Premchand said that when a man tries to become a woman, he becomes a Mahatma and when woman tries to become a man, she becomes a devil. Thus in India there appear a great future existing for some, a little future for many and no future for a great many. The only hope, i.e. from educated and cultured class who carried out reformation in various times today are hopeless. They are busy helping themselves at others' expense with immunity-changing systems indulging in nepotism, casteism, regionalism, prejudices and such negative values. Everything is commercialized- using such technology like cafeteria system, credits, market forces, workshops and adding values in abridged numerical format to suit the American and European formats. The entire psyche of youth during their formative periods is systematically corrupted by disinformation, immorality, cynicism, hopelessness or intimidated.

The Universities, a fountain heads of higher education producing thousands of youth to work in society for society's welfare are bastons of Criminality, nonchalance, apathy, irresponsibility- the faculty and Vice-Chancellors- most of them and increasing with time. A saturation point is feat reaching. While political economic and social crime, are dealt with by electoral and legal forces, the academic crimes go unpunished though often exposed. The country is reeling under the weight of higher education- unaccountable, irresponsible and duty evasive. It is this class that did not caution in advances of ecological writings on the wall and at the same e time use the label of Ecology, Environment, eco-green, friendly for their works, meaning very little for society. Yet they take no stand on any vital environmental issue. Most of their work unreliable, repetitive and sometimes blatantly fraudulent.

22. Summing up the Forebodings

Natural resources are thus exploited or developed through interventions of Science, Technology, trade and politics. Consequently the environment of humankind was differentially effected by depletions and wastes. Thus ecology was affected. All this was done knowing the effects fully well through irresponsibility, escapism, cynicism and a combination of Nimbyism, Nimyltism and Nodmytism. In other words truth is subverted, Morality abdicated equity disregarded and empathy disinherited. In a nutshell ethics is divorced. All these happened and continue to happen to different extents in different societies of the planet more in countries like India. Wherever there careless those societies will survive and flourish. In others, hope can still be rekindled if the systems of higher education, the leaders of society and cradles for future society are not afflicted. Otherwise there is no hope. For most of India and Africa there is no hope, violence and divisions therefore are inevitable. In India disbanding universities the society's predators, pests of human spirit will reduce the cancer from metastasis. And there is nothing wrong with systems and reforms. It is the human material the educators- The lumpen criminal fringe, the genuine ones settling to serve in heaven or hall or marginalized into extinction in India. Though regulations were eased in the economic front, the regulators remain. No matter what is received, planned and done, as long as the unethical, dishonest, incompetent immoral educators remain in the saddle of not flourish, ecology of India, social welfare and future of India and its people will continue to deteriorate. System may be altered but the same humankind run them for their own purposes and others'detriments.

It is horrifying that we have to fight our own government to save the environment.

Ansel Adams

REGIONAL CONCERNS

Chapter - 16

The Kaveri Quagmire

Guidance from Ecological Wisdoms

The failure of discussions on the KAVERI water sharing between Tamil Nadu and Karnataka is not important. What is very significant is that these were held in a regular, friendly, serious, confidential and sustainable manner departing from the acrimonious past and leaving no rancour in the wake of their failure. These seem to me only one step away from a historical success if only a truly Aristotilian spirit is brought to politics, i.e. invoking ecological wisdom.

Kaveri is the most utilized river in India. It has been a part of civilization and culture of Karnataka and Tamilnadu for ages and both need it, as ever before. Tamilnadu being a perennial water scarcity region, enlightened self interest demands practical wisdom. And Karnataka enjoying a natural advantage should not appear to be insensitive to others sufferings. We had enough of antipolitics in the name of politics. It is time to practice politics- as per the original meaning of the word i.e. to bring people together and maintain balances and harmony among them.

Without harping on the data which filled print media in Karnataka and Tamilnadu, I shall approach the issue from the standpoint of ecology.

Enhance Water Availability

The first problem is to increase water availability. Several steps may have to be taken for this.

1. Deforestation must immediately stop at the catchment areas of Cauvery and all its tributaries. Intensive reforestation should take place all along the route of cloud passage from Coastline.
2. Cloud seeding for increasing rainfall should be attempted in these regions with the aid of the Indian Institute of Meteorology at Poona. There should be stations for seeding the clouds at appropriate places. If the seeded clouds pour down elsewhere, they should be monitored and appropriate salvaging steps must be taken.

3. For harvesting the rain water, Auroville, should be contacted. They are a great example for water and soil conservation.
4. Whether the hill streams around the catchments of the tributaries can be maneuvered to contribute to the Cauvery system has to be explored.

Conserve Available Water

The second problem is to save water. Excessive irrigation is often a farmer's privilege. It is difficult to control this psyche. Given the availability of water (Borewell / Pumpset Facility) and the wealth position or access to credit, the tendency to irrigate more lands will always exist at micro level in a free enterprise system.

But this has got to be dealt with.

1 Acreage under cultivation in the command areas should be frozen which seem to have actually increased everywhere.
2. Acreage under crops needing more water should be decreased, especially at the tail end regions of all canal systems.
3. Alternative dry cropping should be encouraged at appropriate places by a generous pricing policy for the yields. Otherwise switch over to dry crop is not that easy in traditional wetlands.
4. The importance of modernizing canal systems to prevent seepage and water logging has been stated before and need no reiteration. Yet the use of this water by the farmers all along the line must be judicious, monitorable and controllable.

Scarce Precipitation

The third is to manage when times are distressful vis-a-vis water. During such times of low precipitation, perhaps the tank fillings should be deferred while canal systems shall continue to operate. Lands dependent on tank irrigation have to switch over to dry land agriculture. Since the upstream regions will still be privileged, Karnataka is expected to help out the downstream consumers with staple food, rice, in our case. Modalities of managing these distress times and trade links are to be worked out. Climatologists and past scenarios of distress are to be the guidelines.

Copious Rainfall

In times of copious rainfall as happened during the last few years and much more this year, temptation to grow water intensive crops must be resisted lest it becomes a habit. The additional water shall be diverted to tanks and work out a strategy to make it percolate into water tables for later use on a dreary day.

Social Responsibility

Last but not the least; the industrialists should be goaded to initiate an eco-political exercise. They can form a consortium and work for realizing the imperatives of the four outlooks mentioned before, thus heralding an aqua green revolution to make kaveri serviceable. Some beginning was made in 1987 in the form of a Kaveri Festival at Erode, but was not continued.

Common Sense

All problems can be solved, at least eased, by common sense. So too this issue. It must be independent of judicial awards and political pressures. This is an ecological issue connecting a natural resource (water) with productivity (agriculture) and linking climatology (precipitation) with human need (consumption). The same Science and Technology or political will that commanded a flowing natural resource into one of total human use, had to be employed to solve this issue, but with absolute ecological sense, i.e. equity.

The entire Cauvery ecosystem has to be assessed considering all links, interactions, interdependencies and chain reactions, i.e., holistically. Experts in Climatology, Agronomy, Hydrology, Economics, Ecology, Geography, Forestry, Trade, Industry, Engineering etc. and institutions such as NIO, NIM, ICRISAT, IARI, For a change, all political parties should exhibit statesmanship and play genuine politics-reserving the gimmickry, melodramas and the like for ephemeral issues - if at all. Confrontational approaches and protest routes to problems of nature and ecology can do no good save increase conflicts, kindle hatreds and unleash violence. Such thoughts divide people irrevocably and break up the fragile Indian polity sooner than feared.

A Silver Lining in this direction already exists in the form of Guhan's book- **"The Cauvery River Dispute: Towards Conciliation"**, undoubtedly the best to have been published on Cauvery. And its time has now come. He was right with his data and therefore courageous and powerful in his assessment of the situation reconciling the genuine interests of all. All political parties intellectuals, social workers and Humanists in both Karnataka and Tamil Nadu will therefore do well to make his book and the pious thoughts expressed in this article by an ecology teacher as the bases for further actions.

1. Tanks, Rainfall, Cropping patterns, Other land uses and users, Other water uses and users, Kaveri Seminar, Erode Festival, Guhan's work, render Politics Secondary and ***Ecological Criteria Primary.*** All disputes are amenable to an ecosystem approach
2. The hawala spirit from Manmohan Rao Dunkel has lead to the installation of a government of mutually checking and balancing partnership of political forces at New Delhi. This welcome mishap heralded KAVERI discussions - regular, serious, confidential and

friendly - another welcome development for all others to emulate. That the talks failed to resolve the issue is secondary. What is significant is the departure from the acrimonious past in a truly Aristotelian spirit of politics.

Chapter - 17

Ecography of Coastal Tamilnadu

A. Introduction

Coastal areas are commonly defined as the interface or transition areas between land and sea; they are diverse in function and form with no exact boundaries and are viewed as an integrated system. Coastal areas are also important ecologically, since they provide a number of environmental goods and services. The dynamic nature is a peculiar characteristic of the coast, which results from the transfer of matter, energy and living organisms between land and sea systems. It is estimated that 90 percent of the world's fish production is dependent on coastal areas at some time in their life cycle. In addition, these areas support large numbers of migratory and non-migratory waterfowl and shorebirds, and endangered reptiles. So these areas are much important for the social and economic welfare of current and future generations since they support key economic and subsistence activities.

Wetlands generally occur at the interface between terrestrial ecosystems, such as uplands, grasslands and aquatic ecosystems. Thus, wetlands are neither wholly terrestrial nor wholly aquatic but exhibit characteristics of each. They also depend greatly on both. The wetland's function as a site of Biodiversity is also valuable to humans, dependent on these ecosystems for commercial, hunting, and recreational uses. The capacity of wetlands to absorb a great amount of water also benefits developed areas. A wetland system can protect shorelines, cleanse polluted waters, prevent floods, and recharge groundwater aquifers, earning wetlands the nickname "***the kidneys of the landscape.***"

The functions of the wetland and the values of these functions to human society depend on a complex set of relationships between the wetland and the other ecosystems in the watershed. The watershed is a geographic area in which water, sediments and dissolved materials drain from higher elevations to a common low-lying outlet or basin a point on a larger stream, lake, underlying aquifer or estuary (EPA 1998).

Physical features of coastal ecosystems, such as reefs and belts of mangrove' are important for the mitigation of the effects of natural disasters, such as storm-tide surges, shoreline retreat or floods. These features also play an essential role in natural processes, such as land accretion, and help to control coastal erosion and other damage arising from wind and wave action.

The environment requires water of sufficient quality and quantity to maintain ecosystem function and biological resources, which in turn plays a major role in maintaining the quality and quantity of water resources (WBCSD, UNEP 1998). Water may be required for various purposes like drinking and personal hygiene, fisheries, agriculture, navigation, industrial production, power generation and recreational activities. Obviously, a variety of wetlands like marshes, swamps, bogs, peat land, open water bodies like lakes and rivers, mangroves, tidal marshes etc., which can be profitably used by man for his various needs and for environment development.

Coastline of Tamilnadu extends over 1000 Kms. The Coastal tracts consist of upland plains, flood plains, Deltaic plains and Coastal plain. The Geographical, Biological, Sociological and economic characters are exclusive on the coast. It contains critical terrestrial and aquatic habitats these habitats together comprise unique coastal ecosystem, which support a rich biological diversity and contain valuable natural resources. These Valuable and irreplaceable assets are subjected to ever increasing pressures. Further the Bay of Bengal is one of the favored regions of low-pressure and is also subjected to the ongoing global change.

Since the coastal ecosystems are diverse and productive. Sustaining coastal ecosystems will require coordinated multi disciplinary efforts involving all levels. In assessing effectiveness, productivity and value, particular emphasis is given to the identification of the on-the-ground outcomes and the status. Thus the coastal stretch of Tamil nadu and its coastal systems were surveyed for its on ground status on wetlands and potable water for its conservation and Development.

Coastlines and beaches of Tamil nadu in particular are dynamic and fast changing systems, which are the vitally important to the local economy. The coasts and their varied users are confronted with many threats including Coastal Biodiversity loss, Shore line erosion, wetland loss, unsustained human access to coastal resources, Developments and pollution. Increasing damage to coastal areas of Tamilnadu will be linked more to the development than populations increase or agriculture.

B. Coast Lines

1. Water bodies: 570 water bodies were recorded from the entire study area of which 155 tanks; 71 backwater areas and 24 sites of mangroves, the status of most of the wetlands are in poor conditions. Many of these wetlands were drained and transformed by Road construction, anthropogenic activities like unplanned urban and agricultural development, industrial locating after road development and resource extraction causing substantial economic and ecological losses. These water bodies play vital role in storing the rainwater and recharging the ground water and the sub surface water (Table 3). Coastal wetlands such as lagoons and backwater help in reducing the effects of flooding by tides and also reduce saline water intruding into the inland. Water which wetlands hold is of greater value, recharging ground water.

Nearly 170kms of Storm drains were along the study area. These are valuable in retaining and detaining water for some time, which allows water to percolate into the water table. All wetlands perform certain function and have different values. Functions of each wetland differ with different degrees from place to place. Coastal wetlands of Tamil nadu are seasonal dependent on the on the cyclonic storms. Failure of monsoons in one or more years will have serious effect on the biota associated.

2. Potable water: The seriousness of the water problems in all the sectors was equal. Almost all the sectors have saline water Intrusion. A total of hundred and three villages were found with increasing water scarcity (Table 2). In the entire region the surface water supplies are often limited by low rainfall, and coastal aquifers are a major water source. Groundwater extraction along the coastal stretch today exceeds the natural recharge rate, causing saltwater intrusion into the freshwater stock. The quantity and quality of water recharge to coastal aquifers are also affected by the drainage of irrigation water and by the construction of coastal roads and their drainage systems. The Saltwater intrusion into coastal aquifers has long-term impacts not only for the sustainability, but also Saltwater intrusion is difficult to handle and even more difficult to reverse.

Large portion of the coastal region of Tamilnadu is of alluvium formation and few regions are Cuddalore sand stone. The dynamics of alluvial landscape and natural sedimentation patterns which determine the nutrient and energy flows in coastal areas are increasingly being modified by human activities. This has higher impact on the vertical accretion by slowing down the accretion and thus aggravates salt-water intrusion and impairing drainage conditions in riverine, delta or estuarine areas. This further also reduces or blocks sediment supply to the coast.

3. Fragile zones: There are around 20 major fragile zones along the East Coast (Fig 3), these are already undergoing stress of different degrees, and further stress on these zones will alter the ecology of the systems and regions. Habitat degradation is the major cause of Biodiversity loss in wetlands about 26% and 44% of the area under mangrove along the coastline is estimated to have lost. The coastal resource system is interdependent with inland resource systems, excessive cutting and clearing of vegetation on coastal areas will result in severe erosion. This enables to carry away the topsoil and cause heavy sedimentation of fish habitats, such as sea grass beds and coral reefs, degrading and ultimately destroying them.

Poor land use practices agriculture activities and clearing of vegetation has increased land derived sediments flowing into the coral reef resulting in their degradation in Gulf of Mannar Increasing demand for water are occurring for three key areas, human needs for safe drinking water and agriculture needs for expanding production to meet population growth and industrial needs to provide more goods and service for the growing population. There is an emerging fourth category need is the environmental needs, which must be meet to attain sustainable development and maintaining the ecological balance. The need for water can't be ignored which is crucial to protect the critical, Fragile ecosystems of the coast. Less fresh water inundation was a reason for the degradation of mangroves and coral reef in Pichavaram, Muthupet and Gulf of Mannar.

4. Developmental activities: The East Coast Road development from Chennai to Kanyakumari is likely to boost up the urban developments. The related impacts of nearly thirty-five coastal towns will be a poisoning threat to the coastal ecosystems, their biological diversity, environmental regulatory functions and role in generating employment and food. The population density along coastal region is higher compared to inland and will tend to increase with development in coming years, will increase the pressure on the coastal resource.

Most of the developmental activities have direct impact on the coastal areas, the impact varies with the structure and type of development, and especially road construction along the coastal zone has variety of impacts. Road crossing are potential barriers to the movement of small stream fishes. Road crossing designs vary from simple, low water fords to massive concrete or earth filled structures, some of these crossing types may act as semi permeable or seasonal barriers for fish movement

Road Development will increase the potential for economic opportunities along the proposed ECR from Cuddalore to Kanyakumari as happening from Chennai to Cuddalore. This will strongly attract immigration from economically depressed rural areas nearby. Larger numbers of younger populations are expected in the coastal areas, which will increase the demand of water, housing, energy, food, employment, goods and services. These will be a higher developmental challenge in future.

A total of 338 Aquaculture sites were recorded throughout the study area of which, the sector-F from Adirampattinam to S.P.Pattinam alone had 183 sites. Sector - G from S.P.Pattinam to Ramanathapuram had 67 (Table 3), concentration of Aquaculture sites on these areas are likely are damaging the existing wetland ecosystem. Further reduction or restriction in the fresh water flow into these wetlands will leave many destroyed. Coastal Aquaculture must be environmentally managed and absorbed into the coastal management planning process.

Large portions of the Coastal areas do not provide unique biological ecosystems from sector H to I. The recreational and aesthetic values of these areas can be utilized for valuable tourism activities, as well as for industrial development and human settlements. Since the Tamilnadu coastal areas are expected to become more urbanized in future the dependence on coastal resource will become more stronger and to meet the needs and raise the economy. Rapid industrial development will emerge; the areas of less biological ecosystem can be utilized with better linking centers.

C. Conservation and Development

Observations on the East Coast of Tamilnadu have brought out variety of issues that has to be put under immediate actions. Lack of potable water and their cumulative impacts due to road construction, degradation of wetlands and other coastal systems are to be studied in detail. All the sectors are in similar stress. South of Karaikal onwards needs an urgent and more detailed work on all the aspects of potable water. Conservation and development of Wetlands have direct effect on the surface and ground water, thus development of wetlands will also improve the potable water needs.

A major enquiry into the water along the East Coast Road is urgent. Since large portions of economic activities, with the exception of agriculture, are concentrated in the coastal towns of Tamilnadu. The Road induced development will bring up more stress on this region, so the coastal systems have to be regularly monitored for its conservation and development.

1. Potable water development: The survey has clearly shown the degrading status of the surface and ground water, strategy for portable water development is highly necessary for the entire coastal stretch. Wetlands playa key role in the hydrological cycle by retaining run-off and slowing the flow of water. The retained water percolates into the underground aquifer and recharges ground water and also filters pollutants to both surface water and ground water along coast. Thus conservation of wetlands is vital to develop the potable water status along the East Coast.

Clearing and loss of wetlands will disturb the hydrological and hydro-ecological balance. If a wetland is cleared, or doesn't receive rainfall for few years the biological systems dependent will vanish disturbing the ecological balance of the region. A detailed study on the ecology of all the coastal wetlands and the related system is necessary for the development of the potable water and maintaining the ecological balance of the critical habitats.

2. Ground water recharge: Though the average rainfall ranges from 75-110 km, which is highly dependent on the Cyclone depression. The number of rainy days is coming down; the discharge of rainfall is concentrated with time in recent years. Fresh water inflow into the sea reduces the length of seawater intrusion into the ground water, still if the withdrawal of ground water is more it brings in seawater into the table, So it is highly necessary to conserve the part of runoff into sea.

Younger alluvial and deltaic deposits cover the coastal track of Tamilnadu. The deltaic deposits have high potential of ground water recharge. Artificial recharge programs are necessary at regions with high potential of recharge to avoid the declining water table. The below said methods will be best suited for coastal regions like Tamilnadu

- Water spreading (spreading surplus water into large tanks and ponds)
- Injection method by injecting water into Aquifer system (Adequate care on the water quality has to be taken)
- Construction of percolation ponds and check dams to allow percolation.

Areas of artificial recharge should be based on the intensity of the problems, the course of action should follow.

- Identifying the coastal aquifers and demarking the area for development and vulnerable areas needing immediate remedies
- Collection of hydrogeological, hydrological, land use, water use and ecological data
- Applying the management technique and monitoring

3. Management Strategies: Coastal zone management is the process of combining all aspects of the Physical, Biological and Social of the coastal zone within a single management framework ***"holistic coastal management"*** to emphasize that careful planning and management of all sectoral activities simultaneously will result in greater overall benefits. Pursuing sectoral development plans independently of one another will result in diverse effects on each other. Since the entire sectors needed inputs in different aspects.

Both wetlands and potable water are interlinked. Since the coastal zones are fragile and diverse special attention is required for managing the resources, for which the resource base of the region has to be increased. The following guidelines can help for a better coastal zone management

- Increasing Resource base
- Land use planning - land use zoning
- Restriction in resource use
- Multiple concept (Wetland, agro forestry)
- Monitoring resource bases and inter related systems

- Identifying constrains and development of management alternatives
- Policy evaluation and revision

D. Acknowledgement

Citizen Consumer and Civic Action Group (CAG), Chennai funded two surveys and Pondicherry University, one. Government officials and Village Heads gave full support.

(in) Symposium on Conservation, Restoration and Management of Aquatic Ecosystems; Center for Ecological Sciences, Indian Institute of Sciences, Bangalore, December 9-13, 2002. p. 108.

NATURE SHRINKS AS CAPITAL GROWS.
THE GROWTH OF THE MARKET CANNOT SOLVE
THE VERY CRISIS IT CREATES

VANDANA SHIVA

Table 17. 1
List of Sectors of the East Coast Road

Sector	***From***	***To***	***Kms.***
Sector- A	Markanam	Cuddalore	50
Sector- B	Cuddalore	Chidambaram	50
Sector- C	Chidambaram	Karaikal	50
Sector- D	Karaikal	Vedaranyam	60
Sector- E	Vedaranyam	Adirampattinam	55
Sector- F	Adirampattinam	S.P.Pattinam	70
Sector- G	S.P.Pattinam	Ramnathapuram	60
Sector- H	Ramnathapuram	Sayalkudi	65
Sector- I	Sayalkudi	Tuticorin	65
Sector- J	Tuticorin	Kulasekarapattinam	55
Sector- K	Kulasekarapattinam	Kanyakumari	75

Total = 670 (Kms Approxunately)

Table 17. 2
Potable Water Status In The Villages Around The Ecr

SECTORS	***IS***	***SA***	***DA***	***SI***	***Total with overlaps***
A	6	9	6	2	**23**
B	10	7	2	5	**24**
C	8	5	0	3	**16**
D	2	14	1	4	**21**
E	2	7	2	5	**17**
F	1	19	2	0	**22**
G	5	8	1	2	**16**
H	10	4	1	2	**17**
I	12	0	0	9	**21**
J	17	0	3	0	**20**
K	30	0	6	5	**41**
Total	**103**	**73**	**24**	**37**	**237**

IS: Increasing Scarcity SA: Seasonal Availability
DA: Depleting Aquifers SI: Salinity Intrusion

TAble 17.3

A Summary Of Selected Data From The Ecr

SECTORS	*VILL*	*CLV*	*BRG*	*SD(Kms)*	*WB*	*TK*	*BW*	*RV*	*MG*	*AQA*
A	30	22	08	26	28	23	02	04	01	07
B	34	61	10	17	34	13	01	01	-	13
C	27	94	12	04	45	16	02	04	-	24
D	25	74	16	13	75	17	11	05	02	25
E	23	74	10	38	98	27	07	01	02	17
F	36	116	21	10	140	17	18	01	05	183
G	08	90	09	06	52	10	18	-	12	69
H	17	48	06	27	53	16	05	-	01	-
I	23	53	05	19	18	11	04	-	-	-
J	16	31	07	09	14	05	02	-	01	-
K	41	74	04	-	13	-	01	01	-	-
Total	**280**	**737**	**108**	**169**	**570**	**155**	**71**	**17**	**24**	**338**

VILL- Villages
CLV- Culverts
BRG- Bridges
SD- Storm Drams
WB- Water Bodies
TK- Tanks
BW- Back waters
RV- Rivers
MG-Mangroves
AQA- Aquaculture

Chapter - 18

Ecological Perils of the Sethusamudram Project

The Supreme Court deferred the hearings on the Sethusamudram Project to mid - July giving everyone another chance to reflect. A Re-look into the documents made available is therefore called for. Prima facie, it appears that the project was rushed through jumping over standard scientific procedures and oblivious of ecological implications and environmental consequences.

Public Domain

The 1000+ pages in the "***public domain***" tells that the current phase of the project began from 1998 when a Steering Group was constituted to oversee future activities. Tuticorin Port Trust was made the nodal agency. National Environmental Engineering Research Institute was asked to make an Environmental Examination, and propose an Environment Impact Assessment after a Project Report was made by L&T Ram boll. An Environmental Monitoring Committee also "***was put in place***" ! Finally a Committee of Eminent Persons was constituted in October, 2007. Thus NEERI carries maximum responsibility for giving EIAs, Environmentally Viable Route, a Techno-Economic Feasibility Report, an Environmental Management Plan and a Post-Project Monitoring Mechanism. By backing the NEERI proposals, the CEP shares its accountability. However, its report is not kept in the public domain!

Impact Assessment

The EIA appears good only on the non-dynamic aspects, which are needed for the channel building. It is weak on the dynamic parameters which affect ecology and its chain reactions. It bestowed only a ritualistic attention on the socio-economic aspects of the coastal ecosystems. This state of affairs is deplorable because EIA consciousness developed since the times of the Growth Models of Club of Rome during the '70s. In fact, some of the EIA

Reports in the world added to knowledge, methodologies and became models for future. NEERI and CEP must be aware of these. Despite the trust reposed on NEERI, Institutions and Academies dealing with Marine systems in India, UK, USA and Australia, United Nations Environment Program should have been consulted.

Ecology

Ecology tells us that this region, an ecosystem consisting of the Indian coast, from Vedaranyam to Tuticorin, 21 small islands, an Inhabited Pamban island, a marine biosphere reserve, a 30-km barrier reef, the Northwest Srilankan coast, Jafna islands and shallow waters, are very interactive. The role of the Reef is yet to be studied. Limiting studies to the Indian Territory only, is therefore unscientific. An effect on any one component initiates chain reactions and impacts at one microhabitat influences others, because ecosystems are contiguous. A channel in this fragile ecosystem, constitutes an intervention without home-work and a risk in the least.

Pollutions

Transport- generated pollutions in future, not only destroy biota, affect the quality of the fish but also cause acid rain from southern Kerala to the north of Vedaranyam.

Marine Silent Valley

The biosphere reserve is a "***Marine Silent Valley***". Data and experiences from other marine parks would be more useful than the exercise on a rare Tsunami which is routine for the Research Institute at Mandapam. Instead, ecological studies on the 30km-Indo-Srilanka Barrier Reef should have been initiated for its role in the fragile ecosystem.

CEP was satisfied with the "***20 Km distance***" between the first island of the Reserve and the barrier reef. This distance is nothing in a sea. Instances of Pollutants traveling thousands of miles should have been known to the committee. Further, irrespective of the type of alignment, the pollution normal in shipping routes and harbours, will impact biota cumulatively and irrevocably. The biosphere reserve will be lost in time. Mangroves on the east coast and between Kilakkarai and Adirampattinam continue to be victims of salt industry and aquaculture.

Biases are obvious pertaining to the dynamic aspects - the physical, biological and chemical parameters which change under stresses. Data on these are to be gathered in an ecosystem spanning between the coasts of India and Srilanka and from Vedaranyam to Tuticorin . Sites are to be selected covering all the representative micro-locations on the basis of statistical requirements.

Sampling has to be periodical based on natural cycles and episodes of storms, cyclones, etc. Sampling duration should be longer because, shorter ones could be questionable.

Home Work?

Though 6 years were available for such a home work from 1998, only 2 years were utilized! Data generated from "***single season***", or "*three seasons*" are unacceptable. Since the ecosystem is tough and data collection painfully slow resorting to shortcuts is a natural temptation, as they started very late. Claims of NEERI, are therefore suspect, and dependence on them by CEP, naive.

Discerning from the memberships on various committees, it is widely felt in the scientific circles that a political decision was taken first, "***internally***", to keep the project "***on fast track***" and initiate "***advance Actions***"

Channel Without Banks

That this is a channel without banks in a fragile ecosystem whose ultimate effects terminate in the coastal societies, is ignored. A project without the benefit of wisdom from a precedent is being pushed at political speed without the backing of sufficient homework. It is still not realized that the adversities of SSCP, if and when materialize, will be borne exclusively by the local people. It is a bad omen for the country if scientists are influenced to bypass standard methodologies and ignore ecological thumb rules. In addition, the Indo-Srilanka barrier reef, like all reefs, will be of significant ecological value, and should not be dented for this very reason, the consequences of which are neither predictable nor reversible.

In this context only a campaign by eminent citizens only can force the government to review the entire project from scientific and ecological perspectives.

University Today, 623: 8, 15 June 2008.

Chapter - 19

Ecological Solutions for Telangana

Ecosystem - Centric Development and Administration

Infogandas

Absurd Info-Gandas in Andhra Pradesh continue to be macabre, propagandas masquerading as information! While the erstwhile British Andhra fears the loss of development spaces, the erstwhile Naizam waits to regain them as "Lebensraum". As the up-man ships and opportunistic games exposed both leaders and parties, the credibilities of institutions and the polity stood eroded. Arguments shouting at each other; people have lost and politicians won. The ideas of India lie dead. Meanwhile, staring at us are the Senghaas ' "Clashes within Civilizations" with Divisions, Conflicts, Hatred and Violence.

Legacies

Legacies held Telangana people captive to the Nijams, Islam, Rajakars, Pakistani Sentiments, Feudals, subsistence levels of existence, isolation and absence of rights and accesses. They are also linked to Urdu, Telugu, Marathi, Kannada, and Ecology of upland, low rain and aridity, irrigation tanks, forests, some wetlands and dry-land crops. The hopes generated by the peasant revolts were nullified by Patel's Military Action, mischievously christened ***"Police Action"*.**

From 1956 neo-feudals reigned in. Mohazirs from Karachi started trickling back. Fruits of pre-1950 peasant revolts disappeared. Resources became re-appropriated. Rural societies were fooled by the Bhoodan Movement. Hybrid Maize made feudals richer; Nagarjunasagar lighted lamps but lamed the children! Most projects were unnecessary and corrupted by the neo-rich. Caste divisiveness worsened. Thus the Naizam districts were lagging behind the British - ruled districts by about 10 generations.

Such regions should have never been merged until this gap was substantially bridged by their own participation. Consequently the divides widened and deepened.

Ecology

Ecology governs the destinies of Earth's inhabitants. Its components - Air, land, soil, water, plants, animals and energy- differentiated Earth into 40-45 ecosystems and constitute resources to carry human settlements and borders with each other, called Ecotones for buffering and balancing. It is these which made all the difference in the long history of the human species.

People yearn only for what ecology provides - land, food, water, shelter and liberties followed by health, education and roads. Instead they are served promises and polemics. Naturally alienations seed, grow into psychic divides and develop into revolts.

Development and Exclusions

Development leads to exclusions. Due to the legacies of a paucity of resourcefulness and absence of ecotones with developed societies, Naijam districts could not use their resources for their development. With the failure of catalysts of the governments and their mechanisms, the primary and secondary stages of development have bypassed them. Primary stage, also termed "***Ecological Development***", involves self-help with local resources and knowledge, fulfils basic needs, improves the settlements and ensures stability in their own ecosystems. With no external exploitation or local appropriation, they experience civil liberties not to migrate. Such a free will readies them for Secondary stage. This also is essentially ecological and rooted in local resources for and by local people but with inputs of science, technology and expertise. Thus stabilized in their own niches with their endeavors, greater external inputs can be made ushering in Tertiary Development.

Niche Consciousness

Leaders and rulers never understood the niche consciousness in human ecology and precipitated economic and social exclusions of people who lost resources first, earning opportunities next, and accesses last. They were rendered Ecological Refugees needing to resist or migrate. A Tertiary Development was thrust on such people which was naturally distrusted. Alienations lead to separatism. Decadence of communists worsened incompatibilities to a breaking point. This has been eloquently discussed in 1960s and 1970s during the Limits to Growth debates.

The Flailing State

Political Parties, Governmental Agencies and Voluntary Agencies are the Agents of Change. They never complemented or supplemented the Ecosystems failing in both transactions and in deliveries. Such states were called "*Flailing*

States" by an Harvard Economist. While Ecology is an evolutionary endowment its resources are the constituency of Welfare Politics. Such Welfare or Ecological Politics dominated the European thinking evolving into the European Union! In India welfare society remained a dream! Because the hiatus at the time of Merger of the two regions widened- with no social ecotones or catalysts, *Status quo ante* becomes a natural corollary, but with some seminal inputs. Otherwise, an already "***Flailing State will become a Failed State.***"

Ecosystem Re-organization

Solutions begin with the re-organizing the new state of Telangana, based on ecosystems - Forests, Hills, Uplands, Plains, Flood Plains, Agrosystems, arid lands, etc, accompanied by a devolution of administrative and financial powers to these ecological units. Development plans need to be based on these Ecosystem principles associated Human Ecologies and their Resourcefulness. Watch-dog Tribunals for Ecosystems, Development, Social Audits of institutions and Ombudsmen should be mandated.

Alternatives exist for all ecosystems. However, Models, Delivery Systems, Transaction Mechanisms, Agents of Change cannot be the same for different ecosystems and for developing and pristine regions. But the human ecologies remain the common denominator. This approach diffuses contradictions and conflicts. e.g., Yelamanchili coastal region by Bhagavatula Charitable Trust, Chitrakoot hill regions by Nanaji Deshmukh, Kunrakudi arid zone by an Adigalar, Auroville experiment by Arabindo's European followers, etc But what have we done with Gandhian Economists like Kumarappa and the concept of Gandhigram Universities!

Mega- Projects

The existing mega- projects are ecologically wrong. with far reaching consequences. Pranahita-Chevella is the chief one. These funds can be used for the revival of thousands of Tanks, Schools, Colleges, Hospitals, Roads and denuded lands. Language can never be the basis of togetherness but it is the Ecology of Resources and Economics of development which also are the roots of cultures and civilizations.

Terrorists and Maoists

Muslim Terrorists and Maoists have same origins; may therefore have same solutions. Laik Ali and Khasim Rizvi's Inheritors and over 5000 Mohajirs and other Pakistanis continue to plague Hyderabad and India. Their black mail will diffuse in a separate telangana. Mukdom Mohiddeen's legacy of struggles against Rajakars, Feudals and Patel's Army, could revive and stem Muslim terroristic psyches. It is this struggle that mothered the Maoists, the ecological refugees, yet dedicated to the Idea of India. With primary and secondary developments the breeding grounds for social exclusions as well as Islamic terrorism would be lost!

Dividing States and Divided People: Unity of India is not endangered by dividing states, but by divided people. The infant-marriage of Telangana with a mature Andhra was a matter of incompatible psyches. If our self deceptions persist, the current flailing state could as well turn a failed state. The present diagnosis on the Naijam-ruled districts is applicable to scores of districts where people were excluded from their ecological niches through development policies and agents of change.

WE DO NOT INHERIT THE EARTH FROM OUR ANCESTORS;
WE BORROW IT FROM OUR CHILDREN

CHIEF SEATTLE

Chapter - 20

Ecological Revival of Tanks

1. Introduction

In these times of escalating water scarcities, revival of the traditional storage systems such as ponds, tanks and similar wetlands is necessary for a fuller utilization of rainfall. This revival is brought about through developmental and conservational inputs for which all the existing knowledge on them need to be obtained. Most tanks occur in Tamilnadu, Andhra Pradesh, Karnataka, Maharashtra and parts of Uttarpradesh, Bihar and Rajasthan. There is no comprehensive data base on them, unfortunately. In fact there is no study available on the State of Tanks in India. Studying these irrigating wetlands on a holistic basis, i.e. ecologically, becomes therefore an urgent need. This is sought to be fulfilled by our two-year long field and lab study on Six water bodies from the North-East to the South-Central.

Tamilnadu covering different climatic and structural conditions. All these constitute a pioneering study on the ecography of tank systems in India which currently account for 3.3 million hectares.

2. Tamilnadu Scenario

There are about 40,000 tanks scattered throughout the State, of which only 7,400 have an ayacut of more than 40 ha. The total potential under this minor irrigation is 1 mha. The mode of water regulation continues to be under the control of the local community itself, though the practices differ from place to place. They have been moulding the rural societies being small farmer managed systems. The area, the associated people and the agroeconomic conditions around the tanks vary only slightly. Hindrances from administrative agencies are also limited. Therefore reviving these centuries' old systems will not be difficult. But these systems are spread all over Tamilnadu and most of them are in a state of disrepair and decay. The larger land owners acquire control of more land by encroachments and more water by sinking mechanized wells in the command areas.

Though 60% of Tamil Nadu is drought prone, only a third of the annual precipitation is utilized. It is therefore imperative to harvest rainfall. That is why the state became a land of ponds and tanks traditionally. Unfortunately, many of these are filled up and built upon. Thus Tamilnadu has been suffering more than its share of water shortage. Tanks should have become subjects of inputs to enhance their utility to local people for they are no longer mere parts of villages; villages are parts of the tank ecosystems.

3. Ecological Perspective

The International Union for Conservation of Nature(IUCN), World Wide Fund for Nature (WWF) and Ramsar Convention(1971-89) recognized the importance of reclamation of wetlands. Therefore the Government of India formed a National Commission on Wetlands in 1989. The various management aspects were to be implemented and managed through four Acts of parliament - The Environment Protection Act (1986), The Forest Conservation Act (1980), The Wetland Conservation Act (1972) and The Water Act (1974). While IUCN has detailed criteria for wetland identification and management, the Government has developed a management plan. Yet no data were raised on the vast number of tanks- the first among the nine types of wetlands in India. Thus it is obvious that there is a need for studying the smaller water bodies - with an ecological approach for economic sustainability of the villages and an economic approach for the ecological stability of the systems - if the tanks are to be used in full and forever.

4. Methodology

A combination of methods has been employed in the collection, analysis and interpretation of data. These include field surveys of tanks for collecting soil and water samples, plant specimens, animal observations and elicit anecdotal information, analysis of soil and water samples, identification of biota and interviewing the officials concerned with tanks. The tanks are: Kaveripakkam, Pillaipakkam, Kaliveli, Ousteri, Mirattunilai and Kambikudi. All data on soil, water, flora, fauna, agriculture, Aquaculture and silviculture are analyzed and summarized for each tank and compared with each other.

5. Habitats

Soil data indicated variations only with respect to a few parameters while most others are rather uniform. Consequently the lands need limited inputs, if the spectrum of crop culture has to be widened. Water quality is also generally uniform irrespective of its longevity in the tank. The tank habitats are deficient in some aspects and excess in a few others. This perhaps is the reason why most flora, fauna and cultural activities are similar among the wetland systems.

6. Biodiversity

The floristic composition varied with the size of the system or its state of degradation. Thus, Kambikudi supports the least number of flora (50) and Kaliveli the maximum (163). Most species occurred on the foreshore and in the ayacut. Even the low richness of plants in Kambikudi appears enough to support many butterfly species and birds. If Kambikudi tank is not improved, many butterflies and other fauna also may gradually disappear - a warning signal to the other tank systems. However there is no relationship between the area and the richness of animal species. Animal diversity perhaps depends on the plant diversity. The species richness of the birds is high, suggesting that their nutritive and roosting sources in the habitat are intact -an indication that all is not lost from the ecological foundations of the wetlands. That some of these have 10-30 migratory bird species would itself warrant their protection as bird sanctuaries. The richness in Kaliveli and Mirattunilai tally well as both the tanks had remnants of natural vegetation till recently.

About 273 plant and 223 animal species occur in the six wetland systems, but each water body harbours only 50-163 plant species and 86-187 animal species. Since the habitat conditions of the wetlands are not very different from each other, there is a possibility that each tank can be enriched biologically.

7. Resources

(a) Land and Soil: The geomorphology and the geochemistry of the land mass, human interventions and natural biological forces determine the resourcefulness of the soil and land. The current pattern of land use in the wetlands suggest under use, over use and wrong use. Bird and animal excreta are fertilizers and energy sources. Most land must be earmarked for agricultural purposes either with or without water. Only a limited part can be left for use as construction material. All the wetland systems have potential to combine traditional practices and modern science. This blending is absolutely warranted as it involves restoration of a million acres of land.

(b) Water: Water availability from rains, supplies and ground, their qualities and patterns of rainfall, run-off, catchments, storage and seepage from the tank, groundwater recharging and irrigation efficiency and finally the economics of irrigation are to be estimated to enable a better stock taking of water resources. It is a pity that the coastal wetlands are not facilitated to rescue the rain fall from flowing into the sea.

(c) Plants: Of primary interest is the diversity of cultivated plants. Traditional cultivars carry special importance, because, many of them possess special characteristics such as resistance to pests and drought. Germplasm of such cultivars of all crop plants should be collected for further use. Many medicinal plants occur around the water bodies. There are other plants useful as hedges, soil binders, fuel, food, fodder and fertilizer, for thatching and weaving, etc. These are dwindling, sometimes giving way to firewood monocultures. Kambikudi has already become alarmingly deficient in plant resources.

(d) Animals: There are many domesticated animals such as cattle and sheep. These consumers of water and plant resources are resources to man and land. Fish are a source occurring naturally, exploited locally and sold out when water recedes - this happens in all the water bodies. Some wild animals are hunted for food or trade. Special mention must be made of the guano excretions of birds, especially during October -February adding great fertility to soil.

8. Human Ecology

There are 5-14 villages around the water bodies with about 200 to 2000 inhabitants comprising landowners, daily wage earners and the unemployed. Some of them encroach upon commonlands. Consequently the current tanks are only a part of their past stature in every respect. Yet agriculture, aquaculture, social and wetland forestry besides direct use of tank soil, water, plants and animals continue around every tank. The people, in addition, have their own perceptions on the socio-economy of the tanks.

(a) Agriculture and Aquaculture: No sooner do rains set in, rice is planted in all the wetlands using channeled water or groundwater. Fertilizers and pesticides follow. This scenario of deltaic culture is unsuitable to these basically semi-arid areas where the accent should be on water economy, soil conservation and equal opportunity in using the common resource of water. In these water scarce societies, water should be retained for a long time in the tank and / or community wells but not privatized through tube wells. There should be dry land farming needing minimum water per unit area which alone increases the arability of land, both by increasing the area available and facilitating double cropping. In such cases it will be easier to develop long-term fertility strategies like composting and biofertilizers and pest control through biological means - plants, birds and others. The water body should be fully utilized for aquaculture by natural means with only inputs of seed and planktonic cultures. The existing scenario is the opposite of the ideal! Wells take away the tank's resourcefulness. Selfishness of the privates pirate common properties and their ecological health.

(b) Social and Wetland Forestry: As for the tree cultures, it should be differentiated into plantations within the tank and outside it. Those within are inundated by water for various durations and serve as animal habitats. Those outside are plantations on wastelands receiving only natural rainfall. The former must be regulated because they yield greater biomass, enrich soil fertility and provide habitat for animals; but cause gradual reduction in tank area which however can be taken care of through engineering works.

(c) Perceptions: The socio-economic status of the local people and their perceptions on the tanks are studied to evolve locally relevant ecological management strategies. Such surveys help understand the impacts of tank irrigation, forestry plantations, dry land agriculture, land use patterns, etc on the local economy. These insights lead to the emergence of a perspective for long-term restoration of the tanks. However the present study was limited to a sample gathering of local perceptions. This has revealed enough on the degraded state of tanks. What they were should be and what is the common scenario, are discerned. It is a sad tale of neglect and grabbing.

(d) Valuing wetlands: Accounting all resources harboured in the tank ecosystem and valuing them on relative and absolute terms may give us an ecological and economic perspective for development options. In the current attempt values for thirteen items are converted to a fraction of 1, the intrinsic value. The present tanks obtained values from 0.9 to 0.6 which means that those in the descending order need correspondingly more conservation inputs. Otherwise they gradually loose ecological resilience and economic utility to the villagers. If the value goes down 0.5 either the tank has to be abandoned or restoration efforts are warranted. The tanks obtaining higher values, on the other hand can, take in developmental activities.

9. Conservation

Conservation has two facets - firstly the resourcefulness of soil, water, biodiversity and enrichment. The second is to conserve the diversity of the habitats - water body, inlet channels, catchments, agriculture, aquaculture, wetland forestry, animal husbandry along with soil, water, plants and animals and above all the villages.

Villages draw sustenance from tank connected resources, which therefore need greater care, so that the human settlements enjoy a longer and richer balance in the resource bank. Unfortunately there are no conservation efforts, leave alone restorative exercises in any of the water bodies. Even the sluices, regulators, supply channels and bunds remain unattended and in disrepair. Biologically rich thickets are removed in of monocultures. But everywhere people want more plants, more water for longer periods and more fertile soil. However fortunately, wetland forestry has turned out to be conservative. The water that normally goes to private properties through the tube wells is consumed by the common forestry which in turn becomes habitat for animals. Together, they gradually enrich the soil. This being the present state in several tanks, conservation of what is still left and restoration of whatever lost can be achieved only under the eign of the local people and their panchayats. A few examples are given.

(a) Land: Identification and quantification of the land use classes should be carried out so that crops for summer and winter, quality soil suitable for construction purposes, can be selected. \

(b) Water: Existing data appear unreliable. Data on rainfall and variations in the run-off, catchment and storage capacity, seepage, ground water recharging, irrigation efficiency and finally the economics of the well and channel irrigation should be obtained afresh before conservation measures are taken.

(c) Plants: Varieties of cultivated and other plants of ecological and economic significance are the objects of survey, collection and preservation. Locally used cultivars carry special importance. The traditional varieties may have been lost, but chances of their occurrence in some pockets cannot be ruled out. Germplasm of such cultivars should be maintained. Many dwindling medicinal species also need conservation.

(d) Animals: Both domesticated and wild animals call for efforts to increase their populations for these have ecological uses, besides the usual economic ones. This is possible only by declaring the water bodies as minisanctuaries or refugia.

10. Development

For development planning, stock taking of all the resources in the wetland ecosystem is a prerequisite. Water dynamics have to be estimated first. The rainfall, evaporation, groundwater and surface water have to be quantified. A qualitative assessment, whether potable or not, need to be made for a few years on a monthly basis.

(a) Resource Inventory: Surveys of land, water, plants, animals and human ecologies around the tanks presented here constitute resource inventories. When these are integrated with other data already available from the national and regional agencies, one gets a composite and reliable data base. There is not much data yet for any single wetland. And there are 35000+ wetlands of different sizes, constituents and characteristics. And these are associated with thousands of villages and several million rural people. This is no small matter but still remains neglected.

(b) Peoples' initiatives: The perennial desire of people is development through the government-not by their involvement. The governmental activities are mostly confined to so -called social forestry, allowing bore wells, making available fertilizers and pesticides and extending the services of the agriculture agencies. Thus some development is seen around every tank.

The villages around the water bodies have been practicing agriculture, aquaculture, and social forestry. If permaculture, sericulture, apiculture, vermiculture are added to these involving grasses, sedges and forbes of perennial nature on a large scale, keeping in mind the low water requirements of plants, the sustainable development will have been achieved. How? Everything will be protected, conserved and yet economic development will be achieved. This should be the ideal for wetlands.

(c) **Governmental Inputs:** Managerial inputs needed by the tanks for their revival as useful wetlands can be made only with the support of governmental infrastructures and extension services, autonomous bodies such as Khadi and Village Industries Board, in association with local Panchayats, NGOs and academia of the region. Formal and informal education of the local people on every aspect of the water body is a prerequisite. Meanwhile declaring these water bodies as protected areas -Resource Sanctuaries or Bird Sanctuaries and reforming the Panchayat Raj Institutions around these wetland systems is an urgent political imperative in our semi-arid tropical region. Or else Tamilnadu's dependence for water on the neighbouring states may become unbearably perennial. This could lead to dangerous geopolitical tensions of which we are already familiar in post independent India. Some managerial inputs are listed in the Table.

11. Perspectives

(a) Tank Ecosystems: The tank ecosystem consists of all the areas and life around the water body, such as catchments, ayacut, villages, channels, wells, agriculture, aquaculture, and animal husbandry, social forestry along with soil, water, plants, animals and supply channels. These systems have some features of irrigation reservoirs, natural lakes, shallow wetlands and riverine shorelines. Consequently subsystems vary from one another among tanks. A few may have even micro-habitats as in the case of Kaliveli. But in all the tank areas, the water body functions as a 'cornerstone' component of the region, drawing the surrounding villages for their water, soil, plant and animal needs. In other words, if the tank is not there, many villages will not exist. This is also borne out from history. Human settlements are known to have disappeared along the ancient Mangamma Salai in Southern Tamilnadu because the tanks seem to have run out of water continuously for some years. Such mishaps can be averted through modern science and technology. The current pilot work reveals the inputs each tank needs based on the size and nature of the ecosystem - water, quantity, duration, ayacut, catchments, associated villages, agriculture, aquaculture, flora, fauna and soil.

(b) Conservation and development: A natural corollary of the above concerns is to quantify the land and water scenarios and the biotic resources.

The imperative is to conserve whatever still exists, restore whatever is possible, so that the system comes to life for future development. The converse of this is to develop the region selectively only in some spheres in such a way that other resources are conserved. In other words, conservation for development and development for conservation shall emerge as the middle path wisdom for wetlands. But this has to be based on accounting, valuing and diagnosis of the system.

(c) Tragedy of the commons: It has been seen that all resources of a wetland are interconnected among themselves and with the rural society as a whole. The tank land and its water are the most important of these common property resources. With the decline in tank rights, duties and managements for reasons of planning and politics, water and land were gradually and systematically owned up.

Water markets: The water residence in tanks, which is anyway ephemeral, is further reduced due to tube well extraction, excessive opportunity usage, etc. The tube wells around the tanks symbolise a privatization of the common surface water. This is extracted mechanically and sold to neighbouring farmers for a profit. And the electrical charges for privately extracting this common water is subsidized!

Land grabbing: In the absence of managerial systems replacing the traditional ones, tank lands are encroached by the neighbouring land owners, local landlords and political forces. Sometimes the government itself "encroaches" on the land in the name of social forestry (which should actually be called wetland forestry.) Consequently many tanks are mere apologies, as they appear no more than temporary ponds and puddles. This anarchy is beneficial to a few and detrimental to most. However there is an interesting variation to this common trend. In Kaliveli, of South Arcot district the offshore is encroached by a large number of villagers, apparently from the lower economic order. And the government has gradually been legitimizing these by awarding 'pattas'. This is expected to be inconsequential due to the vastness of the wetland. But for how long? Socialism through such land reforms at the cost of ecology may also dwindle in the face of the fast completing East Coast Road.

(d) Wetland forestry: Silvicultural practices using leguminous plants in the tank beds improves soil quality and diversity and richness of animal populations. But the production, marketing and managements need detailed inquiries as this is a rare case of doing the best in the worst of times! The economics seem to have been totally ignored wrongly assuming that silviculture in the tank bed is equal to the social forestry outside it.

(e) Ecoagriculture: When water is scarce it is not proper to cultivate water intensive crops. Yet paddy is grown and there seem to be no attempts to study the economics involved. Still worse is the growing practice of cultivating sugarcane just because one has owned up the common tank water through the bore wells. Bore wells seem to be the single most unjust policy input both ecologically and technologically as well as economically for they privatise common properties, spoil ecology, and cause social inequities, by applying a good technology in a wrong place, i.e.around the tanks.

(f) Bird Sanctuary: Tanks are a sort of bird sanctuaries already to different degrees and durations. Birds influence every other life through food chains, reproductive necessities, and nutritional ethologies etc. some of which are of economic significance to the ecosystem. Their excreta contribute to fertility. They bring in plankton from other water bodies. And some of these are predate the insect pests of crops. To ensure a continuity of this it is necessary to regreen the barren areas with permacultures, preferably from the patches of natural vegetation. This results in increased productivity and animal diversity, for, nutrition as a limiting factor no longer exists. The aquatic productivity also gets accelerated leading to an increase of migratory birds and associated fringe benefits in pest control and guanal inputs in agricultural fields. All these are possible only if water bodies are declared bird sanctuaries and given for adoption to local educational institutions. Poaching of the birds should be curbed by appointing local people as guards.

(g) Urban wetlands: Wetlands seem to have become the easiest and popularly justifiable victims of private property and speculative profiteering. In this connection one may recall instances of "converting water into soil" in many small towns by dumping solid wastes for "reclaiming". This owned up land is used for raising concrete jungles. Consequently drainage of rain water was reduced and charging the underground aquifer hampered. What is the gain? What is the loss? And for whom?

12. Imperatives

Having arrived at the conclusion that both conservation and development are to go hand in hand, a framework of actions has to be developed along with the agents of change, the catalysts and ensue cooperation of people and coordination of the governmental bodies. We therefore feel that for the success of ecologically oriented development, governmental bodies, Non-governmental agencies, local people and academicians have to provide complementary inputs.

(a) Empowerment of the beneficiary: Tank irrigation thrived till a few decades ago because water distribution was carried out by local bodies of farmers in the ayacut or the Zamindars. In some tanks farmers do not have any say. Tail-end farmers had to shift to groundwater irrigation. People are not involved in tank maintenance. This can be reversed only by handing over the tanks back to the village Panchayats and rejuvenate the traditional water management bodies.

(b) Policy Inputs: Policies should be based on sound philosophies as governmental programmes and their implementation have long-lasting and indirect effects on agriculture, land use and through them on the daily lives of people .Therefore all decisions have to be based on ecological wisdom of sustainable growth. Agricultural policies are the most important, as they regulate land use through revenue subsidies, extension, training, expert guidance, irrigation policies, etc. Ecological suitability of a crop should be evaluated. Drought resistant plants can be introduced through agriculture and social forestry policies. Water intensive cropping patterns should be prevented through fiscal policies. Long term silvicultural practices are to be given greater priority. Regulation of deep bore wells in tank vicinity, alternative agricultural methods like rotation of crops and organic farming Legal inputs in declaring these water bodies as protected areas is an urgent need.

(c) Actions on the Ground: Strategies for restoration and conservation in non-system tanks and system tanks vary because they differ in water retention periods. The villagers are aware of the importance of tanks in their economy but not their ecology. Only when this is achieved, effective participation from locals in sustainable utilization of the tank resources can be ensured. Sound policies and result - oriented action on the ground are the inputs. Some of them are as follows: Strengthening of embankments and repairing regulators, sluices, supply and feeder channels, desiltation of the water body. Protecting the tank from poaching, giving fishing and grazing rights, removal of encroachments. Regulating the land use in the surrounding areas of tanks so that all rain water drains into the tank and increases its water holding capacity. Providing financial support to the NGO's involved in the developmental and / or conservational activities in the areas around the tanks. Establishing initiatives in favor of organic farming, medicinal plant gardens, ecotourism, wetland forestry, aquaculture, etc.Educating the local people on all the aspects of the tank system so as to motivate them to look after their common property.

(d) Non-Governmental Agencies: Non-Governmental Agencies (NGOS) are recognized as major agents of change. These voluntary bodies have motivated people, built bridges of understanding, achieved goals and most are ecology conscious. Therefore it is appropriate to draw them into tank welfare.

(e) Cooperatives and Panchayats: Most of the governmental efforts go in vain as they fail to win the involvement of locals. Therefore people should be helped to form co-operatives to centralise all operations, especially social forestry. Thus villagers get cheap firewood and a regular income. Later on, such mechanisms can be extended to other areas such as dairying, vermiculture, sericulture, apiculture, floriculture, plant and animal refugia, gardens and above all water sharing. On one hand, people can take up developmental activities yielding produce and income. To sustain this for long, on the other hand, conservation measures of water, soil, plants and animals can be taken up. However villages around the tanks cannot do all these. Even these need inputs from trained extension workers and experts.

(f) Updating the records: The background data on the water bodies given by the Revenue, Irrigation and Agriculture Departments differ from the ground realities for a variety of reasons - natural degradation (siltation, aridity) man made degradation (encroachments), accent on alternatives (wells) wrong policies and practices (reclamation, poaching, deforestation). How much is the rainfall? How much goes to the wetland, to the wells? What is the area of the tank originally and now? How much is encroached and legalized? What is the water spread area and volume of water it can hold? Restoring land and water equity among local people by initiating proper management inputs is a challenging problem for economics and politics.

(g) Panchayati Raj: Tank irrigation requires a strong local authority to generate collective action for tank building, to efficiently distribute the water by optimally allocating it among beneficiaries and over seasons, and to properly maintain the irrigation system. The water dynamics encompass the entire catchment, supply channels, command area, the wells, impediments and adjuncts to these. The villages are intimately connected with this web as a part of it. Therefore both conservation and development have to go hand in hand, Because of a shift from a direct social control to indirect economic hegemony through subtle ownership of water and soil, tanks have degraded severely. Therefore nucleating each tank along with the dependent villages into a unit - An Ecopolitical unit- is a necessity. Empowering this unit with the responsibility to regenerate the water bodies, catalyzed by political power, social consciousness and academic inputs is the chief imperative emerging from the present study. If this happens a progressive chapter, a turnaround, will have begun with far reaching positive consequences all around the wetland country and on the 50,000 villages and 10-20 million people. But can the wetland resources lost during the last 50 years be accounted for, leave alone restored? The concept of values responsible for human progress for so many millennia seem to be sadly out of our dictionaries!

(h) National Tank Authority: The subtle interrelationships among agriculture, social forestry, the land and water dynamics, biodiversity, the socio-economic milieu and above all the political undertones running through the entire system of water bodies have to be realized if they are to be resuscitated. Though this is the story of six of these tanks there is no reason not to expect the same with the rest as well. Since these water bodies are a major group of natural, irrigation and drainage sources, there is a need to change our approach towards their management radically i.e. from an engineering outlook to an Ecosystem Approach, a guiding spirit. This can be done in the interests of the wetlands and their people by the establishment of a National Tank Authority. NTA should be mandated to take care of all wetlands of India. Tank care shall encompass conservation and restoration of resources and development on an ecologically sustainable scale.

Chapter - 21

Resurrecting a Tradition

Once a part of Indian tradition, tanks today are disappearing from the face of the country. The tank system - which was used to store rainwater meant for subsequent use by locals -thrived during the rule of feudal lords and declined with them.

Despite their utility for drainage, storage, charging wells and reducing the havoc caused by floods, tanks were encroached in rural areas while in cities they were filled up. In order to conserve tanks, an 'ecosystem approach of management' is necessary. This can be achieved by setting up a National Tank Authority (NTA) which can be entrusted with conservation, restoration, and development of all tanks in India. The jurisdiction of such an authority shall encompass the tanks of Tamil Nadu, Andhra Pradesh, Maharashtra, Uttar Pradesh, Karnataka and Bihar; the coastal ponds of Orissa, West Bengal, and other water harvesting systems like temple tanks, lakes, and urban wetlands.

But, why has the tank system degenerated? Bore wells sunk around tanks and operated on subsidised electricity, have greatly reduced the quantity of water in them. As tank beds remained fallow for greater durations they were gradually encroached upon.

The number of bore wells, scale of encroachments and the level of siltation all depended on the intensity of the monsoons, the geography of the local catchment area and the statuses of the tank mechanisms. Sometimes, bore wells are sunk in tank beds. The owner of the bore wells and lands in the *ayacut* (irrigated area under a tank), and the encroacher of tanks is usually the same person. Thus what till then was a common resource has now become private property.

However, even today there are some 200,000 tanks in the country most of which are located in Karnataka, Andhra Pradesh, and Tamil Nadu. Tanks account for 10 percent of the present irrigated area of the country and therefore need to be conserved by an authority like the NTA.

The requirements for setting up the NTA are academic, financial, and logistic in nature. These cover rainwater harvesting, storage of water, irrigation usage, potability, recharging groundwater; nature of soil, water, plants etc. The other aspects relate to animals, agriculture, aquaculture, social, and wetland forestry. Biodiversity will be naturally preserved and utilised by people residing in the surrounding areas. The history of the tanks and the perception of local people on all these is a necessary starting point.

Therefore, the existing records on tanks should be revised. Rapid evaluation techniques have to be developed. This will lead to the emergence of a value system to classify tanks and earmark them for either development or conservation, and restoration.

Tank irrigation requires a strong local authority to rally collective action. The water dynamics encompass the entire catchment, supply channels, command area and the wells.

The destinies of villages are intimately connected with this web. Therefore both conservation and development have to go hand in hand, which is possible only by Nucleating each tank along with the dependent villages into an eco-political unit under the local self government.

Although stressing on strong local authority in managing tanks and advocating the need for a body like the NTA in the same breath may appear contradictory, it actually may not be. The NTA is visualised as a body for broad policy making and channeling inputs of a scientific and financial nature so as to motivate, monitor and strengthen the local bodies in reviving tanks. The first duty of the NTA is to declare tanks as protected ecosystems and the associated groundwater as public property. The next task will be to conduct a tank census based on ground realities. By resurrecting tanks and regulating groundwater, a source for drinking water which is a major rural need will be greatly fulfilled.

Down to Earth, Vol. 6 Issue: 30/10/1997

EARTHWORMS, THOUGH IN APPEARANCE A SMALL AND DESPICABLE LINK IN THE CHAIN OF NATURE, YET, IF LOST, WOULD MAKE A LAMENTABLE CHASM...WORMS SEEM TO BE THE GREAT PROMOTERS OF VEGETATION, WHICH WOULD PROCEED BUT LAMELY WITHOUT THEM.

GILBERT WHITE

Chapter - 22

Temples, Coasts, Tanks and Ecology

Hybridize Them and Wait For the Harvest

A. Temples : Survival of Ageless Continuity

Temples are the sacred places of worship to Hindus. They are every where and in fact countless. So countless that even the 30,000 temples destroyed by Muslims and deities transferred as foot steps of Mosques in other countries, constituted only a miniscule. Though their lands are swallowed and properties annexed by the feudals and other land sharks, they are yet well to do. Though the tenants of properties do not pay the revenue and keep it for themselves yet temples are not starving and still venerated. Even if the premises are encroached, yet some semblance of worship continues. Though the traditional priesthood is vanishing and many villages may not have Brahmins at all, especially in South India, leading to their closing, yet some ingenious methods are resorted to see them alive.

This is the pluralism within Hinduism -of Gods, temples, properties, deities, confiscations and appropriations. And the Unity is in the institution of the place of worship, its sanctity and allegiance to. This can be exploited for the common worship of the new God - Ecology -as it is synonymous with the safety of nature. How can this be done.

1. Existing Temples: Water and Trees: Develop the temple tank with water - desiltation, drainage and growing lilies and lotuses etc., Keep small fish and frogs. All by the temples' clientele for the sake of God and his children's common welfare. If the tank is outside the premises of the temple, maintain fast growing and slow growing plants and grass. Temple Forestry - in earlier parlance - Sacred Groves. Rain water can be channeled into this.

2. New Temples for Old Tanks: There are about 40,000 tanks in Andhra Pradesh and as many in Tamilnadu. Many of them have vanished. Especially in the Urban, Sub-urban growing centers. The remainder is under our focus.

Construct a temple on the banks of the tank. In the process the tank becomes temple's property and both together custodians of the regions welfare to some extent. Desalt the tank and built its embankment. Plant grass, herbs, shrubs and trees of slow and fast growth based on the local traditional knowledge.

A sacred grove and a sacred tank will thus emerge in due course. Both serve as additions to local people's wealth from nature. A local natural resource development and its staggered usage will emerge slowly.

3. Biodiversity Refugia and Economic Plants: In the sacred groves and tanks the locals can maintain economically important plants and animals such as medicinal plants and rare fishes. These therefore can function as refugia of vanishing plants - biodiversity refugia.

4. Economics: Income, Grants, Expenditure: This is for the local authorities to decide along with the rules of the government, or laws of the land. If necessary a law can be passed to implement this plan - this model for resurrection of the Mother Earth and development of the local peoples' security consciousness.

5. Fringe Benefits: There could be many fringe benefits - local recreation - parks, bird sanctuary, increased fertility, unexpected income from fisheries, tourist attraction, etc in due course of time. It is unfortunate that the village tank of India didn't find a place either in the wet land convention of Ramsar and after nor in the development alternatives in the literature.

6. Modalities and Traditions: Many of these suggested have been a part of our traditions. Therefore the present proposal is a mere update of the local ethos and wisdom. May be this can be a part of Janmabhumi Programme. Involvement of concerned sections, government, parties, monetary Transactions - in short all the aspects of change, have to be in the local self government.

7. Facts From Tamil Nadu Relevant to Andhra Pradesh: Churches continue to exist. Mosques continue to increase and flourish. Temples loose land income, properties, structure and even priests. I have seen scores of such all along the east.

8. Coast of Tamilnadu: Very famous temples once are in dilapidated condition - one solid example is the situation in Brahmadesam near Tirunelveli - the anchor of the popular tele- serial, Annamalai ! The temple at Kambikudi near Madurai, for long protected the tank. This is the best way to inculcate a psyche of respect to the elements to the nature - water and trees here- through the agency of God - however agnostic or atheistic one may be!

In fact places of worship of all religions must develop a sanctum for water and trees in their premises or neighborhood. I saw a mosque doing this on the southern east coast of Tamil Nadu. Sand Dunes of east coast are being stablized and improved for the habitat and people by the church. Churches and christianity have a programme for nature conservation. Why not others emulate?

B. Tanks And Shorelines

During our errands for surveys of the small water bodies, in Tamil Nadu we came across an interesting practice of tree culture followed in many system as well as non-system tanks. We wish to share our experience and perceptions on this.

In general tanks are not perennial. The tank periphery becomes dry even among these gradually. In these spaces are planted native trees, preponderously, the Acacias. These are not interfered with as they are protected by the local people, as a common property.

35,000 tanks - Census, diagnosis, valuing and decisions to continue, improve, restore, or convert is necessary. Proposal on this that appeared in The Hindu as back as 1997 is still ignored! European Union- financed tank rehabilitation is basically flawed. Both Government and Non-Government activities i.e., engineering and social work appear to be self-help / self-aggrandizement programmes as they ignore the ecological aspects - conservation and restoration on long range basis.

Tanks are more important for Tamil Nadu than any other state south of Vindhyas - A scarcity region. Therefore according Tanks and the neighbouring human activities a status of ecosystems and resurrecting them through ecological methods is one of the solutions to ease the perennial problems of Tamil Nadu.

1. Tree Culture and Guano of Birds: Because of trees more birds are attracted, but seasonally. This encourages poaching for local consumption. The excreta droppings gradually change the soil nature along with fallen leaves. It becomes more fertile and biologically active supporting a little herbaceous and ground cover. In a few years the soils become all the more fertile. This soil is often extracted, but sparingly, for use in the local agriculture.

The trees serve as buffers between the embankment and the overflowing water during floody rainy season. The vulnerability of the bund is reduced, longevity increased and strengthening it rendered sustainable. Physiologically they function as water mines. Water is conserved and converted into biomass for ultimate use in the domestic sector of local people. The shade marginally reduces evaporation. Consequently the capacity of the tanks to hold water by these botanical means increases.

The quality of the water changes. With natural addition of nitrates and phosphates water becomes potable. In agriculture chemical fertilizers become avoidable. If such supplements are added there is a reduction of seed yield and increase of vegetative growth. In near by places the ground water becomes tastier, as claimed by some local people in some places.

2. Stake Development: This experience made the local people wiser. The tank becomes a precious commodity to be protected. Private prosperity from this commons is discouraged - pumping water out, sinking bore wells around the tank for geological withdrawal of surface water, etc. Utility - oriented desiltation and understanding of ecological phenomenon seem to have begun. Proper trees at proper places lead to a series of changes in due course of time - Bird Guano - fertility of soil - fertility of water - portability of water - substitution of chemical fertilizers - protection of bunds - prevention of private exploitations - enhanced water holding capacity - etc.

3. Perspectives For Semi - Arid Regions: There could be lessons from these for the country. There are over 200,000 tanks in the country - mostly in Tamil Nadu, Andhra Pradesh, Karnataka, Maharashtra, Bihar, etc. Many of these will be amenable for this practice to be emulated. There are many other types of wetlands all over the country which can be engineered for the above practice in specific locations, especially at the peripheral regions.

Tanks are wetlands as per definition. Needless to emphasize that local rainfall can be best harnessed by this indirect way of routing it to plant storage in the tanks than allowing it to run off to elsewhere, to the sea in the coastal regions. Tanks themselves may become a vested interest with people as it was decades and centuries ago. Then we can say that ecological wisdom has come to life once again! This practice will be more essential for coastal regions and deltaic systems.

4. Conservation Of Common Property: Where the government, academicians or NGOs did not succeed enough to motivate local people to take care of their common property-facility, the legume trees and the birds on them have facilitated. In fact the local people seem to develop a stake in this system gradually. This good augury for the semi-arid regions of peninsular India, where tanks are very vital structures in water economy, will be more essential for coastal regions and deltaic systems.

5. Coastal Systems: East Coast of India receives rain from North-East or South-West Monsoon some places from both. Most of it drains into sea. Even coastal engineering in the form of storm drains and culverts are designed from the angle of drainage. If there is a shift in the focus to storage, some ways to preserve all rainwater can be found. One of them is to develop strings or strips of coastal tanks at geographically convenient places.

There will be problems of salinity and subsurface water and gradients. Therefore geology, geography and botany should combine with the help of local participation. The tanks need to be lined with appropriate plants with fibrous root systems on the bund, salinity tolerant / absorbent plants and leguminous trees on the inner side of the tank just as in case of wetland forests we discussed earlier. Specific aquatic plants - Nymphaea and Nelumbium should be maintained. We found positive relationship between potability of water and growth of Nelumbium. Also these waters are always cooler than in the tanks without this plant.

6. Deltaic Systems: The deltaic systems in the east coast pose the problem of water logging, subsurface water owing to good rainfall upstream or locally. Consequently most rain that falls drains into the sea. Tanks here should be redesigned to store this extra water and pipe the surplus to other places where it can be either utilized or stored. Tanks here must have high and stable bunds and water consuming plantations on either side of the bunds. Unless the rainfall in one season is fully utilized, transported and stored, the rain in the next season cannot be conserved.

Based on the ecological characteristics and processes the tank systems have to be reviewed. Keeping the objective of not allowing any water from going into the sea, ecology, environment, sociology and engineering can combine to remodel the existing tanks, not merely rehabilitating, construct and develop newer ones.

C. Water Policy

This is relevant to all states and all sections in scarcity state. Policies should be therefore holistic and shall embrace - tank census, restoration, temple tanks and lands, rain water, salinity intrusion, aquifers and bore wells, river connection, wetlands, sewage sand finally relevant education. The agenda for action has to be multipronged and simultaneous.

1. Conservation and Restoration: Agenda must cover a tank census, diagnosis, valuing and decision / classify inputs of engineering and social work and ecology will vary from one to another type. There should be a part of development. All these constitute to make an Eco development Plan - consume to develop and develop to consume - long term growth - sustainable development - an analysis of the prevent with insights into the part - an outlook encourages for future. About 35,000 in all need focus using all approaches: technical, societal and ecological. First or second or both are employed peace meal. The third is neglected. MIDS report on this should be studied. The EU financed tank rehabilitation programme is flawed in the nucleus itself. Its operation is a farce at Pondicherry Union Territory at least.

2. Temple Lands, Temple tanks: Both these need restoration. Since these are the times of remembering Gods and religions, Government should use their revivalist psyches for using their resourcefulness for people. Encroachments in Tanks - Money, Caste, Union, Powers. But need also drives poverty to go for the inviting monsoon tank bed.

3. Rain water harvesting: The agenda here include coastal streams where water and soil going to sea - village tanks - house top collections. A map of water availability state/ district level. A map of measures suitable to each area. A map of ecosystems based on water an availability and man: water ratio and man: land ratios.

4. Desalination: This is relevant whenever other means are not available such as Coastal Areas. Non-entry of salt, Mangrove regeneration and control of salt pans. **Subsurface, shallow aquifers**, deep aquifers. Ground Water charging, percolation ponds etc., Bore well policy should be based on the dynamics of surface and underground water - the location and extent of UG aquifers. Otherwise one will recharge and others will extract, consume and sell.

5. River Connection: Will not be possible unless Jingoism and Xenophobia are controlled and naturally by realization. No one will allow its water to go to the neighbour instinctually. Local river connections are what TN can do; Others do not depend on us. If not use them as tanks. SYL Canal, IG Canal, TG Canal should be lessons for us.

6. Sewage and sludge policy: As a civic environmentalism - reuse of both by municipalities - tilization directly or after partial purification, is of a current relevance.

7. Wetlands: Are natural charges of underground water while supporting or consisting of a kind of surface ecosystem with highest productivity rate in coastal areas. They are the bulwarks against the salinity ingression. Their protection and restoration are necessary.

Talents from all over country, especially neighbours to grow local talents. Ethical and human conduct of "***Knowledge affairs***" will generate and sustain a germplasm bank of intellectual properties.

Chapter - 23

One Step Forward and Two Steps Backward

Ten million people became refugees, fore-going their traditional habitats, while half a million were killed while India was partitioned. This greatest environmental deprivation in human history, is the root of most of the violence and divisive instincts in India today. When deprived of resources, grief becomes a fixed star for most, joy a veering vane for many and profit preserved for some. The story of environment is much the same, 50 years later - one step forward, two steps backward.

Indian Scenario

During the last 50 years, the destruction of environ-ment is not as much as is made out. Our annual rain-fall is still the same as it was then; our rivers dwindled because water is harnessed: their pollution is as much by sewage as by industries. Forests dwindled because they are treated as timber mines. Agriculture meets the food needs, but brings with it the pesticide problem. The biggest crunch is of energy. Drinking water is a casualty, along with it the harmony in the human societies. Pitched battles may be fought for this environment very soon. Absence of genuine land reforms has driven rural people to urban systems to become construction work-ers, rickshaw pullers. house-hold helpers, unskilled labour, head loaders etc, liv-ing on pavements or forming slum - the worst environ-mental refugee problem in history. Tanks were filled up in thousands all over the country and built upon. The harm from destroying natur-al ecosystems that sustained people for millennia is far greater than the pollution problems.

However, the most serious of all is the human health. Pol-lution is toxic to humans directly and indirectly through food chains and. even to future generations. Pests, pathogens, parasites and vectors also are mutated by chemicalization heralding newer diseases and resist***ances*** to drugs. These were termed genepidemics by me as back as 1988. A redeeming feature is the Cap-21, now being initiated to gauge our capability to meet the challenges of the next millennium. India has already made many efforts through institutions, legisla-tions, support to environ-mental education and local activists, etc. We have many biosphere reserves, sanctu-aries, national parks. Yet governments cannot do much without people's involvement. They need right information, rote mod-els, honest leaderships, teachers, scientists etc. We have many scientific bodies, aiding agencies from abroad and 250 institutions of high-er learning. Although the problems and fears were articulated from time to time, we are not realising that there are limits to growth whose veins cannot be open ended. Thus the issue of environment is basi-cally one of ethics, both in the economic exploitations as well as conservation.

Pondicherry

Pondicherry is sliding from a position of strength and harmony to one of suffo-cation. This is chiefly due to a big and unaccountable governance for a small poli-ty, and ad hoc development without monitoring the impact and resistance to criticism. And all this despite a successful Auroville experi-ment in ecological sanity, continuing eco-friendly French legacies and institu-tions of higher learning.

Water

Our water sources are fast depleting. We should urgent-ly conserve rainfall and tanks. PWD has good plans for tanks, especially Ousteri (1989), but no one seems to bother. Out of the first 100 tanks, less man half exist and are in very bad condition. Some exist only as small channels while most were silted and encroached and legalized for cultivation. Ous-teri is the most important one and was well studied during 1993-96 by a team of workers lead by Prof Sharma of Pondicherry University. The nearby wetland Kaliveli was also simultaneously studied by the same group- which has direct bearing on the underground water avail-ability in the Kalapet region.

But none seem to care for these, including the local environment department the entire biodiversity of plants and animals and soil and water quality were analyzed. Biomigration and conservation for develop-ment are the novel feature of these studies.

Land

Agricultural land is being lost. Land is a scarce resource at Pondy. Often, it is senselessly given away. Two examples may suffice. The University was given a great chunk of land that it does not need. This has affected many small land owners and a future expansion of a school nearby. Also, land is given to Housing Societies as plots. The societies could have been asked to construct only two to three storey flats, saving land for future. The government can still action this as most grantees have not initiated construction.

Industry

The industry is in good shape, some of which even uses waste! Only a few of them pollute. Automobile and sound pollution seems to cross all limits, with the government ignoring the cit-izen's plea. The Central Pol-lution Control Board did a good work. Its successor seems to have done little work This may be illustrated by three examples.

— Distillery: The reasons for its shifting from the beach are unsubstantiated by sci-ence. There was never a research study on pollution, its spread and toxicity. But the place to which it was first shifted would have greatly benefited the locals in so many ways but for unallayed fears.

— Kirumambakkam: The pollution and its impacts on human society at Kirumam-bakkam were sought to be studied with the aid of 4 lakh within less than one year. The interim, as well as full report submitted after some years were found unacceptable.

Migrations

The biggest problem is the migration of people from neighbouring areas render-ing the statehood demand environment-unfriendly. If at all it can come, it should be only by transferring Mahe, Yanam and Karaikal to the neighbouring states and bringing about contigui-ty to Pondicherry region by merging the neighbouring "islands" of Tamil Nadu with it.

Some Efforts

Many environment-friendly efforts were made here. The Pondicherry Gazetteer by Cyril Antony and his periodi-cal 'Ponkamaya' have been doing good work And so are the tireless efforts of M.P. John and the data of the Central Pollution Control Board upto 1990. The indus-trial policy leadered by B.V. Selvaraj, the statehood report by Ramadass and the rejoinder to it by Sharma, Antony and Bhat, recent master plan for development deserve great appreciation. However both the industrial policy and the Master Plan appear to talk too much on too little because they are too late! These are like career devel-opment plans just before retirement!

Institutions

The only school of ecology in Indian Universities is here. This has reasonably good faculty with probably the best curriculum and syl-labi in India. In addition to the university, VCRC and JIPMER have major roles in pro-moting environmental health, if properly approac-hed by the local government.

Perspectives

During the last 50 years of planned development, achievements were made at great environmental costs because the principle of '***soul in the past***' was jettisoned. Therefore, the outlook for India's ecology and environ-ment is pessimistic, but not hopeless. A swadeshi spirit is to be freshly inculcated; morality and honesty should strive on their own to become dominant forces.

Indian Express, 14 April 1998.

DESCENDING PLAGUES

Chapter - 24

Genetic Toxicology and Genepidemics

1. Introduction

Chemicals and radiations in the environment enter organisms including man, directly or indirectly. They often affect the genetic material of the recipient systems altering their characteristics. Such alterations pose long term risks to human health causing teratogenicity, carcinogenicity and heritable diseases. In the ecosystems, stability based on interactions may be affected. Since both the above happen without forewarning, these causative agents, Genotoxins, carry a hitherto unprecedented hazard to human welfare.

2. History

This subject has beginnings whose alarming ramifications were not suspected then. Ionizing radiations were found to cause mutations in plants and animals some sixty years ago. During the Second World War chemical warfare pursuits led to the discovery of mutagenic chemicals in UK, Germany and Soviet Union. The post-war ethos inspired the rapid chemicalization of society which generated several unhappy experiences. Various episodes in Europe and America rendered them suspect as to their long term safety. The relationship between inheritable diseases, birth defects and cancer were ultimately established as originating from similar genetic causes. There is now considerable international activity in this field and is ever on the increase.

3. Agents

The agents in the environment to which humans are exposed are of many types. The quantum and period of exposure however vary depending on the life style of and social classes of the recipients. Many of these are often in mixtures and might even interact in the biological system, complicating an already complex scenario.

4. Exposures

The exposures may happen mostly due to consumption and usage in the daily life or in the occupational atmosphere. Sometimes it is caused by the unsuspected ecological routes in the atomic facilities and chemical industry exposes very large populations to high doses with a short time. The most ominous would be the chronic exposures to low doses of mixtures of unknown composition.

5. Social Consequences

As a result of exposure to the various agents in the environment through various channels certain long term hazards occur in human populations without forewarning. Some of these appear in the same generation, such as cancers. Some might appear as birth defects immediately in the beginning of the next generation - thus exaggerating the initial impact through future generations.

Already considerable information is available on the genetic disorders in humans, popularly called syndromes. All these syndromes discovered spontaneously as due to numerical and structural chromosome variations or gene mutations, may in fact due to hitherto unsuspected environmental genotoxins. About 4000 such types of syndromes are recorded so far.

6. Ecological Consequences

In the non-human organisms also the genetic material is well affected, consequently altering specific characteristic in some portions of the populations. Interactions among these are necessary to set up and maintain balances in the ecosystems. Therefore, changes in them may ultimately upset balances. Some of the characteristics that are known to be affected are pest resistance, host susceptibility, soil fertility, breeding system etc.

7. Cellular Consequences

At cellular level, these agents bring about three kinds of responses depending upon the targets of reach. Some affect only the segregating mechanism during cell division-spindle, centromere and cytokinesis. This results in the chromosomal number in the offspring cells varying from the parent cells i.e. in the number of linkage groups. The second effect is the causation of lesions in the linear integrity of the chromosomes, during cell division. Consequently, the offspring cells may have one or more chromososmes different from the parent cells in the number and sequences of some genes. Finally and more importantly the very DNA-operational units, genes, may be altered in their basic constitution. The offspring cells therefore will have some genes mutated form their parental forms.

The above three types of offspring cells, if continue to multiply, will soon dominate the tissues. They impart to them and their constituting organs, the new characteristics arising out of the altered gene composition or sequences or number.

8. Cellular Indicators

That the agents cause effects on the segregating mechanisms or chromosomes integrity of the dividing cells can be realized by screening the relevant tissues for the end points-the ultimate results within a dividing cell. Frequency of these within a cell gives the intensity of damage along the cell lineages. In case of effects on the individual genes, changes in the phenotypic characters under their control alone would indicate the mutation. Such markers are available almost in all organisms.

9. Indicator Systems

The endpoints of segregational errors, chromosome disintegrity and gene mutations can be monitored in the higher plants, animals, cell cultures and microbes. Among the plant systems, meristems of Allium, Vicia, Pisum and Crepis and the whole plant systems such as Hordeum, Glycine, Tradescantia and Zea are more prominent. Among the animal systems, Drosophila and rodents are more frequently used. Cell cultures of rodent and primate species, including man, are generally favored. Among microbes, Bacillus subtilis, Escherichia coli, Salmonella typhimurium, Saccharomyces cerevisiae and Neurospora crassa are the more widely employed.

10. Types of Assays

The above indicators systems can suggest whether or not the agents are genotoxic, in a direct way, But this approach is inadequate in view of several complications arising out of tissue metabolism and ecology. To account for these contingencies several types of genotoxic assays are in vogue.

An epidimeological approach is followed while dealing with the occupational stresses that may cause a genetic effect by looking for genetic end-points and or disease spectra in the susceptible populations. Human excreta are also tested for mutagenicity to find out if mutagens or promutagens are among their daily diet. In situ environment is screened for Mutagen city either through natural inhabitants such as aquatic animals or terrestrial vegetation or by associating monitoring systems with the habitat. In some assays the possibility of agents undergoing activation or detoxification vis-à-vis genotoxicity, in various tissues, is accounted for. However it was revealed that tissues from individuals with pathological history may respond to stresses differently from the normally healthy ones. Therefore relative propensity of different historical backgrounds to respond to genotoxins has developed into another line of assaying. That in vitro assays should involve the cell types which are the usual targets of the stresses is increasingly realized, e.g. lung tissues for inhaled agents. Finally that lethality may be due to genetic reasons rather than toxicity was fortunately realized right from the early times of this subject.

11. Modulations

The genotoxic potential of the environmental agents is modified by several physical factors such as temperature, light and pH, and by chemical factors such as synergism and antagonism. The functional nature and state of recipient tissues also determine the target cell sensitivity.

The influences may be positive in that the consequences of genetic toxicity are either minimized or altogether eliminated. The negative influences, on the contrary, would accentuate or even induce genotoxicity. Because of this modulation phenomenon, potentially dangerous agents are often neturalized and seemingly innocuous ones rendered sinister. The role of plants which constituted the diet of man, directly or indirectly, is multifaceted in this respect and constitutes a vital factor in environmental mutagenesis.

12. Indian Situation

Mutagenic hazards to society from the environmental agents are more or less directly proportional to the level of industrialization and standard of living. In India, a few sources of alarm from this angle can be identified. Chemicals in the industrial atmosphere such as solvents, heavy metal inputs and let offs are an expected source of hazards. Formaldehyde and Cadmium are already known in this respect.

Use of pesticides poses a very great problem, for; they could not only expose the handling personnel but also have an ecological effect by contaminating water and food. A mutagenic pesticide or its metabolite constitutes a very dark threat. The problem here is complicated by the improper use of them despite warnings.

Waste constitutes another source of possible long term hazard. While sewage is present in all towns and cities, other wastes have reached menacing proportions in about seventeen major cities of India. While these can endanger the water sources, some constitute a new dimension due to their usage in agriculture and other similar cultural practices. From this point of view, one could expect susceptible or stress groups to exist in every township of India.

Source of energy production, conversion and utilization are another area of genotoxin emission and spread. This source pervades from the phytomass use in the rural kitchen to the nuclear source in the Atomic Power Plants. The most common and continuous, however, are the nitropyrenes in the ambient air released due to fossil fuel combustion during transport. Chemicals in the consumable items -tobacco, betel nut and alcohol, constitute a daily source of hazardous substances whenever their production is industrialized which, unfortunately, is ever on the increase.

13. Current Trends

Regulation of genotoxins is vital in concern for the genetic health of the present and future generations. By present norms, the suspected agent is initially tested in a few assay systems to assess its potential to be genetically hazardous. Based on the extent of human exposure to this stressor, risk to human society is estimated. On weighing the benefits from the agent against the risks, it may be either banned or replaced or tolerated. In some cases, one may have to be satisfied with protecting the vulnerable or by mere issue of a statutory warning.

To understand and standardize the modulation dynamics of an agent, so as to integrate it into the aforesaid scheme of regulation, is a major current trend if international activity in this field. The other activities are involved mostly in perfecting the existing lines of knowledge and enriching the data bases.

Genepidemics

A. Definition

Occurrence and spread of gene related problems, relevant to human society directly or indirectly, in epidemic proportions.

B. Hypothesis

(a) The theory of genepidemics as due to pesticides was proposed in 1987.

(b) It states that gene-based problems of cancers, birth defects and heritable diseases are on the raise In the 3rd world (then) these will be mostly caused by the pesticides due to their improper use. This improper use will also render agroecosystems unstable and agricultural products unsafe. Both these trigger many other consequences.

(c) Currently the hypothesis is revised. Besides the agroecosystems and human society, two more - aquatic systems and communicable diseases are added.

(d) To pesticides are added the drugs that may cause resistance mutations in the infectious parasites and pathogens. The other chemicals are a wide variety of pollutants released into ambient air and water.

(e) Therefore gene-based problems are generated by chemicals- pesticides and pollutants mostly- in human societies and their subsistence systems such as aquatic ecosystems and agroecosystems and the communicable diseases that pervade all these ecologically. These problems may be in the epidemic form as times pass by.

C. Bases for the Hypothesis

(a) Human genetic Health: North

1. Birth defects 0.5 to 2/5 % (10% 2000 AD)
2. 1% Mental illnesses
3. 4000 syndromes
4. 25% Cancers 50% raise between 1960 - 80 in twenty-eight developed countries
5. Detriment Index

(b) Agroecosystems

1. Resistance: 364 insect pests to 9 pesticides
2. Cotton, Rice etc.
3. 7 Rodent species resistant to Rodenticides
4. Mutations in rhizobia, blue green algae,
5. Pollinators altered, cereal mutations, altered weeds.

(c) Polluted Ecosystems

1. Out of 35 + Studies on animals, 90% demonstrated birth defects and genotoxicity in animals ranging from crustaceans & polychaetes to amphibians and rodents.
2. Out of 20 + studies on plants about 50% of plants manifested genotoxic symptoms.
3. Several plants and animals turned out to be bioconcentrators
4. Many animals exhibited cancers in their aquatic habitats.

(d) Communicable / Infectious Diseases

1. Many vectors developed resistance to pesticides (mosquitoes)
2. Culex - 41 species - resistant to Pesticides
3. Anopheles - 35 species - resistant to DDT, Dieldrin
4. Parasites developed resistance to drugs - Isoniazid (TB) Chloroquin (Malaria) Hycanthone (Shistosomiasis)

(e) Pesticide Genotoxicity Trends

1. Genotoxic to insects, microbes, Plants and exposed humans.
2. Production, Usage, malpractices, Improper usages are on increase

(f) Others

Polymorphism in some plants and many animal species, endangering some of these to different extents, may be noted in the niches under pollution caused by development activities.

D. Existing Scenarios

(a) Industrialization and pollution are on the increase in the developing countries.

(b) More pesticides are being used in agriculture.

(c) In the developing countries the above will only increase and in a corrupted form without appropriate checks and balances due to the ethos of consumerism

E. Perspectives

(a) It is feared that genetic diseases, cancers and birth defects in man will increase not only due to agents exposing them through usage, consumption and occupation, but also through Ecological channels.

(b) The infectious diseases may resurrect because vectors and parasites/pathogens are Resisting the chemical oppressors. These also directly add to human miseries.

(c) The agroecosystem problems arising from increased resistance and reduced fertilities.

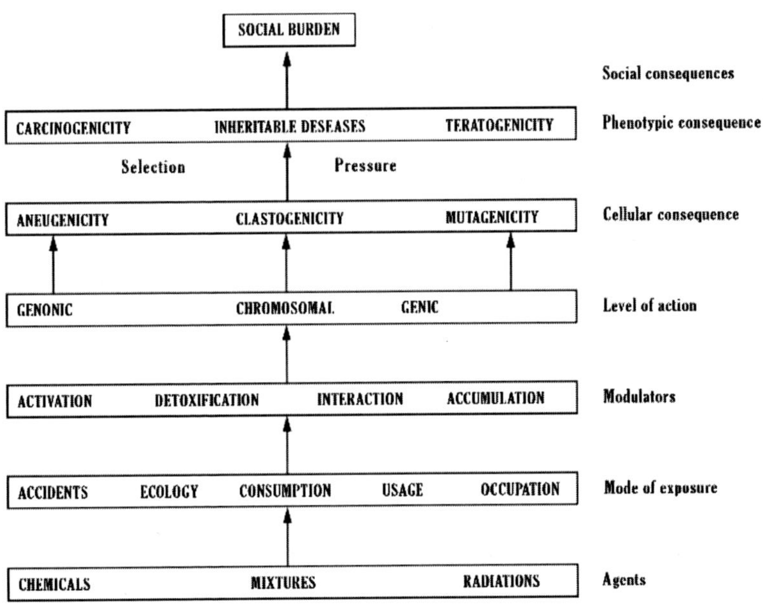

FIG 24.1- GENETIC TOXICOLOGY

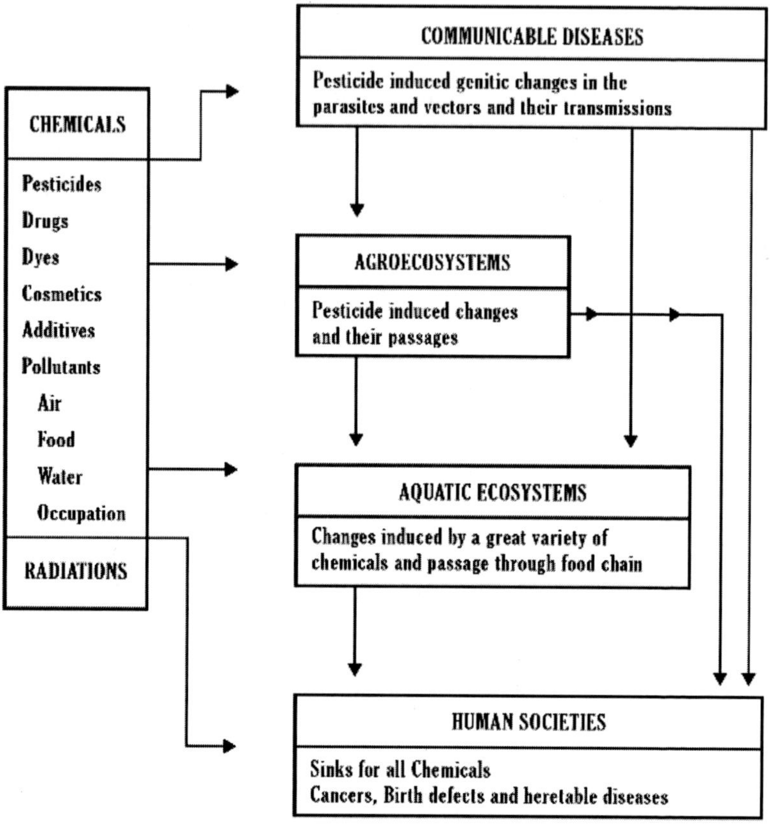

FIG 24.2 - Genetic Taxicology: Pestilence from Environmental Agents

Chapter - 25

Genotoxic Ill - Health from Pollutions

A. Introduction

Alma Ata conference a few years ago set a goal for the world Governments -Health for all by 2000 AD- probably by controlling infectious diseases and malnutrition. Whether or not this can be achieved they have ignored the effects of chemicals and radiation on human genetic systems and their consequences, i.e. genotoxicity. The objective of the present paper, therefore, is to focus attention to these health problems by seeking answers to some questions- what is a genetic system? What changes are forced here, by what and how? What are the impacts on society? How did the leaderships measure up to these challenges and finally what next?

Genetic system of an organism consists of genes and chromosomes present in all its cells. They control all life processes and finally transmit the characteristics to the next generation. When changes are affected in the structure of genes (mutagenicity) or chromosomes (clastogenicity) or the number of chromosomes (aneugenicity) cancers, birth defects and heritable diseases occur. Such a genotoxicity in pests, vectors and parasites of human relevance enhance the impacts, multiply costs and complicate problems. The human genetic effects are irreversible, without alarm signals, incurable excepting a few cancers and heritable except cancers. And the heritable diseases multiply with each generation.

The genotoxic impacts are globally on the raise for some decades. The chemicalization of society is also increasing unabated due to consumerism and industrialization i.e. in life styles as well as in environment. Thus the genetic syndromes increased from 3000 to 5000, cancers from 10 to 20% and chemicals from 5 to 10 millions during the last 10 years in the west. The developed countries are therefore making determined bids to manage these problems. As the developing countries also follow the same or similar models -socially and economically, ill-health due to genetic insults from chemicals and radiations stand out rather innocuously ominous. Focusing these issues, hither to unattended, is in the national interest, for at least, a debate may be generated.

B. Historical Aspects

The historical roots of the problem of gene based diseases can be traced to the beginnings of this century. All the three types of genetic effects were discovered in plants and animals during the 1920s. That chemicals can cause all these was established by mid-1940s. During 1960s many alarming facts on cancers and birth defects involving drugs, foods pesticides and natural toxins came to light which lead to the concept of environmental mutagenesis. Rapid strides in human genetic research in 1950s and 1960s correlating clinical symptoms with genetic effects, greatly strengthened this theory. This subject came to be known ever since, as genetic toxicology. Though great strides are made in understanding these poignant problems of human destiny in the developed countries, they remained largely unrecognized in the Indian Public.

C. State of the Science

1. Agents and Exposures: It is estimated that about 10 million chemicals exist in the world-solids, liquids and gases - 2% of which in daily use to different degrees, in the developed countries. These may be naturally occurring or man made- industrials, dyes, pesticides, drugs, pollutants in the air and water etc. Many of these occur in mixtures. The proportions in which they are present are not known. But it is generally agreed that most chemicals are released without adequate toxicity testing. Testing for genetic toxicity was not included as a requirement until a decade ago. Human populations are exposed to these chemicals singly or in mixtures in several ways- by direct consumption, or through food, water or air, by usage i.e., contacting at home or work place or due to accidents. The quantities taken and the duration of exposure greatly vary. The intake may be through skin or respiratory or digestive systems.

2. Effects on Genetic Systems: Genotoxic agents affect the cells of organisms in three ways. Firstly, they may alter the number of chromosomes. This process is called aneugenesis, and the agents, aneugens. Secondly they may alter the structure of chromosomes by breaking them. The process is called clastogenesis and the agents, clastogens. The third one involves modifying the genes -most prevalent and therefore most important. This is called mutagenesis and the agents, mutagens. A very large variety of chemicals are known to cause these effects through a variety of mechanisms.

3. Consequences to Humans: Whether the cells are altered genetically through aneugenesis, clastogenesis or mutagenesis, the consequences of transformation to the humans are three. The first is carcinogenesis - unending multiplication of tissues forming tumors. These are not inherited. The second is teratogenesis- the effects are seen in the newborns as birth defects. Some of these arise from an impact on the developing embryo in the mother. The others are heritable. The third are incurable and heritable diseases. These are manifest at different times in the life span of the individual. Some of these are seen at birth time and therefore are grouped together with teratomas as birth defects.

When most carcinogens were found to be mutagens and non - carcinogens, non-mutagens, it was realized that the few thousand genetic syndromes and the few hundred cancers and birth defects prevalent in society must have been induced by aneugens, clastogens and mutagens ubiquitously existing in our environment since ages. All syndromes do not however constitute miseries. They vary in their pains, durations, manageability, social burdens and economic costs. The total impact on the human society by all shades of genotoxicity may be arguably grater and severer than the other ailments of humankind.

4. Ecosystem pollution and Genetic Health: The major problems of toxicity probably lie in the pollutions of ecosystems where chemicals exist as complex mixtures. The resident plants and animals respond to these stresses. Thus the toxins enter human systems through water, air and food chains and affect the genome. This issue has been conceptualized only recently, which identifies four routes to monitor genotoxic hazards. Initially, plants and animals living in the polluted habitats are screened for genotoxic symptoms. Alternately standardized monitor systems are maintained in the polluted habitats as sentinels and checked periodically for changes in their marker characteristics in response to the pollution. Sometimes, the air, water and soil are tested in microbes, plants and animals in laboratory conditions. In rare cases, when the organisms concentrate chemicals from their vicinity, their extracts are screened for genotoxicity in the select test systems.

5. Modulations of effects and Consequences: The genotoxicities however are subject to modifications. (1) The genotoxins may be inactivated or innocuous ones may be toxified by the human tissues; (2) the internal effects may be neutralized before manifestation. (3) the manifestation may be prevented from becoming a health risk; (4) The defects may sometimes appear generations later. These are called modulations, enhancing or reducing or nullifying an original effect. These modifications are caused by physical agencies such as temperature, sunlight and pH and chemical processes such as oxidation, reduction, conjugation and a score of others. Further all tissues are not equally sensitive to the toxic properties of chemicals. Such differential behaviors led to the discovery of supermutagenesis and pseudomutagenesis- the former sinister and the later misleading. Deviations from the norm are also caused by excretion of toxins, cellular regeneration, excessive cytotoxicity, barriers to toxin entry, chemical deposition, accumulation, dose-thresholds, excessive gamete production, monthly rejection of some female tissues, abortion of defective embryos and ageing. Some of these increase genetic hazards (deposition, accumulation, activation of chemicals and ageing) while others decrease (detoxification, inactivation, antagonism, excretion, cellular regeneration, barriers, cytotoxicity, male and female gametic dynamics).The latter are predominant.

6. Role of plants: As parts of food chain plants have a significant role in causing as well as nullifying genetic effects. Vegetables contain many substances convertible into genotoxins by human metabolism {flavonoids, alkaloids, glycosides etc)and some capable of destroying genotoxins (carotenoids,chlorophyll,ascorbic acid, vitamin B etc.). At the same time they can concentrate chemicals like heavy metals and pesticides from polluted sites and pass on to humans. The fibrous contents of vegetables drive out otherwise resident genotoxins from the digestive system. Owing to these reasons vegetarianism is being favored in western countries as a preventive measure- a habit India cherishes from centuries.

7. Monitoring the hazards: Since humans cannot be tested, methods involving other organisms and human tissues have been developed to identify chemicals with genotoxic properties. Though it cannot be affirmed whether humans also respond to the genotoxins in the same manner as the test systems, many genotoxins were found to similarly act in several others. Consequently it became possible to connect the cancers, birth defects and heritable diseases with environmental agents. Though over 100 bioassays are available now, involving microbes, lower and higher animals, mammals and human tissues, for testing the putative genotoxins, only a score of these are of greater value, a few of which are given here.

(a) **Ames Test:** Dr Bruce Ames designed a method in such a way that the chemical (a) gets modified by the mammalian liver enzymes (b) certainly enters the bacterium, Salmonella and © reacts with the bacterial DNA, if capable of. The bacterial characteristics are altered as a consequence. This test revealed that most carcinogens are mutagenic. There is no peer to this assay and is therefore the most employed globally. Of course, several modifications are mode in this to suit specific needs.

(b) **Mammals:** Tests involving the genes and chromosomes of plants, insects, mammals and humans- whole body or tissues - reveal all types of genotoxicity. Results from these provided the wisdom to the origins of human syndromes from defective chromosomes. Excessive effects on the genetic system often results in death of embryos. This is monitored in pregnant mammals like rats and mice. Sublethal effects on embryos,i.e., teratogeny are also found here. This is not only the most reliable for linking genotoxicity with teratogenesis, abortions, birth defects, but also the most expensive one, used therefore sparingly.

(c) **Human populations:** Human populations, connected directly or indirectly with pollution- living or working in polluted sites or consuming biota from such sites -are monitored for genetic effects. This is done by (a) medico-social surveys, (b) verifying the health of chromosomes in the blood, amniotic fluid and aborted fetuses and checking the status of male gametes.

(d) **Tier or Battery approach:** For a planned screening of chemicals, generally a group of tests are used following either (a) tier approach or (a) battery approach. The first one consists of a step by step screening, the later steps becoming necessary only if the earlier tests reveal genotoxic potential. The second one involves a simultaneous screening in a group of selected tests and assessing the results by a consensus evaluation. Both have advantages as well as disadvantages. Besides being expensive and time consuming, these exercises encompass the scientist, technician, manufacturer, trader, medical profession, regulatory agencies etc. The vagaries of judging a chemical as a genotoxin or not so as to ban or continue its use, are indeed very complex.

D. Scenario: Developed World

The developed countries are much seized up with the problem of genetic ill-health ever since the foundations of genetic toxicology were laid during the late sixties. Laboratories involving in these problems have gradually increased. Genetic Toxicology became a part of curricula. Environmental Mutagen Societies sprang up in many countries and specialist Journals were launched. So far six international conferences were held in USA, UK, Sweden, Japan and Australia. Both in Europe and USA, the private industry has toxicity monitoring centers. There is at least one country, Hungary, with impeccable genetic health statistics which other countries like Holland wish to emulate. These statistics have given rise to the concept of detriment index -the detriment a genetic syndrome causes to its bearer as a human resource.

Reminiscent of thalidomide teratogenesis three decades ago, a few scores of deformed children are borne recently in the coastal regions of Germany and UK. Similar cases are also found among the seabirds. The inference is obvious. Highest incidences of skin cancer are recorded in Australia and New Zealand (40/100.000). The recent gulf war has already yielded its genotoxic crop of limb deformities and miscarriages. Pesticide connected birth defects were found in Hungary. Electro Magnetic Fields are the current suspects. Uv-ray fallout is expected to be high in the regions of 40-50 degrees of latitude The most widespread genotoxic symptom is the reduction of sperm production per unit by 80% and fall in its quality during the last 40 years.

They have now 10 million chemicals, 10-15% cancers and 4-5% birth defects. OECD found that chemicals were never tested for genotoxicity until 1980 and that there are at least 80,000 of them in daily use and about 1000 most frequently. Testing these important ones alone was estimated to cost Rs.55-60000 crores ($ 20 Billion). Toxic Substances Control Act came into existence. Environmental Mutagen Information Centre (EMIC) was organized and a National Toxicology Programme (NTP) was launched in USA where the budget allocations for genetic toxicology increased year by year. International Committee for Protection against Environmental Mutagens and Carcinogens (ICPEMC), International Registry of Potentially Toxic Chemicals (IRPTC) and International Programme for Chemical Safety (IPCS) were initiated.

E. Scenario:Indian Society and Environment

1. Genotoxic problems: Rapid industrialization, modernizing agriculture and consumerism have made ecosystems and human society highly chemicalized. While the causal relationships between cancers, birth defects and heritable diseases on one hand and the chemicals and radiations on the other cannot be established without doubt, associations and correlations can be. The genotoxic problems in India are due to life styles, disturbed food chains and pollutions. With the development of resistances by pests, vectors and parasites which affect humans very severely, indirectly and sooner, a fourth dimension is added to an already gloomy situation.

(a) **Life style issues :** Indian habits of betel quid usage- betel leaf, lime, areca nut, tobacco -chewing, snuffing, smoking; rural energy usage-dungcake and wood fuels and some food habits have all genotoxic potentials. Processed foods, with a reportoire of thousands of additives, cooking procedures like deep frying, roasting in direct fire, consumption of pickles, sauses, ice creams, usage of cosmetics and electromagnetic appliances, gastronomical niceties, intake of excessive drugs, liquors and chlorinated water entail exposure to genotoxins such as pyrenes, anthracenes, Polycyclic Aromatic Hydrocarbons, heavy metals, Nitrosamines and pyrolyzed amino and fatty acids.

(b) **Pollution Sources :** Improper usage of pesticides, chemicals in food industry, pharmaceuticals, radioactivity, bye-products from water chlorination, industrial pollutants in air and water, food animals and plants growing in polluted waters and soil have all potential to cause genetic effects in human populations. Work atmosphere in various industries constitute a threat to the human genome. Greater UV-ray fall out in the southern latitudes of India pose a cancer threat. The monozite beach sands of Kerala expose people to beeta and gamma radiations. The 50 000 X-ray facilities involving upto a million workers and over-exposed clientele are causing safety concerns to the Atomic Energy Regulation Board (AERB). Kitchen air with traditional combustion and diesel exhausts have Benzpyrene, a very dangerous genotoxin. Several coastlines, riverfronts, urban habitats, Sewage, waste dumps, environs of industrial establishments within and without have similar genotoxic potentials.

(c) **Food chain hazards:** As stated above, many toxins enter humans through food chain including pesticides and fungal toxins like aflatoxins. Flavonoids present in the mushrooms and vegetables are convertible into genotoxins in the human system. Some foods may be adulterated and some may contain dangerous contaminants due to inefficient technology of its production. The effluents from various industries contaminate aquatic habitats with PCBs, PAHs, etc, which enter human systems through food chain.

(d) **Pests, Vectors and Parasites:** In our preoccupation with humans, genotoxicity in other organisms received inadequate attention. Thus in agroecosystems, scores of insect pests and microbial pathogens have developed resistance to pesticides for some decades adding to the social and economic burdens of this sector. Mosquitoes became resistant to many pesticides. Tubercle bacilli, malarial and helminth parasites developed resistance to drugs. One of the mechanisms of resistance is mutagenesis. Thus genotoxicity engenders economic, social and psychological costs also to many sections of people.

2. Health care systems: All our knowledge of genetics, pollution of ecosystems, mechanisms of genotoxicity, methodologies etc will come to naught, if our medical education and profession do not absorb these and health services do not respond to these challenges.

(a) **Education:** Medical education does not have adequate genetic component- human and others. The biochemical component, which alone can elucidate mechanisms, is treated as secondary. The laboratory exercises do not include microbial or animal genotoxicity tests. Illhealth from environment -pollution and its food chain ramifications do not constitute a significant part of medical curriculum. When microbes cause diseases, chemicals cure them. When chemicals cause diseases? The issue becomes poignant when the health of a sacred endowment like DNA is spoiled. It is irrevocable. Without education, how can a profession face new challenges and serve the society?

(b) **Information and Statistics:** Data on health status of people are available with the Indian Council of Medical Research (ICMR) but are unreliable in popular opinion save perhaps the cancer registries initiated a few years ago. Even among these, historical aspects are often ignored. Consequently our data are not up to the desired standards of developed countries, despite the availability of ready-made models to follow. This is our India! Theoretically, every Medical Practitioner, Nursing home, hospital and medical college shall maintain records of all their patients. But there were no attempts to pool the relevant data in these places. Many health surveys were carried out in India on children, women, industrial workers etc for various ailments, but not on genetic diseases. Even the National Health Survey ignored the genetic syndromes and birth defects. These surveys need a careful planning following a thorough understanding of conceptual and measurement problems all of which are sadly lacking.

(c) Health services: The medical profession can prevent some genetic syndromes by scanning pregnant uterii or through amniocentesis. It could caution people on the genotoxic agents in our daily lives and the role of diets and other factors in increasing or decreasing the consequences. The department of social and preventive medicine should be able to monitor community's genetic health through epidemiological exercises. There are several drugs in common use which are genotoxic, but retained in use because the benefits overweigh the risks. In India, we have not addressed to this issue at all. Furthermore there are two serious issues the medical profession and services had to contend with, i.e., caring for the genetically defective is very difficult. While the options of euthanasia are still debated in the developed countries China has gone ahead and banned the birth of defective children and regulated the institution of marriage. The Indian healthcare system is beset with shortage of experts and technicians. Instrumentation is poor and their maintenance costly. Of the Rs.10000 Crores meant for health budget, 60% account only for salaries and office maintenance. All these are made worse by traditional ignorance and current apathy.

3. Scholastic activities: Intellectual activities started in India on the western model, during 1974-5 from the Bhabha Atomic Research Centre (BARC) by organizing the Environmental Mutagen Society of India (EMSI). Initially there was a spurt of scientific activity which has now receded perhaps as a passing phase. The EMSI conducts annual meetings to discuss researches on the genotoxicity problems. In 1985 the whole issue has been reviewed through bibliographic labours.This was followed by the constitution of a Pesticide Task Group to assess the genotoxic and allied problems associated with pesticide usage. This did not accomplish its objectives. Only a theory of Genepidemics came from one of its members, the present author, a few years later.

During this time the then President of EMSI tried to develop an All India Coordinated Project on Genotoxicity in India in collaboration with National Environmental Engineering Research Institute (NEERI, Nagpur) and Department of Environment and Forests of the Government of India for which the present author laised between Dr K.Sundaram, Dr B.B.Sundaresan and Dr T.N.Khoshoo the concerned chiefs of EMSI, NEERI and DOE. With their departure from the scene, the proposal never took off. Similarly, the present author tried in 1991 to assemble the peers in the subject at Pondicherry to review the problems and needs with the help of Late Dr N.K.Notani of BARC. The objective was to generate a treatise incorporating all the work done so far in India and embody a peer-perspective to serve as a guiding spirit in future. But this was sadly torpedoed by the patron's *volte face.* If these efforts were allowed to continue the genotoxicity situation in India would not have been so bleak.

The unhappy aspect of the Indian scene is the neglect of local problems, solutions and needs, to the preference of Western agenda or mere fancies. We have not been able to even copy the role models at all. The negligible interest of medical profession in genetic toxicology save cancer treatment as a routine responsibility is indeed puzzling. There is no genuine leadership in research, education, monitoring and regulation. Industry is expectedly very indifferent. There is not a single study in India concerning genotoxicity problems on the basis of a hospital or a habitat or an occupation. And we have the second largest scientific manpower in the world. There is not even one journal in this field while the newsletter is obscure. Most of the early promises and efforts have slackened, while the problems are on the ascendant.

Fast chemicalization, adulteration, ignorance, paucity of reliable databases and social watchdogs despite consumer protection Act, inadequacy of health services including the apathy of medical profession, inadequate or insincere regulatory activity by industry and Governments, anti-social practices of the profit-motivated enterprises and academia divorced from values compound the problem in India. And this is the same story in all the un-developing countries

F. Perspectives And Imperatives

From the foregoing it would appear that genotoxins and genetic diseases are everywhere. In reality this is averted by the modulations which allow only a fraction of the potential dangers to materialize. Secondly cancers are in far greater frequency than the teratomas and heritable diseases. This is due to barriers for the entry of genotoxins to the reproductive tissues and elimination of affected cells prior to manifestation. As a result, cancers in the west are four times more frequent (20%) than the other genetic defects (5%).In India Cancers are expected to be about 6 million (0.6%) by the turn of century. But with the consumeristic growth in this decade expected to equal the same in the preceding five decades the cancer frequencies will be many times greater than the current projections. The other genetic defects will also increase correspondingly. The net frequency of the genetic pestilence can be anywhere between 1 and 5% and most sufferers are likely to be from the upper and middle classes! So what next? Conceptual changes from the existing mindsets and implementing a programme of action by the leadership classes seem to be the only choice left because hoping to change the development model became a lost cause.

1. New Paradigms: Detriment Index: The 4-5000 genetic syndromes documented may manifest at different stages during the life time of the bearers from birth time to a very late stage of life. Their effects on the bearers also vary from nil to death through minor inconveniences to major suffering. Their frequencies in the populations also vary from extreme low to high. Thus every genotoxic symptom may not inflict suffering on the human and when they do so, not to the same extent. In other words, the detriment the syndrome imposes on its bearers differ greatly. Based on this and with the aid of Hungarian Statistics, an envy of all the developed countries, it was considered that 60% of people suffer some genotoxic detriment though only4-5% only carry birth defects and 25-30% develp cancers (Shankarnarayanan,1985 Annual meeting of the EMSI, BARC, Bombay). Thus the human resource is detrimented considerably from insults to their genetic material. The causes of these are the products and bye products of the development. In other words, by following a pathway of consumerism which the west seeks to abandon, refusing to learn lessons from their sufferings and jettisoning native wisdom to care for the future generations the present leaderships are actually planning a detrimental policy and not a development policy.

2. New Paradigms: Genepidemics: The genetic detriments- cancers, birth defects and heritable diseases have spread all over the globe unlike the communicable ones which are essentially local and controllable. Further the genetic systems of unfriendly organisms- pathogens, pests, parasites and vectors- keep changing under stress, thus becoming uncontrollable. On integration, all these constitute genepidemics with enormous costs- personal, social and economic.

3. Challenges from economic liberalization: The globalization of economies forced a liberalization of policies in India which encourages industrial growth. The foreign investments accelerate this process. Inconvenient industries may gradually be shifted to India. But these may be needed by us for the obvious and comparative advantages they offer. All these have implications. We do not have resources to tackle the pollution problems which will increase at the workplace and in the ecosystems and affect the lower echelons of society. The products may do the same to the upper classes because of increasing consumerism. The substances banned in the developed countries from food, pharmaceutical and agriculture industries may enter markets under the spell of advertising blitzes. Despite having the lists of banned chemicals, outmoded and abandoned processes, industries and technologies in the developed countries, the government will be helpless due to the compulsions of market goddess.

Consequently greater quantities and wider varieties of chemicals are slated to enter human systems accelerating genotoxicity as the most dominant feature of ill-health, detriment and misery by the turn of the century. For example Seveso, Three Mile Island, Love Canal, Thalidomide may join their greater cousin, Bhopal. This bleak scenario is likely to go worse with time unless there is a rapid development of ethics in the human society. A natural imperative of economic liberalization is a strong morality in public and private. But very little is heard of this ethic. Perhaps these values are consigned to the New Education Policy for it is the knowledge systems that monopolize the most unaccountable leaderships in India!

4. Agenda for attention and action: Some activities need urgent initiation and others strengthening. These can be modeled on American and Europian efforts. Having adopted their social growth models, why cannot we follow their ethos of social concern, academic pursuits and regulation? Along with what is convenient -the consumeristic appetites- let some inconvenient duties are also shouldered. Some of them are here.

(a) **State of the Art:** The most urgent task is to take stock of the Indian situation on the occurrence and frequencies of all forms of genotoxicity, genotoxins in our daily life and environment and responsiveness of the academia. These should be compared with the scenarios and experiences in USA and Europe to draw appropriate lessons in making an action programme. Since we are following the westerrn models and probably irrevocably their experiences during 1950-80 may have to be undergone afresh by india if the academic leaderships do not exhibit some hindsight wisdom. The entire genotoxin research in India has to be reviewed and compared with the western efforts- How far is it in tune with national needs and relevance, is to be assessed. Then only any meaningful policy inputs can be made.

(b) **Database on genotoxicity:** The urgent need is to build a reliable data base on the frequencies of cancers, birth defects, spontaneous abortions, heritable diseases, sperm abnormalities, mental illnesses in various sections of Indian population. The data on all these should first be segregated from the existing records of hospitals and medical Institutions. These may then be classified based on familial, environmental and occupational histories. Simultaneouly fresh data should be collected taking the hospital as a basic unit and compiled and analyzed by the medical institutions. By comparing these with Hungarian Statistics, the best in the world, a better system can emerge for following in future. Doing this alone for cancers, as being done now is inadequate, for, it leaves the other genotoxicities. This will not be a formidable task for the experienced and concerned medical scientists in an All India Coordinated Project.

(c) **Database on genotoxins:** Information on genetically hazardous substances, their occurrence and spread are available with National Institutes of Health (NIH) National Institute of Occupational Safety and Health (NIOSH) EMIC and NTP in USA, International Agency for Research on Cancer (IARC), IRPTC, ICPR, ICPEMC etc. These have to be culled and summarized for Indian relevance to serve as a reference source and widely circulated to all hospitals and medical institutions. The objective is to assess the status of these genotoxins in India. Next is to monitor the ecosystems- Occupational workers, Human settlements in and around polluted sites, Plants, animals, water and air that are specific or peculiar to our country. This work also has to be carried out by experienced scientists in an All India Coordinated Project. From all this will emerge a Registry of Potential Environmental Mutagens and Carcinogens. We already have Atomic Energy Regulation Board to deal with radiations. With these two India will have fulfilled the first requirements of the ICPEMC (International Committee for Protection against Environmental Mutagens and Carcinogens) and ICPR (Intenational Committee for Protection against Radiations)

(d) **Industrial Licensing:** All industries should be required to specify the names of all chemicals and radiations they have do deal with from beginning to the end. The product package shall include production details, chemical constitution, toxicity data as in the case of some drugs. The consumer therefore, is not kept in dark. If the chemical is banned in the western countries this shall be mentioned along with the reasons for using here. All industries must be required to have an occupational health unit and open to public scrutiny. With the economic liberalization these few checks are essential. All these should be incorporated in the licensing policy. In fact industries should be the base from where the database for chemical registry and thence the genotoxins may have to be prepared. Should we not know how many of the 10 million chemicals globally accounted, exist in India and in occasional and regular use?

(e) **Consumer protection Act:** This otherwise exhaustive enactment should have covered the possibility of genotoxic chemicals entering our daily lives by borrowing the relevant provisions and regulation mechanisms from American and European models.

(f) **Fresh look at Medical and Environmental Education:** There is a necessity of including more practical genetics in all medical curricula. Medical institutions must be required to collect genetic data from occupational and epidemiological studies. They shall be required to be familiar with genotoxic syndromes and genotoxins- putative and real- and their assays. Considerable inputs on environmental degradation and ecological balances are also warranted. Needless to suggest that the syllabi of Pharmacology and Toxicology have to be updated on par with the ethos in the developed countries and paradigms advanced here. In many Biology curricula genotoxicology is not included though genetics is a continuing fixture. In the curricula of Environmental Sciences, operating in about 60-70 universities, though toxicology figures in most, genetic component is sadly missing. Since genotoxicology is trans-disciplinary involving physiology, genetics, chemistry, medicine, agriculture, molecular biology, biochemistry, sociology, economics and finally the developmental processes and human psyches, it must be given a wider role in education and better public relations.

(g) **Organizational needs:** Genotoxicity research has to be made a thrust area of all agencies-DOEF, DST, CSIR, ICMR-all these have relevances to this field as it is a transdisciplinary subject. Pollution and environment comes under DOEF's jurisdiction. CSIR has NEERI and ITRC, the most relevant institutions for the study of environmental toxicity. ICMR deals with human health, the ultimate concern of genotoxicity knowledge. The entire science and technology that created the problem first which therefore must also resolve, is the responsibility of DST. All these organizations may have to review their objectives, *vis-a-vis* genotoxicology. On the contrary an independent research organization may be mandated with all the research agendas and programmes connected to genotoxic ill-health eliminating the habitual *ad-hocism* and arbitrariness present in this field where benefits are uncertain, risks heavy and costs inestimable.

5. Human Resource Development: The foregone account underscores the deterioration of humans as resources due to genotoxins by (a) rendering a section invalid, (b) imposing a detriment on another section and (c) inflicting a psyche of fear, burden and misery all over. The causes of these are the products and bye products of the industrial development. In other words, by following a pathway of consumerism which the west seeks to abandon, refusing to learn lessons from their experiences and jettisoning native wisdom of caring for the future generations, the present leaderships are actually planning a detrimental policy and not a development policy. The academia is not far behind. They have not given adequate importance to these problems despite increasing incidences of genotoxic diseases in Europe and America which are likely to repeat in India during the coming decades. This situation is rather strange and speaks of the absence of genuine leaderships in our knowledge systems also where hedonistic practices seem to have taken strangle hold over the altruistic ccerns for peoples' welfare.

Table 1
Some Genotoxic Syndromes in Human Populations

Genomic	Down, Patau, Edwards, Kilnefelter, Turner
Chromosomal	Bloom, Buckling, Cri-du-chat, Philadelphia
Genic	Lou-Gehrig, Tay-Sach, Pentosuria, Polycystosis, Retinitis pigmentosa, Multiple exostoses, Spinocerebellar ataxia, Cystic fibrosis, Muscular destrophy, Ataxia, Xeroderma pigmentosa, Sickle Cell anemia, Huntington's disease, Alkaptonuria, Hemophilia, Phenylketonuria, Chapter - Galactosemia, Thalessemia
Others	Dwarfism, Cleft lip and palate, Sperm abnormalities, Limb and skeletal deformities

Total Syndromes : 5000 Frequeny of each syndrome :1/100,000 to 1/1000 Total Cancers : >200 Frequency : 2.5 to 5% in the west Monitoring : Birth or population Frquency : 10-15%

Able 2
Some Genotoxic Chemicals

	Types	*Examples*
1.	Mutagens	rmaldehyde, Chloramphenicol, Hydroxylamine, Styrene, enz(a)Pyrene, Sodium zide, Ethylenimine, Methane Sulfonates
2.	Clastogens	Mitomycin C, Actinomycin D, Isoniazid, Benomyl, Chloramphenicol, Ethidium bromide
3.	Aneugens	Ethylene glycol, Griseofulvin, Nitrous oxide, Lindane, Chloral hydrate, Ethoxy Caffeine, Ethidium bromide
4.	Promutagens	Cycasin, Saffrole, Vinyl Chloride, Morpholins, Olifins, Piperidines,
5.	Antimutagens	Vitamin B,Vitamin C, B-Carotene, Theophylline, Rotenoids, Biometals
6.	Supermutagens	Sodium Azide, Ethylenimine, Methane Sulfonates
7.	Carcinogens	Saffrole, Sodium Azide, Benz(a)Pyrene, Cyclamates, Furfuryl Furamide, Arsenic, Benzene, Chromium, Nickel, Aflatoxins, Nitrosamines, Epoxides
8.	Teratogens	Thalidomide, Hycanthone, Methyl Mercury, Tetracycline, Chloroquin, Mimosine, Bisulfan, Nicotene, Caffeine

Table 3
Sources of Genotoxins in the Human Environment

S.No.	*General Types*	*Specific Classes*
1.	Naturals	Alkaloids, Glycosides, Flavonoids, Others
2.	Foods	Colours, Sweetners, Sterilants, Additivs, Processing
3.	Habits	Mastications, Inhalations, consumptions,
4.	Pesticides	Insecticides, Nematocides, Fungicides, Herbicides, Acaricides
5.	Drugs	Modern, Traditional
6.	Cosmetics	Natural, Synthetic
7.	Dyes	Consumptives, Nonconsumptives
8.	Industrials	Pharmaceuticals, Rubber, Textiles, Petroleum, Paper fertilizers, cement, explosives, Leather, Photography, Glass & ceramics, resins, sugar, Paints, Metallurgy, Iron & Steel, Plastics, others
9.	Energy	Nuclear, Petroleum, Biogas, Biomass (Production, Conversion, usage, Transport)
10.	In Situ	Radiation & Chemicals
11.	Occupational	Heavy metals, Aromatic Hydrocarbons, Halogenated & Aromatics, Nitroso Compounds, Combustion Products & Ambient Air
12.	Water	Domestic and Industrial Effluents
13.	Soil	Pesticides, efflluents, Others
14.	Food Chain	Residues of Toxic Pollutants
15.	Metabolites	Synergistics, Degradations, Activations
16.	Others	

Table 4
Some Genotoxic Stresses to Indian Populations

Agents	*Occurrence*	*Exposure*	*Effects*
B&G rays	monozite sands of kerala beaches	thousands of people for variable durations	Birth defects detected
X-rays	50 000 medical facilities	a few lakhs and for many years	impact reports under scrutiny by AERB
Uv-rays	Fall out in southern most India	millions and for variable durations	spurt in skin cancers feared
EM-rays	electro magnetic appliances at homesteads	millions and for variable durations	Effects suspected
Flavanoids etc	natural in vegetables	millions of consumers and life long	convertible into genotoxins effects suspected
Alkaloids	Tobacco-areca nut-betel leaf quiddings	millions of habituals for brief spells	conform association with cancers in a few thousands of people, more suspected
Aflatoxins	fungal toxins as contaminants in cereal grains	Uncertain size of consumers for limited durations	confirmed carcinogens
Pesticides	contaminant residues in the vegetable foods	millions of consumers for long spells	many confirmed genotoxins, birth defects reported.
Nitrosamines	variety of vegetable and animal foods, beers, wines and liquors	millions of indiscrete consumers	confirmed genotoxins, victims suspected
Nitropyrenes	reprography, diesel exhaust,combustion of biomass, tobacco smoke	millions of occupational workers	BaP is a confirmed genotoxin, victims suspected
Many dyes, Nitrates, AF2 etc	in fast and canned foods, soft drinks, ice-creams and others as colourants, preservatives, flavourants, etc	a few millions of modern gastronomical consumers	many are confirmed genotoxins. victims suspected
Deformed aminoacids, pyrenes and anthracenes, aromatic hydrocarbons	deep fried and pyrolysed foods	millions of consumers and cooks	confirmed genotoxins, victims suspected
Chlorinated Compounds	Purified municipal water supplies	All direct consumers	some are confirmed genotoxins, victims suspected
Drugs	therapies against infections (malaria,helminth,tubercle etc) cancer	a few million patients for brief spells	some drugs are confirmed genotoxins, victims suspected
Heavy metals	Occupational environment of many industries, cosmetics, diesel and cigarette smokes	millions of workers in many industries,cosmetic users,smokers etc	confirmed genotoxins, victims suspected.
.Styrene and SO2	Occupational atmosphere of many industries	thousands of workers for specific durations	confirmed genotoxins, victims suspected.
Arsenic	ground waters of Bengal	millions of consumers	effects expected
Others	40 atomic energy establishments with possibilities of radiation exposures; 9 coastal zones with polluted aquatic food chains; 16 urban centres where most genotoxicity may occur; vegetarian diets contain genotoxins as well as antigenotoxins; Indians are after the same genotoxic practices which the westerners try to abandon		

Chapter - 26

Genetic Toxicology of Pesticides in Higher Plant Systems

A. Introduction

Since the mid-1940s, synthetic chemicals have become an integral part of modern society. Among these, pesticides are the most notable because they are basically designed to be toxic to man's enemies in the environment, at home and in agriculture, embracing a wide range of target organisms. The use of such chemicals is not new. Legend has it that the Romans destroyed Carathage in 146 B.C. and ploughed the land with salt so that all living things would be destroyed. During 70 A.D., Pliny the Elder recommended arsenic as an insecticide, and the Chinese are, in fact, reported to have used arsenic sulfide as an insecticide in the late 16th century.

Today, pesticides are considered an essential part of the current agricultural production technology and have been instrumental in reducing losses to crop yields from insects, pathogens, and weeds. They are also indispensable in insect and disease control in forestry, home gardens, lawns, the clearance of roadsides, and the maintenance of right-of-ways. The use of pesticides against the disease-spreading vectors, and hence their indispensability in public health, is also well established.

A brief account of pesticides may be in order. Kurinnyi and Pilinskaya estimated that there are about 800 pesticides. Khilchevskaya estimated that 900 basic chemicals constitute about 100,000 varieties of formulations all over the world. The Pesticide Index of the Tropical Pest Management also lists nearly 800 chemicals, including growth regulators. Ridgeway et al., on the other land, counted 1500 chemicals, including growth regulators. Ridgeway et al., on the other hand, counted 1500 chemicals, including stimulants, emulsifiers, and solvents, as base chemicals for 50,000 products. The latest estimation of these has reached nearly to 2500.

Concerning the global estimates of the quality of pesticides used no reliable data available. However, in the U.S. until 1984, about 1 billion pounds of pesticides were used annually for pest control, costing about \$2.2 billion, resulting in a groww return of \$8.7 billion in the form of crop savings.

The crop losses in the rest of the world were expected to be over nearly 10 times than the U.S. The inference that pesticide inputs will be constantly on the rise in inescapable. Although it is believed that the U.S. Alone accounts for half of the world's production and consumption, quantities of agricultural usage vary from country to country. Japan consumes 10,000 g, Europe 2000 g, the U.S. 1500 g/ha. India may be cited as representative of developing countries, where more than 100 pesticides have been in use, putting usage at approximately 400 g/ha.

B. Usual Defects in the Work

There have been many undesirable practices in monitoring pesticides, and several prerequisites and precautions have been ignored. This may be chiefly because most of the earlier studies were simple cytogenetic investigations which had no intention of being built into bioassays. However, a brief summary of these defects may be warranted. Many pesticides were tested more than once, sometimes in the same system by different workers (e.g., root systems). Occasionally, the endpoints scored differed. Quite often, however, the same endpoint was scored. In many publications, wrong photographic evidence was presented as effects (e.g., cytotoxic symptoms). Dose responses were rarely monitored. Sometimes there was no basis for the selection of the treatment doses and durations. In many cases, even recoveries were not carried out. Consequently, not only has there been an enormous duplication of effort, but the effort itself was improperly made. Variation in methodology has added to the confusion. Due to a lack of appropriate recoveries across cell cycles, whether the cellular effects are transmittable has remained an unattended aspect of most studies. In these circumstances, only an empirical and inferential summary of the clastogenicity, aneugenicity, mutagenicity, and other effects of pesticides could be given .Sometimes there were no negative controls, solvent controls, or even positive controls in many studies.

For all these reasons, the issue of specificity of pesticide action and its manifestation, consequences and structure-activity relationships were completely clouded and therefore became indiscernible.

In light of these flaws, the following suggestions to remedy the situation may be in order.

1. Test Agents: Background information, such as the nature, properties, mode of entry, mechanism of action, metabolism, and persistence of the pesticide, is a necessary prerequisite of a planned approach to any experiment. General impacts on cell cycle such as acceleration of retardation or inhibition, mito-depression, partial or full repair of the damage during recovery duration are also information required for a proper interpretation of later data. These two aspects may influence the causation, nature, and quantum of effects in any system vis-à-vis the dose employed. All such information on the candidate pesticides for testing should be an essential prerequisite in genotoxicology studies, lest the dose-response relationship become meaningless, even among structurally related pesticides, as far as plant systems are concerned.

2. Controls: The systems under test must be standardized as to their growth conditions, cell cycle duration, background frequency of clastogenicity and aneugenicity, if any, and responses to standard clastogens and aneugens. In other words, positive, negative, and solvent control should be maintained alongside the test agents.

3. Delineation of Effects: It is surprising that no serious effort has so far been made to delineate cellular aberrations into clastagenic and aneugenic symptoms when all deviations from normal cell division can fall only under either of these two categories - micronuclei being a common one. Such delineation is now warranted - cells with symptoms like bridges, fragments and deletions, connections, exchanges, and stickiness among chromosomes should be termed clastoants and those with laggards, arrested metaphases, forwards, scatters, multipolarity, binucleation, and bimitosis should be termed as turbants following

Brogger's term "turbagenicity". Understandably, one may have only the frequency, in the final reckoning, relegating the individual endpoints to the background of details. Whether clastogenesis and an eugenesis will occur together or can be differentiated by dosage manipulation should be evaluated for each test chemical. The endpoints mentioned above will of course increase in meiotic cells.

4. Dosages: It may be desirable to fix an upper limit to the dosage regime under test on the basis of cytotoxicity and progressively dilute to the minimum. This is necessary to fix the minimum effective dose as well as dose-response behavior. This can be done by first determining the concentration of the chemical that is cytotoxic as indicated by clumping, pycnosis, and mitotic delay in the tissues under test; half the concentration of this cytotoxic threshold can serve as the upper limit, i.e., CT.

5. Procedure: By conventional reckoning, many pesticides may appear to be clastogenic, but this may well be a half truth. In some cases, the clastoant frequencies may be comparatively more after longer recoveries, or may even totally disappear. Such an occurrence or retention after a long passage of time is the index of the viability of the pesticide impact. Therefore, only such pesticides are likely to cause similar effects in other systems. This kind of approach lends credibility to the meristem assays in genetic toxicology and plugs several loopholes in the existing state of the art. IN such an experiment, where recoveries cover more than two or three cell cycles, the affected cells contain only the ultimate cellular consequences of clastogenicity, i.e., deletions, exchanges, micronuclei, and aneuploidy far above the control frequency. Obviously, the controls - negative, solvent, and positive - are of paramount importance. It may be that the dose that causes twice the frequency as in the negative or solvent controls, if at all caused, should be considered as the "doubling dose"-a starting point in dose-response evaluation. However, this has its limitations since some of the agents have a very narrow range of response. Furthermore, a too-high or too-low initial dose in treatment may include the possible chance for the doubling-dose effect.

Appropriate planning of the experiment in accordance with the nature of the test system, properties of the agent, and theoretical bases of cellular mechanisms, events, and endpoints is an essential ingredient of genotoxicological testing in plant systems. Without enough checks and balances, the plant systems end up in disrepute through no fault of their own.

6. Residues and Metabolites: The other defect in our current work is related to the biological and chemical conversions of pesticides and their combinations in soil, water, plant, and other organisms. These modified substances may be an entirely different type and extent of effect and hardly any work has been carried out along these lines. Furthermore, the residues are generally in miniscule amounts in nature. What effects do these low doses have on the genetic system expressed over long durations? The food chain magnification of these low level residues may pose a further problem - do these gradually increasing residue levels entail a correspondingly increasing genotoxicity in the biota at the higher trophic levels? A load of "mutation pollution" could then be a natural corollary to biomagnification in the ecosystems polluted by pesticides if they are genotoxic. What then should be the protocols to assess such ecological situations? Many more such questions remain unanswered and appropriate test systems to evaluate such effects need to be developed.

C. Perturbation of the Agroecosystems

If the various criteria mentioned in the previous section are applied, the number of pesticides actually genotoxic could be far less than claimed. Prior to such a check, a few considerations may be relevant in the context of current ethos in pesticide testing with specific endpoints in higher plants. They are (1) what happens to other aspects to crop life, (2) how do other biota in the agro-ecosystems react to the pesticides around, (3) what are the implications of the present knowledge to the whole agro-ecosystem, (4) what is the current scenario in the agro-ecosystems under pesticide stress in the developing countries, where the usage is ever increasing, and finally (5) what could be the future ecological portents in the agro-ecosystems due to genotoxic pesticides?

1. Effects on Crop Systems: From a genetic point of view, mitosis in the shoot meristems, meiosis, seed set, and mutations are the other aspects of importance besides the usually tested meristems of the root. We have seen that all these are subject to pesticide stress. In a comparison, effects on meristems and effects on other endpoints seem to have similarity to varying degrees. 90% of the pesticides producing positive effects on meristems also cause the same effects on meiocytes, 60% reduce seed set, and 20% cause mutations. Obviously, the compounds affecting the meristems will have similar effects on other endpoints too, albeit to reduced extents Theoretically, impacts on all these endpoints, by themselves, may cause developmental variation. Meiotic aberrations may lead to sterility and affect recombination and segregation of characters. Also, a reduction of seed set per inflorescence, being the biological basis of grain yields, may ultimately result in the reduction of crop productivity. Mutations, whether or not scorable, may involve serious characters, including development of susceptibilities to newer strains of pests and pathogens

2. Effects on Other Biota in the Agroecosystem: The karyotoxicity and mutagenicity of pesticides were also studied in a wide variety of systems. From the perusal of these data emerges a picture of concordances between plant responses, on one hand, and those of microbes, fungi, insects, and mammals in vivo and in vitro, on the other. Therefore, the likely similarity of responses between plants and other inhabitants of ecosystem is a cause for concern. A few examples are detailed.

Among the fungicides, benomyl, captafol, captan, carbendazim, fenaminosulf, and mancozeb are geotaxis in plants, microbes, fungi, insects, and mammals notwithstanding a few exceptions Among the herbicides, atrazine and paraquat are similarly genotoxic in plants and fungi (Among the insecticides, dichlorvos, Malathion, methyl parathion, oxydemeton methyl, phosphamidon, and trichlorfon are genotoxic in plants, microbes, insects, and mammals.

These fourteen pesticides are therefore the most hazardous to the agro-ecosystem from a genetic point of view in that they have the potential to subject not only crops, but also pathogens, pests, and others to mutational stresses. The ultimate consequences of these mutational loads to the whole system can only be surmised at this point of time.

Since pesticides are mostly used in agriculture, their genotoxicity to a wider variety of biota may have implications depending upon healthy interactions within the community; interactions depend on the characteristics of the populations of organisms; and the characters are genetically controlled in the individuals. It is these genetic components, not only of the crop but also of the weed, the pest, and the pathogen and therefore of every other member of biota, that pesticides threaten to affect.

3. Implications to the Whole Agroecosystem: At this juncture, it may be appropriate to ponder the likely targets of mutational stress in the agro-ecosystem.

It has been estimated that out of 45 fungicides, 80 herbicides, and 60 insecticides screened in plant systems so far, only 25 fungicides, 20 herbicides, and 15 insecticides have been screened for genotoxicity in other systems might not have been tested in plants.

Among the 25 fungicides screened in various biota, only twelve possess mutagenic potential in both fungi and higher plants. In other words, the crops receiving these twelve fungicides may develop mutations for susceptibility to new pathogens. Similarly, pathogens may develop mutations for resistance to the very fungicides used to control them. In the realities of the agro-ecosystem, this would mean that pathogenic fungi and their host crops may have to be under surveillance for genetic changes whenever these twelve fungicides are in use.

Among the 20 herbicides screened in various biota, all possess the potential to be mutagenic in higher plants. In other words, the weeds receiving these herbicides may develop mutations for resistance to the same herbicides used to control them. In the realities of the agro-ecosystems, this would mean that several weeds and the crops may be gradually increasing their genetic variability.

Among the 15 insecticides screened in various biota, only four possess the potential to be mutagenic both in insects and higher plants. In other words, the crops receiving these four pesticides ma y develop mutations for susceptibility to new pests. Similarly, the pests may develop resistance to the very same insecticides used to control them. In the agro-ecosystem, therefore, one should be looking for newer pest strains and more susceptible crops.

In the long run, crops may develop susceptibility to new pathogens and pests. The pests may develop resistance to pesticides due to their genotoxicity; such compounds, by the present reckoning, only number 16 out of nearly 200 screened for genotoxicity. Even this is only an empirical inference - a whole field of scientific investigation awaits evaluation.

4. Agroecosystem Realities - the Current Scenario: It has been well established that the agro-ecosystem, consisting of crop, weed, microbe, insect, fungi, invertebrates, vertebrates, mammals, and humans, functions in balanced and productive way due to the conservation of genetic material and stability of each of these components. Recently, there have been reports of the erratic behavior of these components, although pesticide-implicated ecological imbalances have been known for some time. For example, ploidy variations, chromosomal aberrations and gene mutations were also known to occur in some of the biota relevant to agro-ecosystem. Furthermore, insect and other pests developing resistance to pesticides through genetic alterations, human workers showing chromosomal changes, and innocuous pesticides often becoming genetically hazardous have been on the increase. This information and other historical and circumstantial evidence led Sharma to propose the theory of agro-ecosystem instability due to pesticides.

5. The Ecological Portents - the Instability Theory: This theory emphasizes that pesticides mutate various biota in the agro-ecosystem directly or indirectly, suddenly or gradually. Since crops are the initial as well as the ultimate recipients of pesticides, it is inconceivable that the biota ecologically associated with them could escape the mutational stress of pesticides. It is further postulated that pesticide mutagenicity could alter the nitrogen-fixing capabilities of bacteria and blue-green algae. Crops may suddenly become sensitive to traditionally minor pests. Insects could become resistant to pesticides. Habits and habitats of useful organisms may be altered; breeding systems may be modified. Since many pesticides may stay in the system as residues in the same or modified form, there could be more severe long-term impacts in the food chain.

All these together have the ingredients to threaten the stability of the agro-ecosystem, which is the very basis of crops productivity.

D. Further Research Of Social Relevance

1. Genotoxicity Testing: In genetic toxicology, testing chemicals for mutagenicity, carcinogenicity, and teratogenicity and applying the results to humans forms the basic ethos. Since the human species cannot be directly subjected to testing, mammalian species form the all important link between the experiments and the ethos. The question now arising is how far are the results of the plant as says applicable to mammals? In other words, do these two systems respond similarly to the pesticides? If the answer is "YES", the credibility of plants as mainstream test systems is assured. If the answer is "NO", the utility of the systems would be restricted to agro-ecosystem only.

Kihlman first attempted this comparison between meristem and mammalian cell responses. Subsequently, Grant made more successful attempts in this direction. Almost all the same time, a workshop at NIEHS, Research Triangle Park, U.S., found at 99% concordance between selected plants and mammalian systems. Subsequently, Sharma found a 73% concordance for genotoxicity and 75% for karyotoxicity between these two systems.

In the present assessment, 25 pesticides have relevant literatures for comparison, i.e., 25 were tested in the mammalian and plant systems. Of these, 15 pesticides elicited similar karyotoxic responses in both. This works out to a concordance level of 60%. Mutagenicity could not be compared because the plant mutant indicators such as chlorophyll deficiency, etc. are morphological endpoints some of which are complex polygenic systems while the mammalian and human indicators were mostly in vitro monogenic biochemical markers

2. Specificity vs Nonspecificity: Most pesticides appear nonspecific in their activity at the cellular level. In other words, only very few can be definitely rated as aneugenic, clastogenic, or mutagenic. Even after all the checks have been followed, a considerable number of pesticides may be found to cause more than one effect. However, some specificity is found with regard to phragmoplast in inhibition leading to bimitosis. Only three pesticides out of about 40 tested so far are known to cause bimitosis. The reason for this situation is the lack of investigations to correlate structures and properties of pesticides with activities.

3. Effective Doses: Similar karyotoxic effects may be produced by varying ranges of doses of pesticides. Some pesticides are effective at very log doses while others are ineffective even at very high doses. These variations could be due to several reasons. Factors such as he chemical nature, permeability metabolism, mechanism of action, and persistence of the pesticides and their impacts on cell cycle, and partial or full repair of the damage during recovery regimen could influence the causation, appearance, and frequencies of effects produced in a system vis-à-vis the dose administered. The minimum effective doses are also likely to be far lower than those obtained hitherto because (1) most of the work was not done in a reproducible manner, (2) many workers used commercial samples which contain many accessories to the pure technical grades of pesticides, and (3) some of the effects may be secondarily caused by pesticide metabolites whenever the interval between treatment and assaying is long, depending upon their half-life periods.

4. Field Applications and Genotoxicity Rating: Despite the innumerable studies on the genotoxic responses of plants to pesticides, the role of field dosage and mode of application were never considered while assessing a compound for genotoxicity potential in the field. The experimentally determined genotoxic dose ranges of the pesticides should be checked whether or not they fall within their prescribed field application dosages. The inferences is that only such pesticides are likely to cause genetic effects on crops standing in the field, in due course of time.

Another limiting factor regarding the relevance of experimental results to the field is the mode of application. For example, seed dressings may have more initial effects on the root systems and embryonic shoots which may be subsequently passed on the adult plant. Soil application poses the problem of low-dose chronic impact on all soil-borne biota. On the other hand, foliage sprays cause the adult plants to retain the pesticide and translocate it to the flowering shoots before getting washed down to the soil. While all three models have an impact on non-target aspects, seed dressing and soil application appear more consequential at the level of development, while spraying has greater impact at the level of reproduction and productivity. It would appear therefore from the present survey that many of the positive effects in the experimental systems may not hold true in field situations.

5. Effects on the Female Germline: Thus far, there has been no work on the effects of pesticides on the meiosis in the female germline due to the obvious logistic difficulties in handling megasporogenesis. An indirect indication is however offered by the reduction of seed set, even when there is no pollen sterility under pesticide stress. It may be noted that studies on female meiosis itself are generally meager.

6. Indirect Effects: Slow and gradual effects can indirectly occur in the plant systems from three sources. first, pesticides may be metabolically modified in the plant system into genotoxins which may exert influence that the parent pesticide did not possess. Several herbicides are known to behave in this way, releasing yet unidentified mutagens into the host plant. Second, pesticide residues, harbored in miniscule quantities, may slowly affect genetic material of the crop. However, such concentrations at the residue level have not yet been demonstrated to have any genetic effect even in the meristems, the simplest of plant systems. This possibility, however, cannot be ruled out. The third source is limited only to developing countries where grains are often stored in both the farm houses and market places under heavy pesticide lacings and sprays. This improper practice subjects the seed stocks to unusual pesticide stresses which may be revealed when the seeds are germinated to raise new crop. Many so-called spontaneous variations in crop plants may, in fact, be due to this source. Hardly any work has been carried out on such effects.

7. The Pesticide Situation in Developing Countries: Pesticide inputs into the agriculture in developing countries has been on the increase and this trend is expected to continue. Yet only fraction of the arable land is covered by these chemicals. The quantities as well as varieties of compounds per unit land is far less in developing countries compared with the developed nations. However, within this small fraction used, the input is great because of overdosing generated by the "***overkill***" syndrome of the farming community. In addition, adulteration of the pesticides with inks, kerosene, varnishes, dyes, etc. is also rampant. In developing countries, the genetic effects are thus limited only to that fraction of land (e.g., about 10% in India) where the crop systems are exposed to pesticide usage and the misuse of pesticides in protecting consumers i.e., lacing the containers of grains, flour, and other similar consumables with pesticides- a prohibited practice- as a marketing method. Obviously, the problems of genetic toxicology of pesticides are more serious in developing countries than in the developed world in spite of a higher dosing norm.

"AN ATTITUDE TO LIFE WHICH SEEKS FULFILMENT IN THE SINGLE-MINDED PURSUIT OF WEALTH - IN SHORT, MATERIALISM - DOES NOT FIT INTO THIS WORLD, BECAUSE IT CONTAINS WITHIN ITSELF NO LIMITING PRINCIPLE, WHILE THE ENVIRONMENT IN WHICH IT IS PLACED IS STRICTLY LIMITED."

ERNST F. SCHUMACHER

Chapter - 27

Genetic Toxicology in India

A. Introduction

The importance of genetic insults from environmental agents to human health was recognized during the 1960s and paved the way for the organization of the Environmental Mutagen Society in U.S.A. Concerted efforts following this have resulted in the screening of environmental agents by using appropriate assay systems drawn from the pre-existing literature in genetics. This message reached India when the Environmental Mutagen Society of India (EMSI) was given birth in 1975, chiefly due to the efforts of scientists at the Bhabha Atomic Research Centre (BARC) led by Dr K Sundaram. Like the fraternal bodies in a few other countries, EMSI has been striving to organize efforts to ultimately provide the public with information on the genetic toxicity of their environment, from where there is no recourse to normalcy, once set in.

B. Historical Aspects

While the investigations on cellular responses to chemicals were first carried out during 1890s, well documented results appeared only around 1940. These included such chemical groups as pesticides, alkaloids, drugs and even routine laboratory reagents involving celebrities such as Kostoff, Eigsti, Auerbach, Robson, Oehlkers and Rappaport. It is interesting to note that one of the early studies on pesticides by Kostoff was published in an Indian journal, Current Science; a survey of the Indian scene in this discipline of prime human concern may be worth while.

C. Early Investigations

The earliest papers in India were by the great botanist, Prof Mehra, in 1946-49 he studied the effects of colchicine and sulfanilamide in Ephedra pollen. He also pioneered similar investigations on pteridophyta which today command great appeal for in situ employment. Spermatozoan abnormalities in fern prothalli in response to chemical stresses were also assayed by him for the first time. Incidentally, the first review in this field was also by him. During this period foundations were also laid in some allied areas. Responses to Gammexane, Ethylene glycol, Carbon tetrachloride, Vegetable oils, Insecticides Urethane were documented in systems such as Corchorus, Vicia, Arabidopsis, Allium, Drosophila, Poecilocerus by Rao (NS,SRV) Seshachar, Bhattacharya, (s) Swaminathan, Natarajan and Kundu. Prof Seshachar was perhaps the first to recognize chromosome segregations as a source of genetic errors.

Works of Prof A K Sharma from mid-fifties are a class by themselves dealing with a wide variety of chemicals, which were major trend setters in the cytogenetical research in India, ever since. During the close of this period came the most significant contribution i.e. the genetic toxicology of the monozite belt in Kerala using on the site vegetation and Tradescantia staminal hair system. This perhaps is the first in situ investigation in the world, for the kind of which much stress is laid presently. Publications from Delhi and Hyderabad have just begun to appear from Natarajan, Kaul, Kesavan, Rao, Reddy, Subramanyam etc. Also appeared at this time was a review by Sharma (CBSR) on plant systems. Investigations being fragmentary, it is apparent that the progress has been devoid of direction.

D. Later Studies

When the EMSI was organized as an affiliate of the International Association of the Environmental Mutagen Societies, efforts were made to provide a direction to such investigations. Annual Conferences, Seminars and Workshops were organized and a Newsletter distributed periodically. As a result scores of workers all over India are today engaged is assaying genetic stresses in model systems. Three major centers can be identified currently which however had independent origins.

E. Major Groups

The foremost among such centers is the one led by Dr Sundaram at BARC whose work was recently reviewed in the Newsletter of the EMSI. This group has not only organized the EMSI but also nurtured it to be one of the best scientific societies of India. A large number of agents-drugs, food colors, dye; industrial chemicals, radiations etc were studied by them in microbial and mammalian systems. No wonder that the Department of Atomic Energy was the first agency to recognize the genetic toxicology as an area for funding.

Osmania University is system intensive possessing expertise in plant, microbial, insect, mammalian in vivo and in vitro systems. Most publications of Professors Nizam, Reddy (OS and GM) Subramanyam, Rao, Ahuja, Polasa and their colleagues have so for dealt with drugs, natural products, pesticides and others. Though their interests are wide ranging involving an equally wide variety of endpoints more emphatic studies were made on drugs.

The most remarkable is the current pace of agent-intensive activities at the Prof A K Sharma, Prof A Sharma and Dr G Talukder. Having initiated chemical screening far earlier to the pre-EMSI period, the center is on full steam with scores of workers screening a great variety of agents such as drugs, naturals, industrials, pesticides in plant and mammalian in vivo systems.

F. Other Groups

Of lesser magnitude and pervasion, but often greater in depth are a few smaller groups spread across universities and research establishments. Significant contributions were made from these locales mostly specializing in specific systems and or specific chemical classes. Among such are the groups led by Kaul, Sharma, Dikshit, Grover Sobti, Singh, Nigam, Kesavan, Marimuthu, Nagarajan, Gopinath, Bhide, Bhattacharjee, Vaidya and Manna from Jammu, Madurai, Lucknow, Amritsar, Chandigarh, Ahmedabad, Varanasi, Madras, Delhi, Bombay, Calcutta, Poona and Kalyani.

Only some centers of activity are presently described though there are many individual engaged in genetic toxicology labours. Therefore a list of Indian work is included. It may be noticed that this bibliography is pregnant with indications that more active groups are likely to emerge in future.

G. Trends

The literature search is very suggestive. A perusal of the publications following this write-up and the appendices on journals, systems and agents reveal past achievements-some of which are wasted and current trends-both disquieting and healthy.

Some of the early achievements in India pertain to the (a) use of fern systems and spermatozoan abnormalities, (b) early recognition and use of systems such as Drosophila, Poecilocera, Arabidopsis, Lycopersicon usage of on site vegetation and Tradescantia to monitor in situ radioactivity for the first time in the world and (d) establishing micronuclei and segregational errors in plants as endpoints incidentally by some zoologists! (Mehra, Bhattacharya, Jain, Rao, Gopal- Ayengar, Seshchar, Shah and their associates).

More recently pesticides received preferential attention from groups led by Kaul, Singh, Sharma and Dikshit employing plant and mammalian systems. Drugs were mostly screened by the groups from Osmania, Calcutta and BARC, the latter infact covering a whole array of agents that human populations are likely to be under exposure. Naturals were stressed by Dr Subramanyam and the Calcutta team.

Much however cannot be said concerning our contributions to the development of test systems in spite of historically significant publications.

The great tradition in plant cytogenetics has contributed for the wider expertise in plant system compared to that in others. Unfortunately this was not exploited appropriately and mammalian expertise therefore appear to be deeper, stronger and more reputed. Yet it may be pertinent to remark that some of the foremost genetic toxicologists in the West who are Indians happen to be from botanical background!

H. Disturbing Aspects

The exemplary work carried out by the pioneers seems to have been ignored for long. Otherwise it is difficult to explain the disrepute some of the later works have acquired. Even today, two disturbing trends are discernible from the literature which may upset the efforts to enhance the credibility and utility of research labours in genetic toxicology.

(a) Even after the concepts of environmental genetic toxicology were developed where criteria for the usage of test systems, protocols, dosage, patterns are well established, sampling, scoring and inferences continue to be erroneous in many papers. And this unhealthy trend seems to attract many a new crop of workers! Consequently, over the years there appears to be no significant increase in the reproducibility of the work from certain quarters. This hopefully may be a passing phase. Nevertheless is should be noted that many papers on irradiated foods, pesticides and drugs, especially dealing with mammalian, and plant test systems are increasingly cited in the western works, considering this as a measure of our credibility.

(b) Same work often appeared in different journals with different titles. Some chemicals are tested many times ignoring the precedents even from India. Essentially single piece of work is often split into several publications. Data on a single chemical from different systems and from a single system on different chemicals were published in multiple forms. Some publications do not reveal the identity of either the test system or the test agents. Reasons for such a state of affairs are not far to seek. Some of the journals are of doubtful circulation, reputation and even integrity. Probable tendency of publishing unreviewed papers on a large scale may also be another reason. Excessive haste in putting into press as many titles as possible, dealing with the widest spectrum of agents to gain historical precendences, if not reckoning, is also a factor in the gradual erosion of standards in the mutagenicity research. Often this develops into notoriety!

In order to gain in credibility something has to be done to stem these disquieting trends. However it should be noted that the scenario is not that bleak, but is replete with major contributions. A few selections are detailed.

I. Achievements

Based on the nature of effects, that some chemicals can be used in chromosome techniques were first recognized by Prof Sharma. Also that growth substances and natural compounds may have a role in the mutagenicity was also recognized by him. The efforts by him, his wife and their group in this and allied fields were unparalleled in the history of Indian science in their sheer sweep. No wonder that the book by them on chromosome technique carrying information on chemical effects has become a major reference source all over the world. And scores of workers in India try to emulate them in their research efforts and scientific ethos.

Relationship between cell cycle and time schedule of harvesting cells for aberration scoring, the importance of micronuclei (Shah and Rao), interactions between radiations and chemicals (Kesavan) and between chemicals and DNA (Podder), utility of the Indian Muntjac as a test system (Sharma T) and chiasmata as a responsive endpoint (Sharma) were among the major areas of contribution from Indian workers. That the associations of chromosomes during segregation could be influenced, which has relevance to centromere synchrony and turbagenicity, was discovered by Dr Natarajan long ago. Cellular pictures in his publications and in those of Prof Mehra and Dr Subramanyam are probably of the best of standards. It is only to be expected that some reviews should also have appeared (Mehra 1961, Sharma 1971, Murty 1977, Kaul 1978, Manna 1982).

The contributions from the workers at the Bhabha Atomic Research Centre are exemplary. Internationally reputed and nationally torch bearing work on aspects such as in situ radioactivity irradiated foods, drugs, dyes, risk estimates, radiation equivalence of chemicals and on systems dealing with Ame's assay, Barley, Arabidopsis, Tradescantia, Micronuclei, dominant lethal test, clastogeny assay in vivo and in vitro and on all aspects of yeast were done/ have been continuing here. (Sundaram, Murty, Chauhan, Bhat, Bhattacharya, Chaubey, Aiyar, Notani, Gopal-Ayengar, Rao, Bhatia, Pradhan, Bhattacharjee, Aravindakshan).

J. Perspectives

With the historical experience and the established expertise behind us attention could now be selective so as to accelerate the natural process of development. (1) Identification of hazard sources (2) building up of relevant data bases and (3) generating an ethos of credibility in our efforts and reproducibility in the investigations, are thought to be among the most primary. Among the issues, the following are suggested for consideration.

(a) The in situ situations such as sewage, effluents, ambient air, waste-fills, occupational atmosphere, and home-stead environment have to be investigated for their composition both in the urban as well as rural human settlements. It is indeed in the home front that most pollution seems to occur!

(b) A more imponderable situation is posed by the food chains. Genotoxins can pass through vegetables, milk, meat, and the processed foods. Many consumables are often adulterated adding insults to injuries. Further, the consumables are increasingly reared on effluents. Sewage and other wastes, sometimes surreptitiously, compound the hazards.

(c) India is famous for the great variety of native medicines, culinary delicacies and gastronomic recipes. In view of their complexity and considering the large populations involved, investigations in this virgin field may be worth while.

(d) Various biota of the agroecosystem may experience mutations in an unprecedented scale from pesticides and their modified versions, given the realities of modern agriculture, upsetting the existing balance of forces. Such a scene has seeds for disastrous consequences due to mutation pollution. An ecosystem approach in genetic toxicology of pesticides is therefore warranted.

K. Conclusion

A co-ordinated programme appears to be necessary for each kind of hazard-source. Thus Air, Water, Consumables, Agrochemicals etc., can be investigated by teams representing the various test systems coordinated by one among them. Assistance of all agencies relevant to environment has to be obtained. This new aspect was found only as a passing reference in a recent hand out of the Department of Science and Technology! Worships to train personnel in the methodology are essential and happily there were two so far in 1974 and 1982 at BARC. Since credible schools are operating in this country with expertise in various systems one can take effective steps to minimize fragmentary reports based on expedient research. If there is one area where team efforts are necessary to avoid costs, duplication, and waste of man power, it is probably here!

An estimation of and investigation into, the genotoxic threats in India through consumption, usage, occupation and/or ecological routes may be best served by an objective assessment of our own strength to gauge the enormity of this problem. If this thought is shared by others, the pains of our bibliographical labours would be without a hangover.

Chapter - 28

Urban Genetic Hazards

Pondicherry is one of the small heavens in Indian comprising the elites of French heritage, Aurobindo Philosophy of action, Bharathi School of literature, Eco-conscious Aurovillians an International Brotherhood, several average institutions and a large goody-goody middle class. And above all it has a comparatively good Government - Political parties as well as administration All these confer on Pondy a plural culture and a cosmopolitan lifestyle. Its is also a haven of consumerism, entrepreneurship, neo-rich, commerce and trade-offs, tourism and traffic. It is a miniature India dominated fortunately by better things of our polity. However I am plagued by the spread of lurking dark forces - the quiet chemicalization of our daily lives. This may enfeeble our society by genetically affecting the human populations, aquatic systems, agroecosystems and ecology of communicable diseases directly and indirectly. Toxicity of Pondicherry's environment to the genetic health of humankind and relevant biota is a concern to be shared by the commons, policy makers and implementers. A brief summary of the subject may however be necessary, for looking at this problem in a proper perspective.

Pollution and Health Risks

Pollution is no pollution unless it is toxic. Toxicity by chemicals is of two types - genetic or nongenetic. Sometimes it could be both. Toxicity to non genetic systems can be mitigated by separating the agents and the unintended targets. Thus neuro-, respiratory-, dermal-, renal-, hepatic- and other toxicities are reversible if the suffering humans are removed from the causative agents. On the other hand genetic toxicity is irreversible and leads to permanent damage in the form cancers, birth defects, infertility - sperm abnormalities and/or abortive fetuses - and heritable diseases. Hundreds of cancers and birth defects and a few thousands of heritable diseases are known in literature, such as Dwarfism, Mongolism, Cleft lip, Sexual abnormalities - half male, half female, super males and females hemophilia, muscular dystrophy, nervous disorders, etc. All these exert enormous social burdens, besides causing misery to the bearer as well as to the family neighbours and health services.

The environmental agents affect the genetic systems of other organisms as well. For example, in the case of communicable diseases vectors develop resistance to pesticides and parasites to drugs confounding the medical profession. In agroecosystems, pests, pathogens and weeds develop resistance to pesticides complicating the quest for saving crop yields and safe foods. In aquatic systems on the other hand, chemicals accumulate, modify and magnify through the food chain and may ultimately cause genetic effects in the human consumers. Consumers of agricultural products too face this possibility.

Status of Genetic Toxicology

The science of genetic toxins, popularly known as Genetic Toxicology, is about 25 years young. It is estimated that there are about 4.5 million chemicals in the world - 100,000 is regular use, belonging to about 15-20 broad categories of usage and scores of chemical classes. Humans and biota are exposed to them through consumption, usage, accidents, occupation, food chain, ecology etc. and through all orifices. These affect the genetic systems of reproductive as well as nonreproductive tissues sooner or later bringing about changes in the basic nature of the organisms. In case of humans, it is established that 25% of us may be affected with cancers and about 10-15% with congenital defects - teratomas and heritable diseases. These were estimated at 1.5% in 1954 and predicted to raise to 5% level but crossed the 10% mark by 1985. Unfortunately all genotoxins have great latency before the affects are manifest - from a few months to several years - and there are no warning signals. This is the scenario in the developed North. It will be same in the South too, as we have chosen the same development model.

With this background we may examine the situation in Pondicherry. Environmental justice demands that people should be cautioned rightly and in time.

Foci of Concern in Pondicherry

There appear a few major problems in Pondicherry - a Complex urban - cum- rural milieu, where every conceivable activity elsewhere in India presents itself.

1. Communicable diseases: Let us take mosquitoes, malaria and filaria. Pesticides are used to control the vectors. I believe not only they developed resistance, but also developed into newer forms by accumulating other mutations. And the parasites are known to become resistant to drugs i.e. altered biologically. Therefore these communicable diseases confound us from two fronts - mosquito and the parasite - both genetically altered by chemicals. While resistant pests are a legion in literature, similar changes in parasites are reported for tuberculosis, shistosomiasis, amoebiasis, etc. only recently.

2. Agroecosystems: In agriculture, pesticides are used to control pests, pathogens and weeds to prevent crop losses. For the same purpose the stocks too are laced with pesticides exposing both the stockists, applicators, handlers and consumers of vegetables, fruits and grains to unnecessary hazards. The cattle feeding on those too may pass them to the consumers of milk and meat. This is the greatest environmental danger in Pondicherry to my mind. With about 40 pesticides of undisclosed quantity in use. In this small territory may be something of a record for India. Small wonder that no one seem to realize this, perhaps because of poor public relations or obliterations by false problems and false alarms. By the time truth reaches next door, falsehood covers the whole city - all hotels, societies, journalists and media.

3. Aquatic Pollution: The third problem concerns the water bodies. The seashore has several sewage outfalls Which may contain anything besides homestead effluents. The marine animals are suspected to carry genotoxic contaminations which may pass on to human consumers. We have in fact pioneered research efforts on this issue taking molluscans (mussels) as Indicators. We would have got valuable data by now but for hijacks and piracies. Nevertheless, certain ponds receiving effluents and sewage were found to be genotoxic. Yet there is no cry in the media on pollution toxicities from these water bodies. Obviously truth is not an expert in advertising nor attracting attention.

4. Human Exposures: The fourth is the human exposure through occupation, usage, consumption and ecology.

We have already mentioned that people exposed to pesticides as users, stockists and as consumers. A glance at the 38 pesticides in use in Pondicherry (22 insecticides, 2 rodenticides, 10 fungicides & 4 weedicides) shows several confirmed genetoxins. (See box). Unless the actual quantities used, exposure durations, places and purpose of usage are known it is difficult to estimate the magnitude of this hazard. Hardly there are any grains, fruits and vegetables In Pondy without carrying pesticide sprays In order to protect them from deterioration from microbial action.

The potable water is chlorinated. This process of disinfection is known to release several species of chlorinated compounds, some of which are genetoxins, specifically leading to cancers.

Pondicherry is gastronomically famous. The pyrolytic products of all meat surfaces, several food additives especially the dyes like yellows and the saffrole are well known genotoxins. However we are not clear on the whole range of chemicals used in bakeries and other food processings.

Occupational exposures to toxins is one of the direct routes of developing genetic defects. Reprographies, X-ray operations, fossil oil handling, pesticide formulations, dyes and ink usage some textile processes, etc are some of the areas where workers may be exposed to toxins. Most chemical Industries may pose similar risks. Unfortunately these are yet to be realized.

Automobile exhausts contain Benz(a)pyrenes which are genotoxic. Almost no one on our traffic joints is exempt from inhaling these, though to varying extents and durations.

There may be other unspecified sources of genetoxins - sewage, industrial effluents, changes. Agriculture and Public Health are transporting genotoxic pesticide into human systems solid wastes, drug formulations, hospital wastes, etc. To what extent they operate here is difficult to envisage.

Human beings are exposed to several of the above. Do they cancel out each other or enhance is another question. The latter is more probable.

Research Efforts on These Hazards

The question arises whether any efforts were made to understand and estimate these genetic effects to human populations. Yes. Initially, we concentrated on reviewing the genotoxicity of pesticides and monitoring sewage and aquatic pollution through Molluscan Mussels, Amphibian Frogs and Water Hyacinth among others. The few water bodies under our study revealed genotoxic potential. However further work could not be carried out due to fascist trends, corrupt habits and fraudulent practices engulfing honest scientific endeavors. The arrogance of power went to the extent of removing this subject from the courses of study! This is the tragic side of our scientific temper!

What Needs To Be Done?

I had stated above that mosquitoes, pests, weeds and pathogens were developing resistance to pesticides through genetic directly and indirectly. Several usage and consumeristic practices and habitats pose genotoxic threats to local populations. All these tend to bestow on Pondicherry the status of a most hazardous place vis-a-vis the genetic stresses. This perception of an epidemic like situation needs verification. Therefore records of hospitals, nursing homes and medical practitioners should be surveyed for the incidence of cancers, birth defects, genetic diseases, spontaneous abortions and other similar phenomena. This should be complemented by a well-designed social survey. To my mind pollution research loses meaning, excepting for propaganda purposes, unless toxicity is first established. On the contrary, genetic toxicity may be manifest even when causes are not apparent. The problem becomes all the more poignant due to the small size of the place, denseness of population with a craving for everything under the sun. And the price that we people may have to pay for the apathy of the government today will be very high because the genetic defects multiply through generations. When they become common, it will be too late to do anything. Forewarned is to be forearmed.

Table 28
Killers Used In Pondicherry!

	Insectides				
1.	BHC	8.	Fenthion	15.	Oxydemeton methyl
2.	Cypermethrin	9.	Fenvalcrate	16.	Phorate
3.	Dichlorvos *	10.	Malathion	17.	Phosolone
4.	Dimethoate	11.	Methyl Parathion *	18.	Phosphomidon *
5.	Endosulfan	12.	Quinalphos	19.	Allethrin
6.	Ethion	13.	Monocrotophos *	20.	Carbaryl
7.	Fenitrothion	14.	Nccm based Insecticides	21.	Carbofuran
				22.	Chloropyriphos
	Fungicides				
1.	1.Captan	4.	Manocazeb*	7.	Zinab
2.	Carbnendazin*	5.	Sulphur	8.	Ziram
3.	Copper Oxychloride	6.	Thiram	9.	Ediphenphos*
				10.	Kitazin
	Weedicides				
1.	Butachlor	3.	Benthicard		
2.	Atrazine *	4.	Anilophos		
	Rodenticides / Fumigants				
1.	Aluminium Phosphide				
2.	Zinc Phosphide				

* These pesticides are potentially very dangerous.

Ponkamaya 2(3): 8-9, 1994.

"A NATION THAT DESTROYS ITS SOILS DESTROYS ITSELF. FORESTS ARE THE LUNGS OF OUR LAND, PURIFYING THE AIR AND GIVING FRESH STRENGTH TO OUR PEOPLE." **FRANKLIN D. ROOSEVELT**

Chapter - 29

Genepidemics, Genepandemics and Darwin

An Ominous Possibility Emerging from a Reconsideration of Darwinism in its 150th Yar

A. Introduction

1. Changes in the genetic systems of organisms including man are caused by chemicals and radiations. These mutations multiply with time and with each generation, horizontally due to the continuation of exposure to such agents and vertically due to the possibility of transmission. In human populations they cause cancers, birth defects, heritable diseases and minor mutations with varying detriments to human performances. In the ecosystems, they cause varieties of mutations in plants, animals and microbes. They include food animals, vectors, pathogens, pests, parasites, infectious as we as useful microbes. Individually or collectively they impact human welfare adversely.

2. Such consequences may be termed Genepidemics and Genepandemics. The former are limited to populations sharing same environmental components in a limited space say around polluted habitations. The Later could occur any where in the world where the populations are exposed to the same kind of chemicals or radiations.

3. Both the above mostly occur in the developing countries due to relative illiteracy, awareness, knowledge and higher levels of needs and wants.

B. Premises

The evidences are considerable - direct, indirect, circumstantial from human societies and ecosystems- both managed and natural, terrestrial and aquatic. Though we tried to integrate all these into a new field of investigations 15 years ago, combining ecology and genetic toxicology, when I first suggested this possibility in a conference in Malaysia, it could not happen. This remained a tragedy of my scientific pursuits, all my life.

C. Evidences From Facts And Portents (Circa, 1995)

(a) Human Ecosystems

1. Diagnosis and drug Administration: Resistance caused among the causative pathogens/parasites to various drugs can similarly be suspected to be one of the side effects of the drug. And some drugs are known to be genotoxins.
2. Resistance of vectors such as mosquitoes to pesticides can also arise from the sub lethal action of toxic chemicals either because it is adulterated or it is not uniformly applied or the instructions of the pesticide companies are not followed in toto.
3. Scenarios: Birth defects varied from 0.5 to 2/5 % Mental Illnesses were around1% 4000 syndromes were already recorded. 25% of people were projected to be afflicted by Cancers ultimately and minor ailments may cause detriment to varying degrees. 35 species of Anopheles and 41 species of Culex became resistant to Pesticides. Parasites causing Malaria and Shistosomiasis also developed resistance to drugs

(b) Agroecosystems

1. Agro-ecosystems: Resistance caused among pathogens and pests such as fungi, insects, nematodes and others to fungicides, insecticides and other pesticides could be mostly due to genetic effects of sub-lethal doses.
2. Scenario: Resistance was recorded in about 370 insect pests to 9 pesticides, including. Those of Cotton, Rice etc. Seven rodent species became resistant to Rodenticides. Mutations were noticed in rhizobia blue green algae, weeds and ollinators

(c) Polluted Ecosystems

1. Aquatic Ecosystems: Pollutions enter animals which could move along the food chain to human societies- whose consequences cannot be easily discerned nor monitored
2. Trans-Boundary Ecosystems: The most ominous and consequential aspect pertains to Ornithology. Birds cover aquatic, terrestrial and aerial environments of local, regional and global pervasions and are recipients of an un-definable array of chemicals -gaseous, liquid and solid particles. The regions they visit temporarily as migrants can be at the receiving end of alarming developments, such as dropping of pathogens through Guano (This is other than such acts of using them for aerial photography of sensitive installations before blowing, as in Bharatpur.)

3. Scenario : 90% of studies on animals - crustaceans, polychaetes, amphibians and rodents demonstrated birth defects in their natural habitats. Many animals exhibited cancers in response to the aquatic media. Several plants and animals turned out to be bioconcentrators. A rare case of speciation in Molluscs within a few generations in two very distinct geographical zones was also recorded in the British West Coast and Californian Coast in response to pollution.

All the above variations are due to Mutations in soma or germ-line. And all these are projected to increase in time. Further all those recorded in developed countries are expected to repeat in the developing countries which opted for the capitalistic mode of development in the name of Globalization.

Genetic Toxicology involving agent-induced genetic changes and their consequences of muatagenicity, carcinogenicity and teratogenicity to human systems now becomes Environmental Genotoxicology- a hybrid of Ecology and Genetic Toxicology embracing other organisms, their ecosystems their ecology, food chains, movements, etc. This naturally takes us into the fascinating world of nature.

Incidentally this is the 200th anniversary year of Darwin's birth and 150th anniversary of the birth of Darwinism and is termed Darwin year in the west. Wide ranging celebrations through Seminars on the futuristic implications of Darwinism from diversity and extinctions to Social Darwinism and exclusions of people.

D. Darwinian Implications

1. It is well known that biodiversity has been dwindling due to a variety of reasons. One of them is the pollution of air, water and soil by chemicals. Pollutants not only cause variations or mutations in the extant biota but also select the suitable ones. The ultimate impact of these forces is the acceleration of genetic and ecological diversity in the gradually decreasing taxonomic diversity- an aspect that is not taken seriously yet, rather strange! This situation is diametrically opposed to the scenario existied upto post-colonial times. What are the implications of such a changes which the Developed countries only are conscious?
2. It is not widely appreciated that the stability of GM Crops will depend on the basic genetic system of the crop and methodology employed for its modification. In the former allelic natures-multiples and pleiotropy, Polygeny, position effects, trasposons, sex determining mechanisms and such special mechanisms are to be considered. The method of evolving any GM crop will be a trade secret. Agroclimatic conditions are another important matter. The concern here is that pollutions causing mutations in other biota could cause the same in

GM crops, affecting their stability and therefore, yields. Food security is directly under threat.

Further all these may happen in a defined region, as an epidemic or in several smaller enclaves sporadically, as a pandemic, both confusing and confounding every one!

Environment, Occupational and Life-Style Concerns - A Trans-Disciplinary Approach - Environment and Health - Beyond. ROCHS-NIOH-ICMR-Complex, Bangalore, 16-19, September, 2009.

THE ECOLOGICAL CRISIS IS DOING WHAT NO OTHER CRISIS IN HISTORY HAS EVER DONE - CHALLENGING US TO A REALIZATION OF A NEW HUMANITY. **JEAN HOUSTON**

THEY ALSO SERVE

Chapter - 30

Phytomass for Energy

1. Introduction

A. Energy Crisis: That energy is one of the threatened resources of man every where was realized for long. The crises arising out of strategies. Thus the likely exhaustion of our fossil fuel supplies in the not too distant future and its potentially effect on the life styles of the manking has stimulated a new field of explorative research for alternative energy sources. The instinctive choise obviously is the ageold and peerless resource, the plants, Energy from plants is regenerative and virtually inexhaustible since they are capable of storing the radient energy of sun fixed in the form of chemical energy, which may take several forms secondarily.

B. Energy From Phytomass: Any plant material can be called as phytomass. Phytomass is organic in nature generated by photosynthesis, the primary product of which is carbohydrate. This may secondarily give rise to other products. Though having other uses, utility as energy sources is the major one from the current reckoning of priorities. That man used plants for his energy needs, from times immemorial may be remembered.

C. Extinct Phytomass: It is well known that all our coal, petrol and natural gas reserves are generated by the fossilized phytomass by itself a product of photosynthesis basically. Each year plant photosynthesis fixes about 2 X 1011 tonnes of carbon with an energy content of 3 X 1021J. This is about ten times the world's annual energy use and 100 times of our food consumption. The total amount of proven fossil fuel reserves "below" the Earth is only equal to the current arboreal phytomass "above" the Earth. The amount of carbon stored in biomass is approximately the same as the atmospheric carbon, and carbon as carbon-di-oxide in the surface layers of ocean. Destruction of forests and soil humus is thought to release about 4-8 X 109 carbon to the atmosphere each year, equivalent to that released by burning fossil fuels.

D. Photosynthetic Productivity: As stated earlier the key of phytomass energy is photosynthesis. The net primary photosynthetic productivity of the Earth which is 510 million Km!! 2!! in area is 152.2 billion tons.This is a sum of productivities of all kinds of ecosystems-Forests, cultivated lands, oceans of energy, never depleting due to the ever replenishing photosynthesis though only a fraction of solar energy is harnessed

E. Present Phytomass Energy: In the industrialized countries where wood is predominantly used as a raw material, 10-12% of it only is used for fuel purposes. However a worldwide survey shows that more than 50 percent of global timber cuts are used as fuel wood. In the poor developing countries this share may reach even upto 80 to 90%. The per capita consumption of wood in the developing countries is estimated to be between 1.3 and 2.1 cubic meter per year.

F. Phytomass Resource Base: Phytomass for energy purposes are of different kinds.Whole plants, plant parts and products, waste arising of food crops, human use, plant based industries and so on, which vary geographically. These plant products may be composed of lignocellulose, isoprenes, glycerides, waxes and alcohol. Already the whole world has been greatly dependent on phytomass for its energy needs for a long time. The imminent exhastion of fossil fuel reserves and acceleration of agroindustries bestow on phytomass a value hitherto unmatched as a source of energy. Therefore there is an urgent need for rapid and extensive studies to explore this source in energy part of the world.

G. Conversions: Phytomass is bulky resources, suited as it is to limited contingencies only that too accompanied by considerable wastage.It cannot meet many of our current energy demands for running the mechanical devices. Therefore there is a need to convert plant stocks to a more usable and generally high quality fuels. There appear to be three main types of conversions of the lignocellulose phytomass wet, dry and chemical yielding different products depending on the constitutents of the lignocellulose feedstock, and the type of conversion process. Other products such as carbohydrates, terpenoids, waxes and glycerides have very specific and important uses despite much narrower base, that that of lignocellulose. Mustering all plant sources and employing the conversion processes available, solid, liquid and gaseous fuels and electricity can be obtained to mitigate the energy problems of the human society.

It is therefore imperative that basic knowledge of natural plant resources is prerequisite to meet the urgent of exploiting judiciously the available phytomass. It will also serve as a guide in comabting the losses of plants and their gene pools without man ever knowing their utility in the energy explorations. The existing information together. Such an effort could be an end in itself as a pository of basic information on bioenergy or a beginning for selective and proper exploitation of the sources. To serve both these pruposes is the objective of the present efforts.

2. Types of Phytomass use for Energy

A. Direct Combustion: Phytomass combustion is the burning of the plant materials with sufficient air i.e. the presence of oxygen, for use as a direct source of heat in the form of radiation resulting in the complete conversion to carbon dioxide, water and inorganic ash. Such a phytomass may have different sources based on which their uses and characteristics markedly vary.

1. Sources: Plant materials for direct combustion have four sources. They are the vegetations on farm and non-farm ands, Wastes and byeproducts of differnt processes and Industrial, Forestry and Agroresidues

(a) Nonfarm lands yielding phytomass for direct combustion are two types- Forests and Marginal lands. The former constitutes a natural wealth of any country differing in their nature depending upon climate, latitude, altitude and other factors. These have other uses besides firewood utility. Based on the forest types, the plant species yielding firewood would vary. However a few tree species from these forests may be considered comparatively more useful which therefore find application even outside the forest niches. The legal status of India's forest land is types.

(b) Marginal lands lie outside the scope of forests and farming.Plants grow as weeds naturally on waste lands, beach lands, margins of roads, rail tracks, canals, rivers, embankments of tanks, reservoirs and similar non-agricultural environments. Such lands are often reclaimed by raising plantations for exclusive use as fuel. Since these lands are nearer to the human society and the plantations are means to suppliment some traditional forest uses, this practice is popularly known as social forestry. As this practice often is in the pattern of agriuclture the term ago forestry too is used when the land is larger, wholesome and the cultivation intensive. Several species are found more suitable for employment in social forestry purposes all over India.

(c) Feedstock from farm lands are of two origins

  (a) Crops specially cultivated for their energy usage and are characterized by

    (i) High annual yield

    (ii) Means of vegetative propagation

    (iii) Short rotation period

    (iv) Hardy and

    (v) High energy content,

  (b) Crop remains from agricultural practices.

(d) Lumbering and Carpentry are based on plant materials. Such activities obviously yield byeproducts some of which can be used as fuels and wastes with no apparent utility. Even such wastes of late are recognized to carry potential for use in direct combustion (Table 6). Saw dust can be compressed at a pressure on one metric ton per aquare centimeter to form saw dust briquettes. This is named as ogalite in Japan. They have heating value of 4700 Kcals/Kg with a specific gravity of 1.1 of 1.2

(e) The other sources are the dry fallen wood, trimings from felled trees, sized timber from thining operations, residual wood from saw mills, residual wood from wood processing, unwanted wooden artifacts and wood resulting from clearing operations.

It should be remembered that plant material have many other uses. Therefore the most effective employment or utilization of this source would be to use it primarily as a raw material for other purposes and finally as a energy source.

2. Characteristics of the Phytomass: a. Calorific value or the heat of combustion is an important attribute of the phytomass. It is known that one gram of oxygen content of the phytomass liberates 3349 calories of heat. Therefore the energy content of the phytomass is related to its state of oxidation. The energy value of the feedstock enormously vary depending on a number of factors. Consequently the forest trees, shrubs, wood vegetation, agricultural crop remains, byeproducts and wastes from plant based industries vary in their calorific values. Even within the same plant, each plant part has different value at different times. This all important component of the fuel phytomass is influenced and limited by moisture, ash and organic contents. b. Moisture content: Water content in the plants is a prblem in energy conversion because increasing moisture content lowers the heating values. It is estimated that 3.5 to 4.5 mega joules of heat energy is required to reduce one kilogram of water. Generally it is considered that moisture content of the feedstocks for direct combustion should be below 15%. The influence of moisture content in the heat energy output is well known which is our day to day experience.c. Ash content: Ash content of the plant material depends upon the mineral content which in turn varies with the plant and soil type. Ash coastes problem in handling by way of erosion and corrosion. Therefore ash content must be less in fuel wood materials. In an ideal firewood it is usually less than two percent. d. Organic content : Carbon containing compounds constitute the organic content of the materials. They are in the form of cellulose, lipids, hydrocarbons, henolic compounds, carbohydrates and proteins Wood is mainly composed of basic polymers namely cellulose, lignin and hemic cellulose. In general, hardwood contains more cellulose than hemicellulose and lignin in that order while in the soft wood lignin and hemicellulose are almost the same quantitatively but less than cellulose. Moreover xylans from the main hemicellulose of hard woods and mannans of the softwood. Therefore hardwoods contain more carbon. The abundance of hemicellulose and the compactness of woody material determine the density and specific gravity with contribute to the weight and volume of the sample. The specific gravities of some of the feed stocks are given

3. Ideal Firewood Species: From the aforesaid information it is evident that the energy content of the feedstock differes according to the sources and depends on several factors. The ideal firewood tree should further be fast growing, able to grow in a wide range of environments excellent coppicing, unpalatable to grazers, be erect without much branches besodes high calorific values as mentioned earlier. Further more a high yield of phytomass is necessary which is usually obtained by a high density planting of the plants concerned, which is popularly known as energy farming. In India the average hectare of forest land produce no more than 0.67 M3 per year.Using central Forestry Commission dats,Warner (1982),estimated 0.4 M3/ha/year on all 75 million hectares and 0.75 M3/ha.On about 40 million hectares of 'real' forests : a more recent estimate (Government of India, 1983) is 0.5 M3/ha/year. Most Indian forest hectares would grow at least 2M 3 /ha/year if trees covered the area.

4. Uses: Direct combustion is very useful in cooking and heating of the house. For this, a furnace is required to collect and distribute the heat generated. It is also resorted to in boilers and for other industrial usage. Wood is combusted and from that heat electricity is produced. In Hawai island 50% electricity is from phytomass. In Tamil Nadu Prof Seshadri produced electricity from phytomass. Trains, tractors, cars, river boats were wood powered during Second World War in U.S.A. However they became out of fashion owing to the preference to coal, oil and the advent of natural gas.

B. Chemical Extraction: There seems to be a high amount of biocrude in some arid land plants and cultivated oil seeds (Wickens 1984) due to the presence of reduced photosynthetic products such as isoperne polymers, triglycerides and waxes. These are rich in hydrogen and carbon, but poor in oxygen and the energy contents is in the order of 40 to 50 X 106 Btu's per dry ton whereas other photomass show 14 to 16 X 106 Btu's per dry ton whereas other phytomass show 14 to 16 X 106 Btu's per dry ton. Further more vegetable products i.e.polyisoprenes, triglycerides and waxes can be upgraded to high grade transportation fuels by catalytic cracking using Hobil's shape selective Zeolite Catalyst.

1. Isoprenes: The idea of utilizing isoprenes has Italian origins. In 1938 Italians in Ethiopia had plans to produce gasoline like liquid from Euphorbia species. They called it as vegetable gasoline and the plants as petroleum plantations. Recently the idea of obtaining hydrocarbons from plants struck Prof Calvin again while waiting for petrol during the oil crisis, who optly considered as the father of this discipline.

Isoprenes are long chain hydrocarbons occuring in natural terpenids. Molecular formula of a terpenoid is (C5 H8)n - C5 H8 where being an isoprene molecule. According to the 'n' number in the order of 10,15,20,25,30,40 and above 40 they are named as monoterpenoids, sesquiterpenoids, diterpenoids, aesterterpenoids, triterpenoids tetraterpenoids and polyterpenoids respectively. Thermal decomposition of almost all terpenoids give isoprene as one of the products. Monoterpenoids and sesquiterpenoids present in the essential oils are volatile. Tetraterpenoids are present in the carotenoids while polyterpenoids constitute rubber. Diterpenoids and triterpenoids present in latex bearing plants.

At present Calvin and other scientists advocate hydrocarbon rich plants as superior phytomass energy sources. Many of these plants have the added advantage of good adaptability to lands that are semiarid, roughly contoured and not good for food production. Many isoprene rich plants are studied for their commercial exploitation. Therefore plants with terpenoids can also be utilized as isoprenes can be extracted from these too as possible sources of Phytopetrol

Plants having polyterpenoids are used for rubber production . If we intend to use these plants for isoprene extraction then it might amount to improper use. Therefore these plants though highly potential for isoprene extraction are limited only to rubber industry and therefore aptly called as rubber plants. Other terpenoid rich plants which are not useful in rubber industry may be called as Petro-plants. Some rubber plants and petroplants are well studied especially in the ways of cultivation methods, raw material harvesting, tapping the latex and chemical composition study etc. Since novetly is attached to the idea of oils of the living plants used like the oils from the fossilized living systems, description of a few selected plants may not be out of place. These can be broadly classified into two categories.

2. Glycerides: From very ancient time man know the energy value of vegetable oils and used it for lighting purposes. Now in recent times vegetable and seed oils are recognized as a diesel substitute and various experiments are underway to establish this idea.

Vegetable oils derived from seeds and nuts are chemically identical with fat. They are composed largely of glycerides of fatty acids. Glyceride is an ester of glycerol and fatty acid in which one or more of the hydroxy groups of the glycerol have been replaced by acid radicals and so they contain upto three different acid groups. Depending on the number of acid groups present it is named as monoglycerides, diglycerides and triglycerides. The fatty acids are chiefly of oleic, palmitic, stearic and linolenic as (a) Rubber plants (b) Petroplants. The more hydrogen the oil contains the thicker the oil becomes. The term fat usually refers to triglycerides specifically, whereas lipids includs all fatty acid esters. The general formula of triglyceride is CH2 (OOCR1) CH2(OOR2) CH2(OOR3) where R1, R2, R3 are usually different fatty acids.The oil esters are identified as a promising diesel substitute for stationary engines and automobiles. Vegetable soils shows more heating value and less costing value than ethanol. When diesel engines were tested with vegetable oils 33% vegetable oil usage as fuels is ruled out because of viscosity and engine cocking problems. Oil esters and microemulsions of vegetable oil in diesel or alcohol seem promising. At present multiple sources of soils are under consideration as diesel substitute.

A large number of plants are known to be rich in glyceride which could by exploited further (Table 14) oils are categorized depending on the type and functions. Based on the use of the oil to man they are classified as edible and non edible oils. Based on the type of the oil they are classified as drying oils, semidrying oils and nondrying oils.

The oils can be extracted at high temperature and pressure i.e. above 60‼ o‼ C and 10,000 to12,000 Psi,the solubility of oils in carbon dioxide changes dramatically with small changes in pressure. Oils can be extracted using hexane as a solvent. Crushing the materials physically in mills is also a device to extract oils

3. Waxes: Waxes are known for their use in lubrication of engines. Chemical composition of wax is similar to fats and oils except that it contains no glycerides, but other esters of fatty acids and alcohols are classed under liquids. Wax has low melting point and water repellent. Wax has high molecular weight and soluble in organic solvents only. Waxes are combustible in nature. Wax has low melting point and water repellent. It has high molecular weight and soluble in organic solvents only. Waxes are combustible in nature.

C. Alcohol Production: Predominantly Carbohydratic phytomass is converted to alcohold by biological fermentation. Such an alcohol could be a useful source of heat and energy through Oxidation.

1. Sources: The chemical nature of the phytomass that can be used for alcoholic fermentation must be carbohydratic in nature i.e., sugary or starchy or cellulosic. The cellulosic and starchy materials have to be hydrolyzed before the conversion process. The three major and popular sugar crops are ***Saccharum officinarum, Beta vulgaris and Sorghum bicolor***. There are however a few other sugary raw materials. The fruits of ***Vitis vinifera*** is widely used in ethanol production. In fact any edible fruit can be used for alcohol production. Starchy Phytomass such as tubers, nuts and grains Tapioca tubers and maize grains are the main sources for this purposes. Wood is an important source for alcoholic fermentation because of its richness in cellulose and hemicellulose. Any source of wood and pulp including wastes can be used for this purposes.

2. Products: The principal and product of fermentation is normally expected to be ethyl alcohol. As the production of several others is also similar, acetone, butyl alcohol and isopropyl alcohol which are flammable liquids are also produced. In the three of the cases a gaseous fuel hydrogen also is produced. The properties of all flammable liquids are well established.

3. The Processes: (a) Saccharification: The cellulosic and starchy phytomass has to be saccharified through acid hydrolysis, whereby the polysaccharides are hydroloysed to monomeric sugars with the aid of sulphuric acid and steam. This biological process has two problems. One is the crystalline nature of cellulose and another is the presence of lignin. Both these resist degradation which could be overcome by preheating.

(b) Fermentation: Though anaerobic fermentation carbohydrate can be reduced to alcohol. However to get the other desirable end products in the fermentation particular specified microorganisms should be used

(c) Direct Process: Cellulose and starch can be directly degraded anaerobically to ethanol by some thermophilic mocroorganisms.

4. Advantages And Limitations: Most of the wastes, otherwise difficult to dispose can be used most successfully creating wealth from wastes. There are no transport problems because the fuel is the liquid.

Directly fermentable sugar crops and starch rich grains are very expensive to use them as feedstock for alcohol production. In cellulosic feedstock presence of lignin is highly undesirable. Further, alcohol in itself is in high demand for industrial usage and human consumption. Therefore usage of alcohol for fuel purposes will necessitate great changes in planning priorities in agriculture and land use. Keeping in view such limitations the Research Association for Alternative Development in Japan recommended EUCALYPTUS wood chips as the best source for alcohol production while the lignin fraction can be used for coke production. On the otherhand Brazilian scientists stress the importance of studies on lignocellulosic conversion. In India efforts are beind made to develop sugarcane specially suited for alcohol generation. However the continuous fermentation techniques converting phytomass directly to alcohol using thermophiles may hold great promise, atleast in terms of costs

5. Uses: Alcohol became a popular fuel for lighting 1830 A.D. when it replaced malodorous fish and whale oils. In 1880 Henry realized the value of power alcohol for automobiles. In 1930s 75 percent alcohol and 25% ether was consisting in automobile fuel. By present projections 100% ethanol cars are expected on the road very soon.

D. Anaerobic Digestion: Anaerobic digestion is a biologically mediated transformation of organic material to a combustible gas. This combustile gas is called as biogas which is composed of methane and carbon dioxide in 211 volume ration.

1. Characteristics Of Biogas: The biogas released from the digester contains 60 to 70 percent methane, 30 to 40 percent carbon-di-oxide, traces of hydroges, nitrogen water vapour and hydrogen sulphide. A typical biogas sample containing 68% methane, 31% carbon-di-oxide and 1% Nitrogen gives a calorific value of 5871 kilo calories per cubic meter. The quality and the heating value can be improved by reducing the carbon-di-oxide can be reduced contacting the biogas with monoethanolamine or with lime solution. Hydrogen and particulatles can be removed by ferric oxide and iron filings. Pressurized water purification system called Binax system is another purification method. A biogas sample with 80 percent methane has a calorific value of 8660 kilo calories per cubic meter. The typical Binax gas contains 98% methane, less than 2% carbon-di-oxide and 4 ppm hydrogen sulphide. Its other characteristics are all well established. As compared with hydrocarbon fuels biogas has a slower rate of flame propagation and larger delay period.

2. Sources: Any wet phytomass can be an appropriate substrate for anaerobic digestion. The wet biomass can arise from four sources. They are (1) Aquatic plants (2) Terrestrial plants (3) Wastes and (4) Forest sources.

(a) Aquatic plants: Hydrophytic plants, both submerged and floating, which are considered to be weedy in nature are a very important source of anaerobic digestion. For example EICHORNIA is capable of producing 0.9 to 1.8 tonnes of dry biomass of weed per hectare per day.

(b) Terrestrial Plants: Weeds on marginal lands or elsewhere may be collected and used for anaerobic digestion.

(c) Wastes: The refusals arising due to different activities of man can be a source of phytomass for anaerobic digestion. The wastes are of different types as (1) Agricultural wastes (2) Agroindustrial wastes (3) Municipal solid wastes.

(d) Forest Sources: Forest sources are very rich in organic materials. Some such sources are forest litter, any green plant matter including felled wood. The yields of biogas of different plants are known to vary.

3. Conversion Process: (a) Process: Anaerobic digestion is a conversion process of wet phytomass of complex into methane and other gases. This process is a biological process carried but by living organisms in the absence of oxygen. The reaction may be expressed as below. Organic matter Methane bacteria + Carbonidioxide + Hydrogen sulphide water + stabilized effluent The containers of feedstock i.e. digesters keeping the feebstock isolated from the air can be of two types batch type and continuous flow type. The Phytomass, containing polymeric substances is gradually hydrolyzed to soluble mono or dimeric units such as glucose, xylose, meltose, amino acids, and fatty acids. These are degraded to give acetic acid, propionoic acid, hydrogen and carbon dioxide. In the final stage the acetic acid, hydrogen and carbon-di-oxide are converted to methane and carbon-di-oxide, the biogass.

(b) Methanogenic Organisms: The bacteria belonging to the Genus Streptococci, Bactericides, Clostridium, Escherichia, Methanobacterium,

Mothanobrevibacter, Mothanobacillus and Mothanosaricina are the principal methanogenic organisms. The methane forming bacteria are sensitive to the environmental conditions such as p^H, (6.6 - 7.6 p^H is optimum) temperature (35°C to 54°C is optimum) carbon nitrogen ratio (30:1 optimum) and solid content (7 to 9% by weight)

4. Advantages: Biogas energy to the rural families reduces the prespsure on firewood supplies. The digested sludge can be used as an organic manure, in agriculture. anaerobic digestion practice improves the environmental health in rural areas and reduces the drudgery of women.In India keeping all these advantages in mind the sixth five year plan envisaged establishment of one million family size plants and hundred community size plants.

5. Limitations and Improvement: The biogas technology need developments pertaining to rectification of corrosion of gas holders, fermentation techniques, diversification of feedstock, elimination of hydrogen sulphide, ligning digestion and lignin biosynthesis, biology of methanogenic organisms etc. Nitrogen fixing plants need better production as they are good source of phytomass for digestion. To increase phytomass production plant breeding and plant growth regulators need thorough study.

6. Uses: Besides for home cooking in rural areas, biogas has been found to be useful for several purposes such as lighting, heating, engine fueling electricity generation and even automobile fuel after liquefaction.

E. Thermal Conversion: Thermal conversion is the decomposition of substrates in the presence of heat. Here the dry phytomass can be thermally converted to energy rich fuel products in the (a) absence of oxygen (b) presence of limited amount of oxygen or (c) presence of hydrogen. These processes are generally called pyrolysis, gasification and hydrogasification respectively. The phytomass feedstock for this thermal conversion myst be dry in nature.

1 Sources: Any dry phytomass can be a source of feedstock for thermal conversion. They can be whole plants like trees, shrubs and herbs or refuses such as agricultural wastes, agroindustrial wastes and municipal solid wastes.

(a) Wood Sources: Wood is a very ideal source for thermal conversion because of its lignin content. Moreover in the harwoods hemicellulose chiefly consists of xylans and in the soft woods mannans. Further, softwood lignin contains guaiacyl propane units which has phenolic groups with emthoxyl group whereas hardwood lignin, in addition, has syringyl propane units. The high content of acetyl and methoxyl groups in hardwoods explains why hard wood is preferable in thermal conversion and especially for methnol production. (b) Terresrial weeds: Any weedy terrestrial plants, all herbs and shrubs, can be used for thermal conversion. (c) Agricultural Wastes: The crop remains of agricultural plants can be used for thermal conversion after drying. (d) Agroindustrial Wastes: Waste arisings of some industries which are based on plant raw materials can also be used for thermal conversion. e. Municipal Solid Wastes: The municipal solid wastes are rich in organic materials like papers, clothes etc. which are all having phytomass origin. So this can be a source of phytomass for thermal conversion.

2. Pyrolysis: Pyrolysis is the decomposition of polymeric organic materials into simpler form by heating in the absence of air. The resulting products are a solid char, a liquid pyroligneous acid and a gaseous mixture.

The solid charcoal which is a carbonaceous residue and is rich in mineral components. The liquid pyrolytic oil or pyroligneous acid is a complex mixture of hydrocarbons, ketones, aldehydes, acids, alcohols, sugar residues, furan derivatives and phenolic compounds.

When the phytomass is heated in the absence of air, water vapour will come out in lower temperature and when the temperature is raised slowly various compounds will be expelled.

The general output is 30 to 40% charcoal, 20% pyrolytic oil and 20% gaseous products of the original phytomass. Yield and heat content of pyrolytic products however varies with the source woody phytomass.

Charcoal can be used for home cooking. Blacksmiths use charcoal for their hearths. Charcoal has replaced coke use. Pyrolytic oil can be used as a diesel substitute. The pyrolytic oil got from the wheatstraw is called as tarry oil. Pyrolytic gas can be used for electricity generation. Gaseous fraction of pyrolysis contains Hydrogen and carbon monoxide and this would enable methanol to be synthesized. Methanol along with alcohol it is added to gaseoline and used as a petroleum substitute.

3. Gasification: Gasification is the conversion of a solid or liquid to gas. The phytomass is burnt in the presence of limited air or oxygen resulting in a low to medium energy gas and a little solid char residue. The gas is composed of carbon monoxide, hydrogen and methane. This gas can be upgraded to get different products like methane, methanol, ammonia etc.

Phytomass is burnt in a much higher temperature and pressure, than in pyrolysis. Gasification can be done in the presence of air or in the presence of oxygen. When the Phytomass is gasified in the presence of limited air, low energy synthesis gas is generated whereas gasification in the presence of oxygen results in medium energy synthesis gas.

Products: Producer gas : When Phytomass is burnt at over 1000° C in the presence of insufficient air and steam gases are produced which contain carbon monoxide as the main combustible constituent together with Nitrogen. Hydrogen and methane gases are present in small proportions. Thhis low energy gas is called producer gas. The calorific value of this producer gas is 1200 Kcal/N m^3

Synthesis gas or Syn gas: This gas is a mixture of carbon monoxide and hydrogen. Traces of nitrogen, carbondioxide and hydrocarbons also present. The energy content of synthesis gas is 7 mj/m^3 Synthesis gas can be used as a pipeline gas. It can be used as a fuel for boiler and internal combustion engines. Synthetic natural gas can be used as substitute for natural gas.

Substitute natural gas: The low to medium energy gas is methanated catalytically to produce methane which is favored by high pressures and low temperature i.e. 250 to 450°C. The energy content of SNG is 38 mg/m^3 substitute natural gas contain 95 to 98 percent methane and its calorific value is about same as natural gas.

Methanol: Through various technological processes the synthesis gas can be upgraded to methanol. Methanol can be used to power phosphoric acid cell and used in vehicles. Methanol along with ethanol is added to gasoline and used as a petroleum substitute. Methanol powdered automobiles, however are under exploration. Electricity can also be generated from gases and used to power engines.

Ammonia: Ammonia may also be formed from the synthesis gas. Carbon monoxide is converted to carbon-di-oxide which is eliminated. If the hydrogen and nitrogen are present abundantly and ammonia can be formed from this.

4. Hydrogasification: The Phytomass can be gasified in the presence of hydrogen or synthesi gas to produce a high energy gas rich in hydrocarbons like methane and ethane, leaving a little char.

Conversion of Phytomass to high energy gas is done by reacting the organic materials with water at or above its critical temperature and pressure, so that critical density of the water is achieved. The desired conversion to high BTU gas can be done either without catalyst or in the presence of a catalyst that promotes hydrogenation, reforming and cracking of the organic material. This high energy gas can be liquified to get a hydrocarbon rich oil. This liquid can be formed directly by Thermochemical liquefaction by reacting the phytomass with syngas under a temperature of 750°F and a pressure of 5000 Psi. The resulting oil can have heating value of 35,200 Btu/Kg The high energy gas is rich in methane and ethane and other hydrocarbons. Its chemical composition is just like natural gas. The oil liquefied from the gas also is rich in hydrocarbons.

The high energy gas can be used as a substitute to natural gas. the liquified gas can be directly used as a diesel substitute.

3. Perspectives

A. Quantum of Phytosources: A recepitulation of the survey hitherto made makes it amply evident that more than 500 species in the least, hold promise for exploitation through one or the other five major channels of eneregy utilization. This is in stark contrast with the popular belief that a few plants will serve as a penacea to the energy crisis. The species and genera related to the present compilation may be as large and awaiting investigation. Many other plants, unrelated to these, may yet be found to be useful from ethnobotanical investigations. It should however not be ignored that only a fraction of these may ultimately befound suitable for different ecosystems which may have to be individually worked out. A whole new world of plants has opened up for introduction to Indian homoclimes and genetic and agronomic imprevoement.

B. Historicism in Plant Selection: In selecting plants for introduction and improvement, it is generally agreed that priority should be given to the local species. In the absence of any such 'sons of the soil' two alternatives exist. First and an easier one is to select the relatives of such non existent sons ! Second and yet unrecognized, though ecologically the soundest, is to choose a past denizen of the now barren habitat. Culturing the same would not only just revegetate but engender regeneration of the lost habitat. Then how to know such plants ? Archives like these of East India Company ancient treatises, early floristic reports such as those of Roxburg, travellogues would help a great deal. Archealogical screening of superficial depths of now-abondoned human settlements, environs of the ancient monuments could reveal plant parts of yore-stems, roots and pollen. These are identifiable and such plants and/or their relatives can be reintroduced which might have vanished under stresses. With such restorative approach, regeneration of ecosystem if possible though time consuming. Such a historicism in afforestation is the most ecologically sould approad. And a whole new field of environmental history could develop. Unfortunately, to the detriment of the ecological health the accent has been laid on exotics.

C. Pros and Cons: To utilize this regenrative and inexhaustible resource, perfection in planning as well as implementation of phytomass schemes is essential. In opting for such extensive phytomass programmes there are advantages as well disadvantages. Plants are good stores, versatile in products, convertability and energy contents. Exploitation is dependent on technology already available with minimum capital input with present man power and material resources. This has a large biological and engineering develop potential, reasonably priced in any instances and ecologically inoffensive and safe. On the other hand every step in the energy farming has problems. These pertain to land use, acquiring new land areas, disturbance to the existing patterns of agriculture, site preparation and crop selection, irrigation and water demands, origin and spread of few pests, fertility, chemicals, manpower diversion, transportation, acceptability, shading and pathogenesis, underground water depletion, humidity and ambient temperatures stockage, uncertainty of costs and initial supply. Yet as if to override all these in the underdeveloped South, considerable barren land and stray man power exist.

Inspite of the theoritial hopes and despairs, whether a phytomass scheme should be implemented or not, depends on several practical factors such as (a) resources, (b) available technology, c) distribution and marketing prospects (d) social acceptance (e) economic viability and finally (f) the political will.

D. Land Use: It may sound simplistic. But the moto should be to grow as any varieties of plants, in as many places, in as much density and as soon as possible. The rest will take care of themselves. Consequently, social forestry schemes were developed where in every strip of available land is sought to be vegetated by useful plants. A tentative selection of plants and their properties are assorted by the Ministry of Agriculture, Government of India. This concept of utility gravitates human mind towards commercial benefits. Automatically the barren land is put to production, no matter where the profits would be reaped. One can understand the wealth of unutilized landmass in India from the targets various states set for themselves to achieve for the 6th plan as against 1st plan. So much is awaiting to be done. The twin purposes of generating productivity from potential habitats and human awareness of the problems especially on the availability of land. Therefore it will be the land use patterns commitments and projections that may determine the extent of phytomass exploitation which can be illustrated by revenue from forest and donar supports The available land will always be less and demands more. The former a natural consequence of the latter. A big rider here is the question of how little is the minimum, that should never be harvested, should be maintained as a reserve, as an insurance against ecological boomerangs ? Can this atleast remain unravaged and untrampled - the practical philosophy would be plant everywhere, so much so, the minimum would still flourish to maintain biota free from ecological backlashes.

In conclusions, it may be recalled that the current quest is to produce fuels akin to the fossil fuels-gaseous and liquid-either to supplement with some or similar efficiency in view of the infinite variety and mass of plant life available. It is crystal clear that the phytomass of plant life available. It is crystal clear that the phytomass of the present can contribute to this objective greatly, just as its ancestors were responsbile for today's fossil, fuels.

Chapter - 31

Ramar Pillai's Petroplants

The Tragedy of Indian Science

A. Introduction

There is a considerable spurt of activity in the media covering Ramar Pillai's claims of converting water into a petrol-like liquid with the help of some herbal extracts. While much has flown under the bridge, the bridge itself was under the threat of a collapse! Therefore I am trying to put on record some informations so as to enable a proper direction to emerge in the discussions and researches pertaining to the petrol magic of Ramar Pillai. A brief introduction to the use of plants as energy sources may serve as an appropriate backdrop.

B. Phytomass and Biocrude

The phytomass, organic in nature, is generated by the harvest of sun, photosynthesis, the primary product of which is carbohydrate. This secondarily gives rise to other products some of which are valuable in developing alternative energy options. These are chiefly composed of lignocellulose. Others such as Isoprenes, glycerides, waxes, terpenoids, carbohydrates and alcohol have very specific and important energy uses despite much narrower base. The isoprene polymers, triglycerides and waxes are said to constitute the Biocrude. These are rich in hydrogen and carbon, but poor in oxygen and the energy contents is in the order of 40 to 50 X 106 Btu's per dry ton whereas other phytomass show 14 to 16 X 106 Btu's per dry ton. Furthermore vegetable products i.e.polyisoprenes, triglycerides and waxes can be upgraded to high grade transportation fuels by catalytic cracking using Hobil's Zeolite Catalyst. In this connection it may be pertinent to recall that all our coal, petrol and natural gas reserves were generated by the fossilized phytomass.

C. Isoprenes and Terpenoids

Isoprenes are long chain hydrocarbons occuring in natural terpenoids. Molecular formula of a terpenoid is $(C5\ H8)n$, where - $C5\ H8$ is an isoprene molecule. Depending on the value of the 'n'- 10, 20, 30, 40 and above the terpenoids are named as monoterpenoids, diterpenoids, triterpenoids tetraterpenoids and polyterpenoids respectively. Thermal decomposition of almost all terpenoids give isoprenes as one of the products. Monoterpenoids and sesquiterpenoids present in the essential oils are volatile. Tetraterpenoids are present in the carotenoids while polyterpenoids constitute rubber. Diterpenoids and triterpenoids are present in the latex bearing plants.

The idea of utilizing isoprenes has Italian origins. In 1938 Italians in Ethiopia had plans to produce gasoline like liquid from Euphorbia species. They called it 'vegetable gasoline' and the plants as petroleum plantations. Three decades later this idea of obtaining hydrocarbons from plants struck the Nobel laureate Prof. Calvin again while waiting for petrol during the oil crisis, who aptly was considered as the father of this discipline of petroplant research. He felt that the hydrocarbon rich plants are superior phytomass energy sources, which have the added advantage of good adaptability to arid lands, roughly contoured and not good for food production. Many isoprene as well as the terpenoid rich plants were studied for their commercial exploitation, as the former can be extracted from the latter together constituting the sources of Phytopetrol.

Plants having polyterpenoids are used for rubber production. Extracting isoprenes from these will be improper. Therefore these rubber plants though highly potential for isoprene extraction are limited only to rubber industry. Other terpenoid rich plants which are not useful in rubber industry may be called as Petro-plants. Some rubber plants and petroplants are well studied especially in the ways of cultivation methods, harvesting, tapping the latex and chemical analysis.

D. Glycerides

From very ancient time man knew the energy value of vegetable oils and used it for lighting purposes. In recent times vegetable and seed oils are recognized by Calvin as a possible diesel substitute. Vegetable oils derived from seeds and nuts are chemically identical with fat. They are composed largely of glycerides of fatty acids. Glyceride is an ester of glycerol and fatty acid in which one or more of the hydroxy groups of the glycerol have been replaced by acid radicals and so they contain upto three different acid groups. Depending on the number of acid groups present it is named as monoglycerides, diglycerides and triglycerides. The fatty acids are chiefly of oleic, palmitic, stearic and linolenic acids. The term fat usually refers to triglycerides specifically, whereas lipids include all fatty acid esters. These are identified as a promising diesel substitute for stationary engines and automobiles. Vegetable oils show more heating value and less costing value than ethanol. Oil esters and micro-emulsions of vegetable oil in diesel or alcohol seem promising. At present multiple sources of oils are under consideration as diesel substitute.

A large number of plants are known to be rich in glyceride which could by exploited further. Based on the use of the oil to man they are classified as edible and non edible oils. Based on the type of the oil they are classified as drying oils, semidrying oils and nondrying oils. The oils can be extracted by several methods including physical crushing in mills.

E. Waxes

Waxes are known for their use in lubrication of engines. Chemical composition of wax is similar to fats and oils except that it contains no glycerides. Wax has low melting point and is a water repellent. Wax has high molecular weight and soluble in organic solvents only. Waxes are combustible in nature. It has high molecular weight and soluble in organic solvents only. Waxes are combustible in nature.

F. Quantum Of Phytosources

Although over 500 species in the least hold promise for exploitation through one or the other five major channels of energy utilization, only a few scores have relevance to hydrocarbon channel. This is in stark contrast with the popular belief that a few plants will serve as a panacea to the energy crisis. It should however not be ignored that only a fraction of these may ultimately be found suitable.

G. Historicism

Some new species of hydrocarbon potential, or unknown uses of the existing species can be obtained from the Archives like those of East India Company, ancient treatises, early floristic reports such as those of Roxburg, travelogues etc. Such a historical approach may lead to the emergence of a whole new field of environmental history. It may be recalled that the current quest is to produce fuels akin to the fossil fuels either to supplement or function as substrates. It is hoped that the phytomass of the present can contribute to this objective greatly, just as its ancestors were totally responsible for today's fossil, fuels. Long dead they served, yet living they would continue.

While all these have the chemicals mentioned, only some may have them in quantities viable for exploitation on an experimental scale which can further be upgraded through the intervention of science and technology.

H. Microbes

There are also microbes which generate hydrocarbons. Perhaps a combination of phytochemistry of the local petroplants and microbial metabolism may harbour the secret of Ramar Pillai's process! A brief summary of the microbial routes to hydrocarbon production may be therefore appropriate. The microbial products of metabolism have many acceptable fuels and /or substrates. The biosynthetic mechanisms may be various. The quantities may or may not be sufficient for commercial exploitation directly or indirectly. Isoprenes and hydroisoprenes of chain lengths ranging between C-15 to C-30 occur in a variety of bacteria - phototrophs, halophiles, thermophiles, acidophiles, methanogens, etc, yeast, fungal spores, mycelia. Some promising ones are: Botryococcus braunii, Dunaliella salina, Pseudomonas maltophylla, Desulfo vibrio, Debaryomyces and several micrococci. All the processes being under genetic regulation, they are amenable for manipulation through biotechnology and genetic engineering.

I. Scientific Culture

There have been claims of innovations from factory workers, caterers, workshop owners and others who have been writing letters to organizations and scientists and governmental bodies from time to time, but without the needed response. Therefore the purpose of this article is not merely to place the Ramar Magic in the right perspective, but also to provoke introspection whether the leaderships in science have begun to develop scientific culture. Both together may induce the plant biotechnologists to review the state of the art in the biological hydrocarbon research, now that Biotechnology has become a major thrust area or an expedient fad to accommodate sons and sons-in -law, covering everything under the sun including conventional fisheries! This is especially addressed to the leaders in Photosynthesis, plant molecular biology, botanists, biotechnologists and Biomass experts of Tamilnadu from where Ramar Pillai and his magic potions hail. Some of these vaxed eloquent that science is their first love but manouvered endlessly to remain in power endlessly and absolutely to the detriment of society and institutions. The question of accountability is more relevant to these Peers and Powers, now a days busy in assessing others and awarding accreditation grades, than to their underdogs. I doubt this kind of research on hydrocarbon-fuel generation is going on anywhere in the country despite having major Biomass research programmes at Madurai University and Lucknow for many years.

J. Inescapable Inferences

Scientific temper demanded a careful and dispassionate scrutiny on the quiet before any publicity is warranted. The responsible people at the helm should have searched the phytochemical literature for isoprenes, reviewed the previous efforts in this direction in India at Hyderabad and abroad in California, surveyed the flora around Rajapalayam and delved into the ethnobotanical lore of the Tamil Country, checked the Maarten Voogd contention of water-petrol compatibility and should have entrusted these jobs to the nearest University at Madurai famous for botanical energy, photosynthesis and biomass research whose leader is pervading the Indian Scientific Horizon, therefore most relevant to the current issue. Instead they resorted to witnessing a show as if it were magic and labeled it dishonest. Lack of an application of mind and home work forgetting that science being laborious and quiet perseverence followed by careful scrutiny must necessarily shun publicity. It is unfortunate that the country has once again demonstrated that it has not developed a genuine scientific culture which made an eminent scientist like Prof P.M.Bhargava to quit all the scientific academies of India calling them Mafias!

In Euphorbiaceae, the genus Euphorbia consists of about 200 species ranging from small herbs to large trees, most of which are rich in isoprenes. They grow well in semi arid climates, but found everywhere else also. Melvin Calvin has selected Euphorbia lathyris growing wild in his native California in USA for his pioneering experiments on converting phytochemicals into hydrocarbon fuels with considerable success during 1970s and 1980s. It was hoped then that the species could be improved using tissue cultures and genetic maipulation mechanisms.

The Oils and Fats Division of the Indian Institute of Chemical technology at Hyderabad, was reported only a few weeks ago to have developed a process to convert the natural substances from the seeds of wild castor, Jatropha curcas into fuel grade hydrocarbons. This plant has nearly 200 varieties. Similar work was also taking place in Thailand and Nepal. Three years ago Maharashtra and Andhra Pradesh tried to cultivate Jatropha curcas on large scale and build a processing plant of its seed crop for hydrocarbon fuels. With about 175 million hectares of wastelands, Jatropha cultivation for fuel extraction is a distinct possibility. The Hyderabad Institute has worked out many details of the Jatropha project. And this plant having known to be of traditional uses in medicine etc is grown easily in arid and marginal lands. Probably this may soon be improved by genetic means.

A decade ago the then National Productivity Council was urged to identify the best candidates for Hydrocarbon production from the expertise available with the American Solar Energy Society who established the feasibility of growing many varieties of plants as a source of renewable energy including Euphorbia, with its 2000 varieties. But the National Botanical Research Institute at Lucknow while confirming the ability of the petrocrops to yield high quality liquid fuel after conversion found that this is economically not feasible because the net dry material yielded was only a mere 8%!.But about genetic and other biotechnological means to upgrade the situation?

A scientific possibility and its probable realization has been blurred by human credulity and temerity to defraud the public. The issue of obtaining petrol like substances from plants can now be examined.

> "WE ARE BEING MADE AWARE THAT THE ORGANIZATION OF SOCIETY ON THE PRINCIPLE OF PRIVATE PROFIT, AS WELL AS PUBLIC DESTRUCTION, IS LEADING BOTH TO THE DEFORMATION OF HUMANITY BY UNREGULATED INDUSTRIALISM, AND TO THE EXHAUSTION OF NATURAL RESOURCES, AND THAT A GOOD DEAL OF OUR MATERIAL PROGRESS IS A PROGRESS FOR WHICH SUCCEEDING GENERATIONS MAY HAVE TO PAY DEARLY."
>
> **T.S. ELIOT**

Chapter - 32

Botanical Pesticides

A. Introduction

Species are distinct from one another due to the genetic characteristics conferred on them during evolution. Such genetically distinct species have to interact in any ecosystem in order to serve a common function. These interactions are the roots of productivity and stability in the agro-ecosystem. Protecting its constituent genetic pool from the pesticide-induced stresses is therefore very vital.

Several thousands of pesticide formulations from a basic repertoire estimated anywhere between 800-1500, are in use. Inputs and acreage covered vary from country to country however. Japan consumes 10,000 gms, Europe 2000gms, USA 1,500gms and India 400gms per hectare approximately. From the considerable literature existing on the genetic effects of pesticides estimated that 120 out of 240 studied were genotoxic in one or the other systems.

B. Genetic Instability Due To Pesticides

1. The Instability Theory: Based on the accrued knowledge a hypothesis was proposed that pesticides threaten the stability of the agro-ecosystems by altering the genetic constitution of the interacting biota In other words, the pesticides are suspected to gradually mutate various biota in the agro-ecosystem directly or directly Such a mutation pollution may alter the stable but delicate balances without which the system could break down. In support of this contention, several evidences are presented.

2. Historical Evidences: In USA, the wartime agriculture was revolutionized by the introduction of high yielding hybrid maize accompanied by fertilizer and pesticides inputs. Subsequently the crops needed an obligate supply of fertilizers to maintain the initial yields. Similar was the experience in India following green revolution in wheat as well as rice. Such a reduced soil fertility may be due to the genetic alterations of microbial populations under the pesticide stress. Further, in USA the pesticide inputs increased 12 times in the last 30 years but the pre-harvest crop losses doubled.

3. Circumstantial Evidence: Mutation in the gypsy moth due to environmental factors is well known. Most mosquito populations abounding effluent channels are likely to be mutants. Natural vegetation on the habitats receiving industrial effluents showed genotoxicity. Similar results were obtained even when tester strains were used as monitors. Genetic abnormalities in the pesticide workers is also well known. In this context, it is reasonable to expect similar changes in the biota of the agroecosystem also under the pesticide stress which is increasing in Afro-Asia.

4. Indirect Evidences: Crop plants are the ultimate recipients and carriers of the pesticidal chemicals to humans through food chain. In Barley, Wheat and Sorghum a large number of them were found to be genotoxic. It is most likely that microbes and mammals respond to pesticides in the same way as do plants. This is borne out from a comparison of responses of various biota to pesticides from the published literature. Nearly 80% concordance has been found between plants on one hand and microbes and mammals on the other It is thus implied that the biota in the agroecosystem are under the genetic stress of pesticide to the same extent as the experimental plant systems. It may be relevant to recall the reported mutations in the nitrogen-fixing bacteria and Myxophyceae

5. Direct Evidences: Plants Considerable evidences emerged directly from the agroecosystem. Ploidy variations, mutations, cytogenetic abnormalities were known to occur in Wheat, Barley, Sorghum, etc. Fertility losses in Triticale, Sorghum and Barley were also recorded Chlorophyll deficient mutations have been reported in barley used as an experimental system Interference with chiasmata and crossing over was also suspected Resistance to herbicides in both the crop as well as the weed plants were known Abnormal weeds due to pesticide history of the soil were also recorded. It is conceivable that many spontaneous isolations of aneuploid lines in several economically important plant could be due to the pesticide history of the soil medium.

Animals: Rodent populations exhibited genetic changes in Vietnam as a result of extensive herbicide sprays Seven species of rodents became resistant to rodenticides including Rattus rattus and R. norvegicus FAO estimated 182 strains of insects and mites to be resistant to pesticides in 1965 which grew now to 428 About 223 agricultural pests developed resistance to nine major groups of pesticides. At least 10 species developed multiple resistances to five groups. Cotton leaf worm alone developed resistance to ten insecticides during 1961-1976 in Egypt. And the genetic bases of resistance are well established).

C. Indian Situation

In India, following the green revolution, several, pests were reported in the media to have become more troublesome. A very minor pest, Karnal Bunt became a major one in wheat, Yellow and brown rusts assumed epidemic proportions. Gall midges, bacterial blight, Tungro virus and its vector the brown plant hopper became gradually major pests of rice. The pests are evidently mutating to new strains or the crops are losing their resistance under the influence of chemical applications. It is known that pest resistance to pesticides and crop susceptibility to pests both gone controlled and therefore inherited

Thus evidences from various directions unequivocally favor the hypothesis that mutation pollution in the agroecosystem does occur and the consequences will appear sooner or later.

D. Consequences

It is difficult to visualize the scenario characterizing the breakdown of the agroecosystem. Crops become more susceptible to newer pests. Pests develop resistance to pesticides resulting in a constant need to look out for newer crop varieties and pesticides. Pesticides-fertilizer interactions may lead to an entirely new generation of hazardous compounds. Natural fertility of the soil due to microbes may be eroded. Breeding system of the crop may be modified. Habits and habitats of useful organisms may be altered affecting the predator-prey relationships. These changes would play havoc with the ultimate productivity, on which rests the human welfare.

To mitigate the existing and future problems due to synthetic pesticides an integrated pest management appear to be the only imperative. Yet the genetic modifications generated by these synthetics cannot be averted unless substituted by naturally occurring ones Therefore a brief account of them is essential for assessing their possible role in reducing the genetic risks.

E. Pesticidal Naturals

About 2000 higher plant species appear to possess principles deterrent to other biota Pyrethrum is reputed to be known from Marcopolo's time. Sabadilla, Ryania and Tobacco are also known for the last one to two centuries in South America and Europe. Use of Neem against pests of stored grains has been a practice from times immemorial in India. At least 40 major ethnobotanical practices against pests are known from India, Burma, China, Iran, Yugoslavia, South and Central American Countries, North Africa, West Asia, Southern Europe and Russia. Information on the herbivore deterrent plants and the active principles responsible for such properties is highly scattered in literature. From this a more frequently credited information is compiled for future consideration important properties of the naturals which may have a bearing in obviating the adversity from the synthetics are summarized herewith. These pertain to their (a) stability (b) residual (c) possible synergistic coupling with synthetics and (d) niche-limitation.

Stability: Since the natural pesticides are known to lose the activity through oxidation, they are environmentally desirable. But prolonging their activity to the extent desired as a pesticide is necessary by the use antioxidants as was done in case of food preservation.

Residues: It is generally agreed that residues of botanical insecticides are less or negligible. This has the following consequences.

1. Minimal chances of mutations in the pests
2. Negligible chances of mutations or toxicity in the predators
3. Absence of long term mutagenicity to soil organisms
4. Absence of chances for mutagenicity along the food chain due to less bio-magnification.

Synergism: Since the naturals loose activity sooner in many cases, they may be used along with the synthetics as synergists. If the natural is found to inhibit the detoxifying enzymes, the synthetic pesticide required in the combine would also be low. On the contrary, if the natural induces the detoxifying enzymes, the synthetics would be degraded leaving no residue to be either left in the soil or biomagnified through the food chain. Therefore the impact of the naturals on the detoxifying enzymes, is vital in this respect.

Niche-relevance: Contrary to the situation with the synthetic pesticides, which are designed and developed for global use, the naturals may have to be employed on a small scale and within the agroclimatic zones. The insecticidal plants may even be grown as companions to be the crops. Geographical spread of resistance and development of cross and multiple resistances are averted by limiting the use of naturals to specific niches. Thus ecologically sound agriculture can gradually develop The advantages mentioned above may however be nullified if the naturals are also genotoxic for they also would ensure the same genetic instability feared from the synthetics. Therefore a survey of the genotoxic potential of the naturals is necessary.

F. Genotoxicity of Naturals

Our present information on the genetic toxicity of naturally occurring substances is unfortunately scarce. Though there are about 2000 references dealing with about 500 chemicals, the information is very fragmentary How many of these are herbivore-deterrents is yet to be ascertained. Most of these were not tested in more than one system which is inadequate to infer whether or not one is genotoxin. Furthermore, the insect systems, the most relevant from agricultural point of view, were rarely subjected to testing. Therefore it appears that there is a great need to assess the research need regarding the genotoxicity potential of natural pesticidal substances, before they can be hopefully employed to feared instability in the ecosystem.

G. Current Needs

That synthetic pesticides pose tremendous genotoxic problems to the environmment, is indisputable though some information is yet to come forth. An alternative is provided by the botanicals, which may also have limitations. Some of these constraints can be overcome only be research thrusts on botanicals as well as synthetics. A few of these are therefore dealt with presently.

1. Further testing of synthetics for Genotoxicity: Testing synthetic pesticides for genetic effects on microbes, insects and plants would indicate the likelihood of similar effects on the pests and the associated milieu in the agroecosystem. This would suggest the likelihood of resistance development in the pests. Thus the mutagenicity test system would serve as a kind of an alarming system. This in turn facilitates an early detection of pest resistance which is environmentally and commercially important. While there is considerable work on this aspect, field studies pertaining to commercial samples are negligible

2. Genotoxicity of naturals: It is possible that the natural pesticides also may possess genotoxic properties. Ascertaining this on all the candidate pesticides, in the microbial, plant and insect systems is necessary.

3. Role of antioxidants in the Genoxtoxicity of naturals: Antioxidant chemicals which may have to be used for increasing the life of the natural pesticides, may induce, nullify or modify the genotoxicity potential of the pesticide. Exploring the chances for such an activity in the insect and microbial systems is highly necessary if antioxidant use should not turn counter productive.

4. Role of Metabolism: The plants, insects and mammals may activate the natural pesticides to powerful mutagens or inactivate them into non-mutagens. If a correlation can be established between insecticidal activity on one hand and pro-mutagenicity or lack of it on the other hand, we may as well develop a new class of substances- the pro-insecticides. Necessity of accent in these lines is obvious.

5. Effects on detoxifying enzymes: Piperonyl butoxide derived from safrol increases the pyrethroid toxicity by its ability to inhibit the detoxifying systems and thereby increasing the longevity of the natural pesticide. There could be several other substances which cane used likewise, both with naturals as well as synthetics. On the contrary, there could be substance inducing the detoxifying enzymes, which could be used along with synthetics to reduce their persistence potential. Positive results in this area will strengthen the utility of integrated pest management.

6. Mutability of Acetylcholine Esterase: There are reports of Ache species with differential tolerance to carbamate insecticides Obviously, micro-mutations limited to one or few chains of peptides are involved. If these isoenzymes can be electrophoretically distinguished both in vitro and in vivo a genotoxicity test which would detect micro-mutations in the AchE locus in response to the pesticides - synthetic and natural can be developed. Such a system, so far absent in the genotoxicology kit of tests, would tell us whether or not the pesticide is a potential mutagen in the fastest possible way.

G. Conclusion

It is clear from the aforesaid that many adverse side effects of the synthetic pesticides will be overcome by the selective use of naturals owing to their inherent properties pertaining to stability, residues, synergism and niche-relevance. However, genotoxicity testing of these naturals and to some extent of the synthetics with couplers such as antioxidants and synergists, with and without the metabolizing systems is necessary. This would yield information that could be further useful in the quest for an environmentally clean and agriculturally safe ecosystem.

Table 32.1
Pests Resistant to Pesticides

Pests	*Resistant pests*
Arthropods	428
Fungi	91
Weeds	5
Nematode s	2

Georghiou and Sato 1983.

Table 32. 2
Plants with Properties against other Biota

Property	*Species*
Insecticidal	974
Antifeedant	219
Fungicidal	92
Nematicidal	55
Piscicidal	47
Bactericidal	37
Rodenticidal	30
Herbicidal	6
Repellant	209
Attractant	22
Sterilant	1
Others	159

Regupathy and Jeyaraj 1985.

Table 32. 3
Some Soil-residual Pesticides with Mutagenic Properties

Pesticides with Mutagenic groups	*Pesticides transformed into Mutagenic derivatives following activations*	*Mutagenic groups*
Acephate	Captan	R - Br
Atrazine	Carbaryl	R - OPO
Bromacil	CDEC	R - SPO
Dichloropropene	Diflubenzuron	Cl = C = C
Ethylene dibromide	Mirex	Cl-CH2C = C
Fenaminosulf	Nitrofen	CSCL2
Methyl bromide		Ar-N-OH
Parathion Simazine		C ON-OH

Based on Sharma 1988 and Sharma and Panneerselvam 1991.

Table 32. 6
Promising Plants with Insecticidal Properties

Acacia Catechu	Maringa oleifera
A. concinna	Marranta aurindinaceae
A. dealbali	Mellettia subserosa
Acorus calamus	M. avirculata
Adhatoda vasica	Mundulea sericea
Agave americana	Ocimum basilicum
Allium sativum	Parmelia tinctoria
Amorphohallus companulatus	Phytolacca dodecandra
Anacardium occidentale	Plumbago zeylanica
Andrographis paniculata	Pogostemmon sp.
Annona procyanidin	Pongamia glabra
A. reticulate	Quassia amara
Argemone mexicana	Rhus coriaria
Azadirachta indica	Ricinus communis
Calotropis gigantea	Ryania speciosa
Carica papaya	Sansveria marginata
Cassia fistula	Sapesidus emarginatus
Chrysanthemum cinerarifolium	Solanum sp.
Curcuma domestica	Swietenia mahagoni
Eucalyptus naundina	Tagetis erecta
Haplophyton cimicidum	Tephrosia purpurea
Kaemferia galanga	T. villosa
Lantana camara	Trachylospermum wilfordii
Madhuca indica	Vitex negundo

Inaddition to the plants mentioned in Table 4. Compiled from scattered literature.

Table 32. 4
Selected Pesticidal Principles from Plants

Names of the compounds

Absinthin	Galamgin	Parthenin
Achillin	Geranial	Patulin
Alkaloids	Glycosides	Phlorizin
Allethrin	Gossypol	Picrotoxin
Allicin	Guaiacol	Pinene
Allyl isothio	Piperonyl butoxide	
Cyanate	Harmine	Procyanidin
Amygdalin	Heliotropine	Protoanemonin
Anabesine	Hypericin	Psilotin
Andromedatoxin	Psilotinin	
Anemonin	Isocumarol	Pulmitoleic acid
Anonaine	Isoemetine	Purpuranin
Arbusculin	Isoflavones	
Aurmillone	Isotubulosine	Quercetin
Azadirachtin	Quinidine	Juglone
Barthrin	Resins	
Betaine	Kaempferol	Ricin
Benzoquinones	Karanjin	Robustic acid
Benzyl	Rotenone	
Isoquinoline	Lapachol	Rutin
Berberine	Lecanoric acid	Ryanodine
Lonchocarpic acid		
Camphene	Lycorine	Safrole
Cantharidin	Saponins	
Cineole	Mimosine	Scabrin
Cismethrin	Mundulone	Scopoletin
Colchicine	Myristic acid	Scopolin
Coumarines	Myricetin	Sinigrin
Sparteine		
Damsin	Niacin	Stemorin
Demecolcin	Nicotine	Strychnine
Dhurin	Nornicotine	Supinine
Diallyl disulphide	Nicandrinine	
Diallyl trisulphide	Tetremethrin	
Oleodipalmitin	Tropital	
Emilien	Oleoresins	
Esculetin	Oleundrin	Vanillin
Eugenol	Ollic acid	Veratrine
Orcinol	Villosin	
Filicin	Viscidulin	
Fulvine	Vulgarin	
Fumarin		
Xanthinin		

Compiled from the scattered literature.

Table 32.5
Ethnobotanical Lore of Pesticidal Plants

Plants	*Locality*	*Uses*
Annona cherimola	Tropical America	Insecticidal seed resins
A. squamosa		
Adina cordifolia	India	Insecticidal sap.
Aesculus californica	California	Insecticidal nectar & seed
Amianthium	U.S.	Insecticidal bulb & lead
Muscaetoxicum		
Amorpha truticosa	Southern U.S.	Insecticidal fruit
Anabasis aphylla	Near East	Insecticidal
Anabasis aphylla	Russia, Asia	Insecticidal leaf & stem
Anamirta cocculus	India, U.S.	Insecticidal berry
Canna generalis	U.S.	Insecticidal leaf & stem
Celatrus argulata	China	Insecticidal leaf & root, bark
Chrysanthemum coccineum	Southern Russia, Iran	Insecticidal powder
C. cinerariifolium	Yugoslavia	Insecticidal,
C. marschallii	Iran	Insecticidal
C. parthenium	Iran	Insecticidal
Cimicifuga foetida	India	Insecticidal root
Cistus ladaniferus	Mediterranean region	Insecticidal , deodorant
Clausena aniseta	Tropical Africa	Mosquito repellant
Croton ladaniferus	Mexico	Insecticidal
Croton tiglium	China, India	Insecticidal seed
Cyanchium arnottianum	India	
Cymbopogon nardus	Tropical Asia	Insecticidal, disinfectant
Delphinium staphisagria	Europe, U.S.	Insecticidal seed
Derris elliptica	Tropical Asia	Insecticidal roots
Delphinium consolida	Southern Europe	Insecticidal
Dryopteris (Aspidium)	U.S.	Insecticidal rhizome
filix-mas		
Duboisia hopwoodii	Australia	Insecticidal leaf & stem
Eupatorium capillifolium	Eastern North	Insect repellant
	America	
Funastrum clausum	Tropical America	Insecticidal latex
Gardenia lucida	Burma	Repellant
Heliopsis scabra	Mexico	Insecticidal root
Helotropium peruvianum	South America	

(Contd...)

Ipomoea quamoclit	Tropical America	Insecticidal
Jacaranda filicifolia	Panama	Deterrent
Jacquemontia tamnifolia	Tropical America	Insecticidal seeds
Liquidambar orientalis	China, North America	Fumigant
Lonchocarpus nicou	South America	Insecticideal roots
L. urucu		
L. utilis		
Mammea americana	Tropical America	Insecticidal
Melanthium virginicum	Eastern U.S.	Insecticidal bulb & leaf
Melia azedarach	Asia	Insecticidal leaf, fruit &
Seed		
Milletia panchycarpa	China, India	
Mucuna spp.	Tropical America	Insect repellant
Mundulea sericea	Africa, India	Insecticidal bark
(M. suberosa)		
Nerium oleander	Algeria, India	Insecticidal leaf
Nicandra physalodes	India	Insecticidal leaf
Nicotiana rustica	North America	Insecticidal
N. Tabacum		
Patrisia pyrifera	Tropical America	Ryanex source
Physalis mollis	U.S.	Insecticidal leaf & stem
Phellodendron amureme	China, Japan	Insecticidal fruit
Rhododendron		
Hunnewellianum	China	Insecticidal flower
Rhododendron molle	China	Insecticidal flower
Sapindus marginatus	U.S.	Insecticidal fruit
Schoenocaulon officinale	Central America	Insecticidal seeds
Simarouba versicolor	Brazil	Insecticidal bark
Solanum tuberosum	South America	Insecticidal tuber & sprout
Sophora flavescens	China, Japan	Insecticidal root
Sophora pachycarpa	China, Japan	Insecticidal root, leaf & stem
Stemona collinsae	Tropical Asia	Insecticidal
Stemona sessilifolia	Tropical Asia	
Stemona tuberosa	Tropical Asia	Insecticidal root
Tephrosia vivginiana	Eastern U.S.	Insecticidal root
Tournefortia	Maita	Insecticidal leaf
Hirsutissima		
Trachelospermum stans	Mixico	Cockroach poison
Tripteryguium wilfordii	China	Insecticidal root

Lewis, W.H. and Elwin-Lewis, M.P.F. 1977.

Chapter - 33

Ecology Burns to Make Way for Energy

A. Introduction

I was informed by several students, a few faculty and some non-teaching employees that the scrub jungle in the campus is being burnt to raise tree crops for energy yielding purpose as a contribution to rural development. This is done under the leadership of Dr Kannabiran, Lecturer in Life Sciences assisted by the Deputy Registrar Mr.Subramaniam and the department of Horticulture. I was also told that some natural vegetation was already disturbed quietly here and there on the campus for the above purpose. After seeing it myself I found the burning and affecting the natural vegetation has gone too long.

As I was involved with the campus development in the past and I have reasonable experiences, with men and matters on the campus, I am venturing to advance the following for the sake of common good. I hope you give the attention that my perceptions and suggestions deserve coming from a socially conscientious and value abiding senior teacher.

B. Plea For Immediate Steps

(1) Kindly order to stop the destruction of the natural habitat and convene a meeting of the Ecology School and all others concerned with the above activities to consider my plea without affecting any body's interests. (2) To facilitate the above herein I put forward my perceptions on this issue, along with some historical facts and alternatives for immediate and future action. With an objective to keep R.V.Nagar open for non-destructive ecological development.

C. Concerns

R.V.Nagar should be developed / ecologically restored conserving all biodiversity, water and soil. This should not be disturbed by clearing the scrub to raise monoculture of trees however fast growing they are. In fact we should take steps to convert the campus as a 'Biodiversity Park' in line with nearby Auroville and in tune with the current global and natural ethoses.

D. Habitat and the Environment

(1) The campus is a degraded habitat. Soil and Water are lost every year into the sea. The vegetation is less but very diverse (Plant species = 150+). Animal life also is rich (Mammals = 20+, Reptiles = 30+, Butterflies = 30+, Birds = 80+, Insects = not accounted- probably a few hundred species). All these having established certain balances through interactions and interdependence - this is what Ecology is all about - need to be conserved. 2. In recognition of the above only, DoEF has been encouraging projects for ecological development of campuses.

E. Efforts of Ecology School

In realization of the above, our School has made many efforts, some as failures and some with partial success.

1. Ecodevelopment of Campus: During 1987-88, I prepared an ecological development plan for our campus in collaboration with the then horticulturist Natarajan and Chief Engineer Arumugam. It envisaged extensive bunding, impounding of water in rainy season, vegetating and turfing, as per natural topography, a network of slab-layed footpaths, narrow roads, park sites, ponds, etc. This involved more labour, involvement of our School in Toto, but less expenditure - instead of the present National Highway type of roads, monoculture, etc. Suggestions to locate the departmental buildings, hostels and faculty housing and even a design for 'Ecology School' building was given then. Why I was involved in all these was a different story. But most of my efforts were disregarded for reasons not known to me.

2. Ecology of the Campus: However considerable field work was done by the students guided by chiefly Dr Davidar. Others such as Dr Rauf Ali, Sundarvel, Dr Parthasarathy, Dr Chapekar and myself also obtained data on the campus through students - as per my information from students, Contour bunding, regeneration of natural vegetation, planting and fencing to conserve soil and water was a major thrust under Dr Davidar's leadership carried out for 2-3 years. Students further carried out flora and fauna enumeration, variations in plants, phenology, associations, frequency and density studies, mapping of the region under my guidance. Faunistic enumeration - birds, reptiles, mammals, butterflies - plant and animal interactions were carried out under Dr Davidar's inspirations. Habitat studies and other ecological aspects were the interests of other faculties.

3. Ecorestoration project: In early 1991, after getting some guidelines from DoEF on a visit to Delhi along with Prof Gnanam, I initiated an Ecorestorative project with two scientists from Auroville (Edward and Ardhendu) and one colleague here (Dr Poyyamoli) and a student (Francis) to conserve soil, water, biota and vegetate in a way to increase soil binding and water holding ability and soil fertility. But this did not continue as the VC was in favor of a major ecorestoration work for whole of Pondicherry under Dr Meher-Homji's leadership and through DST funding to the tune of Rs.50 lakhs. After a couple of meeting Dr Poyyamoli sent me a message of his disinterest through the student! This bad culture and withdrawal remains mystery. Thus my attempts could not succeed. Probably the other bigger one must be going on successfully.

4. Tree Planting: Occassional tree plantings were taking place under the aegis of NSS or Tree Planting Committee (i/c Manivannan (AR) doing Ph.D.,in Eco-Agriculture under Dr Ramaswamy of Futurology) and our Ecology students without disturbing the existing communities.

5. Field works for students: When the annual field trips for ecology students were discontinued by Dr G nanam, he advised me as a substitute, to concentrate on our campus and accelerate the existing works (so that the funds could be saved and spent for the excursions into the past (tribals) of futurology department!).

F. Effects of Your Actions

(1) Habitat: Soil erosion will be accelerated. Microclimatic changes ensue. Scrub forest will be lost. Biomass turnover rate will be reduced. (2) Flora & Fauna: Floristic and faunistic diversity will be reduced. Neem trees regeneration along with Phoenix will be hampered. (3) Ecology: Shelters for amphibians, reptiles, mammals and cursorial birds will be lost, leading to local extinction of those species. Animals interacting with vegetation will lose partnerships. Butterfly diversity and 'undergrowth foraging guilds' will be endangered. Food resources for birds will also be diminished. (4) Permanent loss of data: The accumulated data of seven years will be rendered useless and labours of all these years ill come to naught. (5) Field-trips Revival: Field ecological exercises can no longer be done and the annual ecological tours - have to be resumed. In any way these should be retrieved in preference to Past-ology practices (visiting Toda settlements) of the Futurology Programme. (6) Acceleration of Immorality: In the light of 1987 plan, 1991 project efforts, students and faculty studies on the habitat - disregarding all of them - the present attempts will further accelerate/promote unethical activities.

G. Imperatives

(1) Immediate action: Kindly stop clearing the scrub jungle pending a meeting of all concerned people under the Chairmanship of Dr Gnanam. The suggested alternatives do not affect anyone, save non-science and arbitrary power-yielding. (2) Foresight or Myopia:All the above uniformly envisaged to improve the existing campus, revegetate all empty spaces, maintain soil and water intact - and all forever. constituting a very good eco-tourist spot - development through restoration - perhaps a unique model for the rest of the

200+ campuses in India. It is still not very late. Our school is too desirous to give our energies in the endeavor of converting R V Nagger into a biodiversity park - fitting too well with the diversity we have in our faculty.

H. Why Stop the Present Work

1. Between nature conservation and monocultures, the former is recommended and preferred by all environment-conscious people ("Environment Consciousness does not pass beyond conference rooms" - statement of Prof AGnanam, the VC, in Indian Express). 2. There is no urgent need to raise plantations for

(a) meeting 'energy crunch' in the campus
(b) 'raising income' for the university
(c) Aiding rural development as there is a big local Government for this small place
(d) There are more pressing jobs in the University.

I. Ideal Project Site

(1) It is proper to carry out this project on a really barren wasteland - without the destruction of the existing ecosystem-such as the University land opposite to the Engineering College. (2) Even in the campus itself there are many acres of land without any scrub vegetation, such as the place between the Administration block, Library and the SOM building, where plantation can be attempted.

J. Pertinent Observations

(1) Excellent energy plantation work was done at MKU. These are gardners' jobs. Why to waste time and energy of a Reader (Dr Kannabiran), a precious human resource. (2) However if Dr Kannabiran wants to fulfil a frustrated desire of working on vegetation - 'Neyveli project' along with Meher-Homji way back in 1988-89 - he can as well reclaim that project through the good offices of Prof Gnanam and not destroy our own backyard. Grave injustice was done to him and Dr Homji in those days by pirating their project on plantation undergrowths at Neyveli Thermal Project! (3) It is painful to us in School of Ecology that a senior man like Dr Kannabiran who began teaching a paper on 'Biodiversity' under the cafetaria system, should lead the destruction of biodiversity (either he does or not believe in biodiversity at all or is the subject being marketed by a wrong shop - Bose School instead of Salim Ali School). Either way it is a bad example to students. The credibilities are at stake. (4) A very senior and productive Faculty (because of which promoted from Lecturer or Reader) like Dr Kannabiran should not be under- used for pedestrian works of this sort. He should be used in frontline fields such as Tissue Culture or Plant Biotechnology. At stake will be his own credibility;

people will label him with disparaging honorofics and the VC as promoter of servile labours. Dr Gnanam may be accused of considering Dr Kannabiran as 'fit for nothing else'. All these are uncharitable. (5) If the apparently innocent exercise of burning scrubs of ecological value to raise plantations of economic value is of future research utility to Dr Kannabiran, he can as well initiate ecological research from now itself on the same scrub jungle. We will be happy to help him, though he has started teaching Biodiversity and Environmental Biology trespassing into the domains of Salim Ali School. (6) Putting Mr.Subramaniam, the Deputy Registrar to assist Dr Kannabiran raises a question of impropriety. Not only we cannot provide assistances of this sort to faculty in their research projects, we are putting an officer under a faculty of same or lesser scale of pay (I was told that Dr Kannabiran is in a lecturer scale though designated as Reader). This will be a dangerous precedent. When all faculty are almost secondary to all officers despite academic superiority, how come one officer is rendered assistant to an equal or lesser in the ladder of ranks. Or is there something invisible? People raise these questions which do not increase the credibility or honour of our university and its leaderships. Or, are we using his expertise obtained at the Ponds company from where he leapt into our university service. Or, is he working for a Ph.D degree just as many others have been doing under various faculties on the quiet? Then it is all the more curious!

K. Open To Charges

(1) Recently Indian Express carried a story "*How green is the Campus*". Our VC Dr Gnanam lamented that "environment concern rarely goes beyond conference rooms". What is happening on the ground in his own campus is just the opposite. This contradiction exposes Dr Gnanam and our entire university to various charges related to credibility and hypocrisy. (2) The present plantations may pave the way for a new project - to study the undergrowths of monocultures - after some years. Thus we are creating an issue where none existed - Is this ethical Science? (3) We are committing the future of the campus to this ethics of destroying existing systems and erecting monocultures. If the next VC wants ecorestoration, shall we to pull down these energy plants? 4. Or is the present exercise planned for something else in future - as an insurance - a quiet first move in the grand game plan !

L. Conclusion

(1) I reiterate that University campuses should be biodiversity parks - not stands of monocultures. There are many places for this. The university infrastructural machinery should not be geared up for such unwise purposes. With lesser inputs the existing soil - water - plant - animal continua can be conserved in an accelerated speed. (2) If my informations are right, please verify if these actions are known to Dr Gnanam. If not, it is all the more

sinning. If it is with his approval these acts are indeed star crossed misfortunes, but not too late to retrieve. (3) We must think a dozen times, discuss such issues openly, before a decision is taken on such issues. This is what democracy and ethics is all about. Openness unfortunately has become a great causality. Silence is fast approaching conspiratorial proportions. (4) I request you not to treat this letter as insubordination, defamation or indiscipline. This is merely my moral input in the interests of our university - academic, social and humanistic. It is not meant to cause losses or inconveniences to the executive and its favorite faculty, but to minimize or stop losses to our university y the current exercises. In the process we are assured of long term ecological profits that sustain us all. (5) I am just echoing the local imperative of the ethos that Dr Gnanam has highlighted recently in the conference of Vice Chancellors at Delhi which he shared with me also. It is exactly due to this reason I hope you will do the needful to retrieve our campus for ecological restoration.

SUBMITTED TO THE VICE CHANCELLOR DR AGNANAM FOR CORRECTIVE STEPS AND TO OTHER PROFESSORS TO INTERVENE IN DEFENSE OF ECOLOGY. NEITHER HAPPENED, BUT I WAS SERVED A WARNING!

ETERNAL HOPES IN THE HEART

Chapter - 34

Higher Education in Ecology and Environment

With Special Reference to Natural Sciences

Preamble

In the mid-1970's, three universities in India started Schools/ Departments of Environmental Science in spite of discouragement of the Department of Environment, Government of India. Since then a number of other higher education institutions have introduced programmes on different aspects of the environment. In the present paper an effort is made to take stock of the scenario of higher education in Ecology and Environment in India. The organizational structure, curricula and courses, training programmes and practicals, textual resources, research projects and products and models of environmental higher education are discussed. An attempt is made to highlight the gap between the ideal and real environment and to understand the philosophy behind the identification of courses and the formulation of curriculum. Various suggestions are. put forward regarding faculty, syllabi, evaluation techniques and the role of the university. There is a need to cultivate a holistic philosophy (and approach) towards environment after evaluating the various components in a historical perspective.

A. Introduction

Global consciousness on the importance of environment was so aroused by the Club of Rome (CR) activities, United Nations Conference on Human Environment (UNCHE) at Stockholm and United Nations Environment Programme (UNEP) during 1965-72 that many countries have launched ministries of Environment well before the next UNCHE Conference at Nairobi in 1982. India was one among these, where environmental concerns have graduate into an all encompassing activity after the Rio conference. Added to these the recent globalization of Indian economy and the GATT impacts on our policy have deep environmental and ecological implications. If the challenges are not well responded, the consequent inequalities may- unleash enormous violence that ever before. Even if it succeeds bringing social uplift, and international respectability, environment here will deteriorate depleting natural resources and disruption balances. To preempt and manage these portents new thrusts of activity are warranted. These are best done through education at higher level that supplies all the levers and agents of change for the country. Therefore a stocktaking of the present and past is an urgent responsibility, so as to do better in future, vis-à-vis higher education pertaining to environment and ecology. If we do not improve on this front the ecological and environmental challenges from the anticipated acceleration of economic activity cannot be met, letting down the country.

The Department of Environment of the Government of India initially discouraged the establishment of separate sectors of environmental studies in the Indian Universities during mid 1970s, yet three Universities initiated Schools or Departments of Environmental Science (1975-80) while there is no such science, *sensu stricto*. Everyone found or made to find"relevance" there. Intellectual dishonesty and the faltering steps thus existed right from the inception, rather unfortunate for our country. Faltering continued, trial and error method of learning became the habit of even mighty people as if environmental 'science' is an entirely new field! Consequently pitfalls resulted. The net result is that there are only a few top class institutions in this field today; most others are mere patch work quilts, pushing holism to a backseat. Thus there are nearly 60 educational establishments in India with outfits prefaced "Environmental". While environment and ecology is everything, they became different to different people; some assumed "omniscience" and became 'omnipotent' and 'omnipresent'. The ecological wisdom that has taken deep roots in the developed North is yet to seed fruitfully on our campuses. Our objective is therefore to take stock of the whole gamut of 'higher education in our campuses and suggest approaches to reforms and inputs m the light of hindsight wisdom, ground realities and future needs.

B. Geography of Environmental Education

A number of institutions, spread all over the country, now offer programmes on different aspects of environment, Powar (1993) has described the status of environmental studies in Indian Universities as prevalent in the early nineties. Latest information on the programmes offered by different institutions in India is available in the Universities Handbook, (Association of Indian Universities 2000).

C. Heritage and Traditions

Though the educational efforts started in mid 1970's, research activities on various aspects of Environment and Ecology have been existing for long:

It is appropriate to briefly summarize these as a backdrop and to assess if the traditions bequeathed to us have been followed, for these will have telling effect on the quality of education.

Right from the times of British East India Company information on resources, environment and ology were getting documented. Surveys of flora, fauna, native people, other natural resources like water and minerals have gradually intensified. By the time of political freedom from the British, enormous informations accumulated on our plants, animals, lakes, rivers, etc. All these and the departments of agriculture, forestry and fisheries, had bequeathed very valuable knowledge for their conversion into wisdom to guide the destinies of our country. People like Misra, Santapau, Agharkar, Salim Ali, Bharucha and Mani began their every valuable careers followed by Ganapathy, Panikkar etc. on several aspects of ecology and environment. Following these the institutions of modern India under CSIR, ICAR, ICMR and outside it accumulated great informations. The Surverys -Botanical, Zoological, Geological, Anthropological arid others have laid foundations for all modern investigations. How much of these efforts are reviewed and brought out in print in a state useful for the country's ecological exercises? How much of these informations were culled and crystalized by the academics for use in education? How many standard treatises resulted from our great records barring MS Mani's and Salim Ali's monumental works and the floras by the British pioneers. We have not even tried to review the works - either subject-wise or institution-wise or Scientist-wise for use in education. Indian Botanical society has done this once 30 years ago! Many students have never heard the names of Bharucha, Misra and Mani!

Our track record of research efforts in ecology and environment - our strengths and weaknesses, challenges and responses, knowledge and information - warrant an assessment to function as foundations. Prima facie maximum research outputs have come from Institutions dealing with forestry, agriculture, flora, fauna, etc. Though a very elaborate exercise is indeed involved, universities had to build on these. Current Statuses of or resources and their ecology should have been a major basis for all our efforts since 1975 when the first School of environmental sciences in India started. Did these happen? A modest exercise in this direction, pertaining to a small area of Environmental pursuit was carried out by us some years ago which is in the process of updating (Sharma *et al,* 1985). The importance and relevance of a status report on Ecology and Environment in India in developing a model curriculum as well as regional and local variants is obvious. The other advantages of this exercise are too evident to need an emphasis. Environmental history dealing with the dynamics of all natural resources, interactions among them and their fate in time and space are necessary.

D. Experiences: General Scenario

Though there are many significant contributions by individual scientists, the ecology work of the Botany department of Benarus Hindu University, the Centre for Ecological Sciences of Indian Institute of Science, are perhaps the best - the former with a tradition of over 40 years and the latter, 20 years! The Salim Ali School at Pondicherry University perhaps has the best teaching programmes to start with. Despite good intentions, monetary inputs, topicality, patronage environment/ecology schools did riot click well - probably beacause of an absence of ecological ethos, work ethic and academic leaderships? Let us take a brief stroll into these green paradises fading.

1. Organizational Structure : (a) Appendages of Biology: In many Universities, the unit became only an appendage of the existing biological schools ignoring other and often more important areas, especially geography. This is due to two reasons - ecology of plants, animals and toxicities from pollution were their bother and therefore had a claim.

(b) Schools or Departments: Wherever Environmental Schools were initiated, the experiences were not encouraging. Whoever wanted to join the bandwagon and has the means to obtain the seats on faculty, became environmentalists - learning the subject leisurely by trial and error. Meanwhile students got only the labels and suffered a lack of content. While some made graceful exits, others continued ad-hocism at India's expense!

Sometimes, well planned guidelines preceded the organization of a unit. They were defeated by recruiting people on the faculty irrelevant to the planning. Consequently the content of the education became deformed, to suit the faculty's whims. Such units became refugia for riffraff and bred similar students.

In some units best qualified people were recruited with a fond hope that they will build the excellence in some ares of environmental relevance. But the faculty continued to perform only in their original areas, of little relevance to environment.

(c) Co-ordinating Cells: Some Universities established environmental cells with a skeleton staff diverted from the existing departments with a mandate to organize activities of training and research by associating relevant people from the remainder in the University, the cell often became a monopoly turning into a friends club. More often it is torn by rivalries. In the best of circumstances, the subject is reduced from a glorious holism to a mere sector sacrificing student aptitudes while faculty absolved themselves of responsibility towards society.

It is often found that "***exaggeration for effect***" model was put into practice to cover up deficiencies by faculty. In other words, over emphasis of one's own speciality as constituting the whole and soul of environment education was very common at least until the deficiencies are filled! Meanwhile the students become scape goats of half baked knowledge! Pollution and toxicology is dominating as if it is "***be all***" of education, where incidentally a great deal of irreproducible research takes place in India.

2. Curricula and Courses: (a) All Agricltural Universities: have subjects of direct relevance to environment and sustainable development-soil science, agronomy, plant protection, water technology etc.

(b) A medical institution: in India has yet to reorient its social and preventive medicine to ecological realities, though health effects due to occupational hazards and noise are receiving attention.

(c) Most Engineering Institutions: have some environment -oriented outfits within their civil engineering departments and rarely separately.

(d) Ecology and environment: are taught in the postgraduate departments of Botany and Zoology in all the universities of India as a full paper. In many universities it is also included at the undergraduate level in one or the other from.

(e) Post graduate courses in geography, geology: soil science contain stray doses of ecology and environment.

(f) Sociology: economics, history, politics, philosophy, anthropology, literatures do not have any component of ecology and environment in most of the Institutions where these are major concerns. Similarly the above areas are seldom touched in the curricula and syllabi of courses on Ecology or Environment, having been more loaded in favor of biology. In many places even physics and chemistry are ignored. In other words objectivity is wanting..

3. Training and Practicals: Practicals in the natural habitats, Surveys of manmade systems and prolonged experiences in the ecosystems are wanting in most places. The training of students in most places is highly inadequate in other respects also. This can be offset if tie-ups with the local and regional institutions are organized. Social surveys, field works, lab exercises, experiments are all important equally.

4. Textual Resources: We have yet to have a decent text either in ecology or environmental science with Indian experiences and examples, save M.S. Mani's treatise on biogeography.

Unfortunately a bandwagon approach has dominated several ingredient of environmental education, pertaining to books and other source materials for teachers and students. There are a plethora of books published, rarely text books or treatises dealing with basics or any subject in its entirety. Most are edited volumes. Many publishers specialize in these, price them 6-10 times the production costs, dole out cuts and commissions to push sale these half-baked products to make a quick buck. A veritable racket, this has not happened in any field of education so far in India. Similar is the case with researches also. Anything is passing off as environmental- un-verifiable-therefore liable to be fraudulent. Adding to these are the publicity gimmicks by the professionals in the educational institutions. These trends are bad for a healthy growth of education in environment and ecology. It is a pity that this is not being realized enough any where all paying lip service.

5. Research Projects and Products: Research projects and researches themselves, publications, journals, books, publishers, library policies and practices, local problems and consultancies have their own stories to tell. Prima facie most of them can be assessed from inadequate to downright cheating. Academician - Industrial nexus are already taking the society for a ride. The frauds and scandals on all of these above are a legion.

Enormous number of papers and books are being published by the Indian academics. Many of them are fraudulent. Publishers make an easy money because of the topicality of the subject. Even M. Phil, theses are published as books and Vice Chancellors are glorifying them! Some people are publishing one paper/ month on average and awaiting awards! In the name of consultancies for expertise, futures are being made in collusion with academic authorities, bureaucrats and politicians. Funds overflowing, 5-star living can be seen here and there. Such success stories build an ethos of easy and quick success. This can only be at the cost of environment and ecological balances^ This is how we respond to challenges.

Most tragic is the absence of social consciousness and ethics among the faculty demanded by the ecological wisdom. The same lacunae, careerism and exploiting the name of "environment" or "ecology" for selfish ends are bequeathed to the students. Thus the band wagon moves - the great Indian environmental Safari -full of sound and fury, signifying nothing. Environmentalists flourish and environment perishes and students end up as any other careerists!

Having narrated our experiences, it is appropriate to see what should be the guiding principles in organizing and imparting environmental education.

E. Environmental Studies: Ideal and the Real

The environmental study is the Science of Nature as well as of Society culminating in an understanding of human ecological predicaments. Environment will not become a concern unless ecology of its resources is affected nor pollution would bother us unless toxic either to us or our resources. Therefore controversies or debates on these are irrelevant yet a Vice Chancellor thought it fit to debunk a department of his own university recently along with a faulty member - despite both being parties! Queer sense of ethics indeed.

Thrusts of Environmental activity in University systems, as the possible best, should cover the natural resources, "regional institutional facilities and the ethos of the local leadership based on global knowledge to meet regional challenges. Therefore the subject of environment deals with most branched in a University; ideally, there should be only a holistic education on environmental issues through a Coordination of efforts. In delineating the details however the ideals may be supplanted by the local forces of expediency and opportunism. Yet when the subject, as an ultimate outcome of academic exercises, is offered to students it embodies the twin content of Student Needs and Faculty Resourcefulness. Thus a programme on Environment - be it called Studies or Sciences, must be imparted by working out a golden mean between students needs and faculty resources. Both these influence development of the curriculum, the final outcome should not be far away from the initial ideal. As we have seen, University systems are not all that good as one hopes for and therefore must be put to public gaze by adding surveys of and services to local environments in the curriculum by involving and interacting with regional facilities. Keeping all these the following proposals are made.

F. Imperatives

1. Philosophy: The criteria for identification-of courses and formulation of curriculum and programme may be idealistic or opportunistic. Current wisdom may opt for the expediency in the name of realistic solutions are the idealistic ones. Even if everything is lost in society, university still shall keep the asphidistra flying with an objective and hope of retrieving the society. It should ventilate and articulate the eternal social values. A kind of perennial philosophy has to be upheld against the myriad ephemeral ones coexisting even within a single campus. All this is a challenge to the leadership. Despite my experience I am for the idealistic one because I do not wish the intellectuals continuously fail the society. This should be reflected in the most important discipline of educational endeavors-environment.

Social Sciences and Geography have been lately begun taking interest on environmental issues. This development is expected to have a salutary effect on the environment education. For an overall well being in this very vital area of education medical disciplines have also to come off age just as engineering and agriculture. Then alone a holistic subject will develop for a proper comprehension of the problems of nature and their management. Of course this is a very challenging task.

2. General Suggestions: (a) Out of 200+ higher institutions about 150+ offer courses in Botany and Zoology. And all these have an ecology/ environmental biology as a paper. This needs to be revised, updated and perhaps a common paper can be formulated as in the case of Genetics and cell biology. (b) Out of 10.000 colleges in India, most of them offer B. Sc courses with

Biological components in which ecology is included as a part of syllabus of a paper. This needs an upward revision with greater coverage and emphasis. (c) In every department one can have some environmentally oriented components- Botany, Zoology, Chemistry, Physics and so on. Ecology and environment being most holistic in nature than any other subject, education must necessarily reflect this, specializations depending on the trainees choices. (d) Separate syllabi need to be prepared for social sciences and humanities at college and university levels to increase the levels of their awarenesses and make the students and teachers realize who valuable their own disciplines are vis-a-vis the environment Checklists of courses and syllabi are therefore to be prepared based on the four criteria

(a) Natural Resources of the region
(b) Institutional Facilities in the region,
(c) Major facilities in the other universities in the state and finally
(d) the global thrusts. Faculty specialization can always be moderated to suit student needs and regional scenarios. Naturally a tough homework is warranted. A tougher one is to encapsule them within the degrees of M.Sc, M. Phil., diplomas, certificates and building curricula, for each of them with appropriate constituents.

3. Thrust Areas: Geography tells us that there are vast stretches of unproductive lands, aridity, salinity, wetlands, coastlines, forests, mountains, etc. Anthropogenic activities include industries and development projects of the government and nongovernmental agencies. Trans-boundary conflict with the neighbouring polities involving waters, islands and human settlements are many. Artisan, craftsmen and ecological refugees are another 'perennial resource' for studies. In fact all the thrust areas should revolve only on these and the universities (schools/departments/ centers) must focus on these local specialities. The entire planning as well as implementation can thus be innovative, socially relevant and path breaking for the country. Thus a good teaching and training programme is accompanied by appropriate researches.

4. Curriculum: Since environment is all pervading, a student in this needs a basic knowledge of many a discipline, whether or not the teacher has! Only with such foundations one could branch off to a pursuit of Nature - oriented, or Lab-oriented or Society-oriented activities. Since it is not always possible to separate Nature, Laboratory and Society from one another the subjects under these may be grouped as Foundations, Basics and Derivatives - applied major fields and minor specialities. The urgent need is to develop an exhaustive curriculum at M.Sc, M.Phil., and Diploma levels in Ecology and Environment replacing or reforming the existing variety of species as per the local fancies. The curriculum shall consist of (1) Theoretical papers on (a) Foundation subjects, (b) Basic disciplines, (c) Applied disciplines and (d) In-Depth Specialities (2) Practicals consisting of both (a) Laboratory and (b) field exercises and (3) Field surveys on (a) urban and rural societies as well as (b) pristine and disturbed habitats. Each one of these sectors has to be very broad based and structured into many papers.

5. Schedules or Streams of Courses: (a) Primary: This consists of the usual M. Sc, with four semesters - the last semester devoted entirely to dissertation project. Each semester shall have devoted entirely to dissertation project. Each semester shall have five to seven papers chosen from the suggestions given here subsequently. The stream will be manned by the entire faculty within or across the school.

(b) Secondary: This consists of an M.Phil., of two semesters in a major branch of Ecology/Environment. A few faculties can offer such a programme covering their disciplines such as Environmental Toxicology, pollution, etc., this may further be in collaboration with. the extra-university facilities available in the region. Thus M.Phil., in several subjects can be offered.

(c) Tertiary: When the faculty gradually become experts they should be able to offer a Certificate (one semester) or a Diploma (two semesters) Programme of teaching and training. in their specialities.

A school thus can be very active offering M.Sc, M. Phil, Diploma and other programmes serving the regional needs with a global ethos.

As Environmental studies are diverse, some courses cannot be structured by any of the existing methods. Such ones may have to be organized into 4+ or 8+ week courses instead of the customary 16+ week schedules. Fresh looks on all aspects of these including syllabi formats for practicals and surverys are warranted.

6. Models: Approaches to Programming : Over all three approaches of higher education on environment are discernable. In the holistic approach everything is included in the menu of Curriculum. In the Stream approach the holistic programme is delineated into physical, biological and social sciences. The sectorial approach, by far the most common, concentrates only on that part of environment in which the department running it is interested.

The foundations - basic courses, applied subjects, in-depth specialities, in theory, practicals in lab, library and field, surveys in the habitats and society can be structured and offered as curriculum for Post-graduate Programme in four ways. In each model any set of courses can be chosen to from the module. From these very same, M. Phil modules, Diploma modules, Certificate courses can be chosen. Foundation Courses had to be covered first Semester and Basic Ecology, Applied Ecology and Specialities in the 2nd, 3rd and 4th Semesters respectively. Practicals and surveys will be spread out all semesters and Dissertation in the Fourth Semester.

(a) One/Three Stream Programmes: The components of all the four semesters will be so organized that it should mostly suit for students of any back ground with the aid of some foundations. Conversely the components of I, II, III and IV semesters will be different depending on the background of the students. Thus we can envisage three streams each for students with the backgrounds of Biological sciences, Physical and Chemical Sciences and Social Sciences. Practicals and Surveys will be different.

(b) Model -1 Maximum Choice System: In this system students can choose their papers in all the four semesters.

(c) Model-2 Minimum Choice System: In this Foundation, Basic and Applied courses of the three semesters are compulsory. The choice is limited to the elective in the fourth semester and the dissertation with it.

(d) Model -3 Tutorial Approach: During the First semester all foundations will be taken as tutorials and exempted from evaluation. Two Basic Courses will be compulsory. Semester two and three have only compulsory papers. In the 4th semester Electives can be more, than one and Dissertation is compulsory.

7. Mechanisms of Oragnization: (a) Holistic: All branches of relevance to environment are brought together in a single programme. This may be done by a large school or by a small coordinating unit.

- **(i) Unitary:** A full fledged school of environmental studies. All programmes -Research, Training, Extension and Education leading to M. Sc, M. Phil., Diplomas, etc., can be carried out following two mechanisms which are in vague at present. Both are holistic in approach to education can be organized so as to be self sufficient in faculty numbers and competence to carry out all responsibilities including programmes, curriculum, programmes, syllabi, etc. But this is difficult because it entails a large number of faculty - competent enough to claim autonomy for each one.
- **(ii) Federal System:** A nucleus or center for Environmental Studies may be formed with a minimum faculty. All other faculty can be co-opted from other departments. Where ever there is relevance and interest they can run common programmes while each can carry out researches in their own departments. The difficulty here is that common often in spite of tragedies, Accountability may become a casualty.

(b) Sectorial: This consists of dealing environmental education in different branches separately. For decades ecological and environmental subjects were parts of syllabi in Botany, zoology, chemistry, physics, geography, pedology, geology etc. If these are strengthened in the light of current global concerns, the purposes of environmetal education will be served.

8. Faculty: In any University the local leaderships are more important than a check list of recommendations from a workshop of Shakespeares. The leaders are expected to be master of one and jack of few trades and among the best either as a teacher or a Scientist, if not both. It is my experience that most good scientists become so mechanical in their selfish pursuits that they cannot teach. They gradually develop fear to be in a classroom! If the leadership is sensitive and idealistic, in addition, it will just be wonderful. I do not entirely believe in the selection of the "best men" and leaving the rest to them. Quite often these 'best men' developed empires, impregnable forts and even fascist tendencies due to which Science in India did not flourish while scientists did, and do still. This is mainly due to a singular lack of the four Es (***Ethics, Ethos, Equity and Empathy***) and domination by the four Ps (**Power, Profit, Pleasure and Popularity**) in our academic society to climb up social ladder in any name under any guise. One can trace ail the evils of our country to this singularly Himalayan source. Therefore besides the teacher's experience and expertise, dynamism, careerism and philosophical outlook determine the fats of all programmes, they are in situ agents of change.

9. Syllabi: Syllabi are generally brief in most Indian Universities. But this can be misleading to the student as well as the observers and evaluators but convenient to their framers and teachers, They should be elaborate and unit wise. There shall be no scope for ambiguity. Syllabi should be invariably followed by the recommended texts and incorporate besides the global status of the subjects the regional compulsions. Ecology should always maintain a golden mean between latest knowledge and regional needs.

10. Evaluation: The conventional examination system does not hold in toto. The evaluation need to be continuous and involve such wide varieties like library, field and laboratory exercises, surveys of fields, people media and governmental records, question banks. Open book system of answering and even framing and posing questions, aproaches to understand and solving problems have to be tried. Genuine social service o restorative nature in the field also reveal the ethics of the student and therefore reckoned for evaluation.

11. University: The University must empathise with regional problems of environment and people and apply global knowledge by taking up relevant projects for research, development, education, extension, etc. This is the only way to develop an ethos for people oriented educational ethic. The chief thrust of Environmental Studies should be conservation of natural resources and ecology followed by their staggered utilization - i.e. ecological development.

G. Conclusions

1. Pioneering Efforts: Some of the suggestions made here are neither new nor novel as they are advanced or carried out by UGC, AIU, and IGNOU in other fields, a few of which have relevance to Ecology and Environmental studies. But this is the first time that an exhaustive analysis of the problem of higher education on a vital field like Environment is made - assesment of the current scenarios and consequent imperatives for action. The difference between the presently advocated approaches and the beaten tracks lies in the stress on developing a national ethos and interests. Only a patriotic effort on the part of the responsible institutions and senior teachers and scientists can salvage the country's current efforts from the mess on environmental education. My efforts constitute a beginning in this direction.

2. Holistic Practice: Environment appears many things to many people. There will be as many opinions as there are Environmentalists. As in the "elephant and the blind men" the 'experts' may pursue only their personal interests. It is therefore imperative on the leadership to be honest with themselves and the very restive but impressionable youngsters, these present and future leaders have to' cultivate' a 'holistic philosophy' towards the Nature and its components in a historical perspective, lest they fail our country for ever, and never to be resurrected!

3. Leaderships: All said it should be realized that reviewing the status of higher education in India is a very tall task for an article. It just consumes a space that befits a book or booklet. The work done in India is a major aspect that needs to be surveyed. So is the work done by Indians. There are many Institutions especially under CSIR and ICAR that is partly or greatly ecological in nature. A new dimension is added by the gradual and increasing entry of NGOs into the areas of ecology and environment. They influence the education and also influenced by it in turn. As a corollary comes in the role of foreign agencies and foreigners, more so in this age of globalization and remote sensing. Unless a great feeling of self preservation and patriotism pervades the. veins of Indian enthusiasts of environment, India may turn out once again a land of open veins! Adding to this are the ever increasing corruptions in the granting and conduction of research projects, publication of papers and books and awarding of degrees. In this background can we expect correct choosing of courses, or making syllabi, or teaching what to talk of improving them. The culture of practice schools in business management is needed in ecological and environmental practices, in addition to introspection on our duties to wards the country, the people and future generations - nothing short of a spirit that pervaded India under the leadership of Mahatma Gandhi some decades ago can redeem Indian environment for which the need of appropriate higher education in this field is pivotal.

PREPARED FOR THE COMMEMORATIVE CONFERENCE IN HONOUR OF PROFESSOR K.B. POWAR, SECRETARY GENERAL OF ASSOCIATION OF INDIAN UNIVERSITIES, 2000

Chapter - 35

Higher Education in Ecology and Environment

With special reference to Social Sciences and History

A. Introduction

1. Environment has been a subject of study in India for centuries. British East India Company accelerated this tradition by focusing on the surveys of forests, minerals, humans, plants and animals, among others. They continue even today in one or the other form, because of their intrinsic resourcefulness.

2. Although they became subjects of studies of variable importance at various levels of education, they remained sectorial and un-integrated until 1970 in the developed countries and in India until 1978.

3. Due to the Five Limits to Growth Paradigm and other developments, such as the Club of Rome, The Green Philosophers and Activists, United Nations Committee on Human Environment, World Conservation Strategy and finally United Nations Environmental Program, Governments across the world opened ministries of Environment and initiated Education at the highest level, the Universities. India too naturally joined the global mainstream.

4. It is noteworthy that the Indian Institute of Science, Bangalore, initiated a School of Ecological Sciences in 1971-2. At the same time came a plea from Prof Seshachar on the International Biological Program. They were thus in tune with the best thinkers of Europe and USA and ahead of the rest of India by nearly a decade! A few years later Ecologists of India reviewed their subject and identified areas of future emphasis. Alas! All these were totally ignored.

5. By 1978, ignoring the above as well as the cautions from the Environment Ministry, a full fledged School of Environmental Sciences was initiated at Jawaharlal Nehru University and an outfit carved out of the Botany Department at Andhra University. Madurai University made inputs in the curricula in the Life Sciences and in the School of Energy, Environment & Natural Resources by 1981. By 1986, another full fledged School, in Ecology

was given a caesarian birth. The permissiveness thus began was never disciplined.

It should be realized, at this juncture, however, that Ecological and Environmental subjects are common subjects in one form or the other in Medicine, Engineering, Agriculture, Forestry, Fisheries, Botany, Zoology, Geology, Geography, and in all the surveys of India- Anthropological, Botanical, Geographical Geological, Zoological, etc. And yet...

B. Environmental Education

1. From then on, there has been a feverish activity on the environment front both in research and teaching, with sound and fury signifying nothing. Though the thrusts, inputs and spread were great, covering over 300 institutions, the outputs and impacts were controversial in the minimum and scandalous at the worst. Degrading ecosystems and student Morales bear testimony for these. But all are covered under the carpet of prestige the subject enjoys. Every teacher in these areas seem to know everything of the subject and at the same time none knows anything. Typical story of blind (wo)men and the elephant! The subject became so much subjective as no other discipline has become ever ! There are several reasons for this state of affairs. But we shall focus only on one aspect pertaining to the Curricula and Syllabi.

2. Environment Departments, Curricula, faculty, Students, Researches were all monopolized or dominated by Biological sciences, followed by Chemical Sciences. Subjects like Geography and Geology rarely figured despite their holistic value. Despite the fact that the subject of Environment has become vital due to human predicaments, social sciences and Humanities are totally ignored.

3. That the Environmental consciousness was aroused by social scientists first is has dawned on the educational leaderships too late. We come across during the last decade environmentally relevant researches from Sociologists, Anthropologists, Economists etc. But transmitting them to students is rare.

4. Most shameful, however is the total neglect of history forgetting that everything will have a history and those who ignore it are often condemned to repeat its blunders. Historical studies made great contributions to the knowledge of earlier environments and the then influencing factors. Human ecology was always dependent on history.

5. The study of local Environments of any ecosystems, require past history. Teaching of every discipline begins with its historical background and the then methodologies followed. Yet the subject did not unfortunately develop good public relations, as pollution-wallahs could.

C. Good, Bad and the Ugly

1. Exceptions are not altogether absent. Pondichery University did

introduce 3-Credit courses, namely, Environmental Politics, Environmental Philosophies and Activism incorporating Anthropology Philosophy, Sociology, Politics, Economics and History, as dimensions in the holistic understanding of environment and a general Basic paper of four Credits on Natural Resources, underscoring the multi-disciplinary, inter-disciplinary and trans-disciplinary nature of the field of Ecology and Environment. But all these are out-weighed by subjectivity, vested interests and down right crimes against student futures remaining uncorrected for decades! But they are removed with the utmost urgency and rarest enthusiasm, the moment the professor was eased out in the name of superannuation!

2. Environmental Economics is introduced in several Universities. Subjects like Environmental Management, Renewable Energy and Law, figure in the Distance Education programmes of many Universities There is a whole department of Environmental History at Jadhavpur. A very important subject, Human Ecology also figures in some Universities.

3. UGC woke up very late and could bring a wholesome syllabus, but limited to under-graduate level. It also commissioned and brought out a model text. This constitutes one of the well-intentioned inputs by the UGC resulting in good result. It has an adequate component of Social Sciences. A very rare example. They could do nothing at University level !

4. NCERT's Syllabi and books are just excellent. College Lecturers also use them. Some of my colleagues at the University level refer them clandestinely.

5. Several Efforts were made by some Individuals, Institutions as well as AIU to reduce scandals and initiate welcome corrections in the Curricula of Environmental Studies at University level. They come to naught for obvious reasons.

6. A welcome subject Ecological Philosophy was also introduced. But never taught. Skolimowski's Dancing Shiva is the lecture Notes and its contents the syllabus! A course on sustainable development has no one from Economics, taught by two biologists! Environmental Law was taught by a Physicist ignoring a doctorate in law. Many syllabi have only 3-4 lines of contents which is against the rules of the institution. There are hardly any practicals either in the laboratory or in the field. And there are no teachers now a days- only faculty unqualified to impart either knowledge or skills, for they themselves entered the portals through backdoors or bore wells! These guide research in all fields because Environment covers all- Agriculture, Forestry, Wetlands, Mangroves, Energy and What Have You ? Some departments resisted a Board of Studies in Environmental sciences for a neat 14 years. In all these crimes Vice-Chancellors were the onlookers, acquiescing. This kind of India is the dominant India. It needs urgent attention and discipline.

7. With these backgrounds as foundations, three objectives are presently pursued:

To summarize a SPECTRUM OF PERCEPTIONS prevalent in well-informed circles pertaining to Ecology and Environment as a subject in education. This is to tell what thinkers feel.

To re-view the global HISTORY from an ECOLOGICAL angle, where wars are considered as conflicts for ecological resources. This is to provoke a rethinking on history of all events in society.

To list the topics, followed by the author in his teachings, AS A COMPENDIUM, to serve as a base in building a bigger social science dimension of Environment. This is to realize that formulating and integrated Curriculum is an intellectual challenge.

D. Spectrum of Concerns and Perceptions

1. The ethic of ecological limits to growth has emerged chiefly due to the efforts of Club of Rome and the activities of UNECHE and UNEP during late 1960s and early 1970s. India has responded by establishing relevant Institutions and making inputs to higher education. Though many departments of Universities and government have been carrying out select agenda of Ecological and Environmental nature, nearly half of India's 200+ Universities initiated sectorial outfits during the last two decades. Experiences from these can be summarized as follows.

[a] Knowledge of environment exists in Humanities, Social Sciences, Natural Sciences and Applied Sciences. [b] The existing centers are a motley crowd of sectorial outfits claiming to teach a holistic subject. [c] Yet these differ in their modus operandi. d) While scholars in Humanities and Social Sciences took leads in the developed countries, leaderships in India mostly remained as preserves of Biologists. Thus neither the western models were imitated nor innovations are made. A fresh look into all these is an urgent imperative.

2. Role of history in understanding ecology or environment or natural Resources is not a component in the Indian Knowledge Systems at higher levels. Similarly, historical aspects of ecology of plants and animals are totally jettisoned, discontinuing the earlier practices in vogue until a few decades ago. Surprisingly, historical aspects of any of the Ecospasms [10-15] though well documented are generally given no or a low priority. Ignoring the origins and growth of consciousness of Ecology from the Rainbow Warriors to the COR - UNEP - UNCHE- Die Grunen times of 1970s lead to a lack of perspective among the students of Environment and ignorant of the imperatives. This is a sin!

3. Having known the fate of environment and ecology in USA after the Second World War (called good war) and in Europe following the Marshall

Plan, even a mild tremor of *'history-repeating'* in the Times of Dunkel & 400+ SEZs has not entered the campuses (!) because history was totally ignored. Environmental Movements and some NGOs alone offered a few saving graces. Yet neither these nor the role of history on Environment and the influence of the latter on the former formed a part of syllabi in any university, save one and that too for some time!

4. The nomenclature "*Environmental Sciences*" shall be, replaced by "*Environmental Studies*" for, this subject encompasses social sciences, later day history leading to philosophies and Ancient Indian History. Although there are researchers in history, sociology, economics, anthropology, psychology etc, covering some aspects of environment, they are not in the higher curricula dealing with environment. Even Environmental economics is a part of economics curriculum, and not of Environmental Sciences. Strangely, a Central University runs a Program on *Sustainable Development'* sans Economics Department!

5. At Pondicherry University there was a successful attempt in 1992 to integrate - Policies & Politics, Sociology, Economics, Anthropology, History, Philosophy and Literature into TWO papers in the Environment Science Program : ***1. Environmental Politics, 2. Environmental Philosophies and Activism*** - but kept only as optional papers. Yet these were sought after, mostly by the students from Social sciences, Humanities and others. The Environment students were prevented from taking this in a myriad ways! This state of affairs continued for a decade until the teacher retired !

6. Both the papers vanished, supplanted by Two Eco-philosophy papers, in two departments : In the First, there is no syllabus and no teaching. The prescribed text is Dancing Shiva by Skolimowski -its contents make the Syllabus; in the Second, the syllabus is the English Poetry dealing with Nature, ***Thoreau, Emerson, Ginsberg excluded***- full of seminars where the syllabus is completed by the students.

The above example illustrates how ignorant but powerful people dominate serious knowledge workers with numbers. Students, Institution, Subject and the Country can go to hell, which seem to be anyway empty, for all the devils descended on to the Education Band wagon during the last three decades !

7. Ecology, Environment & Resources overlap and shall be studied together holistically. Therefore, this area is not merely multi-disciplinary, but also inter-disciplinary and trans-disciplinary. There are many Examples of interplay between History and Environment, which most historians are aware of (!) but ignored by most in the field of Ecology and Environment - The saga of human settlements and migrations, Red Indians and their Buffaloes, the great Dust Bowl of USA, Resources and Wars, Guano and European Agriculture, Indian Waters, Dams, Tanks and Deltas and conflicts as vested interests, the myths of races, medieval and ancient environments of

India. Most colossally wasted in India are the travelogues and the entire ARCHIVES!

It is the absence of Environmental History in our knowledge systems, in my humble opinion, is responsible for the display of ignorance, fascistic thinking, foolish utterances and ecospastic actions associated with the Sethusamudram Channel Project.

8. The ecological wisdoms are wide spread in the literature of Sanskrit and other Indian languages. The philosophies enshrined in the Vedic literature are greatly ecological in nature- the concept of God, the understanding of the Panchabhootas, the realization of Sun as the source of all energies. etc.

Some religions are more ecocentric than anthropocentric. In fact"Die Grunen" had embraced Marx, Gandhi, Feminism, religious and revolutionary ideas. On the other hand all parts of nature - SWAMPEAH are dealt with by various sciences. These two streams of knowledge systems have been interacting far less than desired to the detriment of all learners. The necessity of synthesis is obvious.

E. Ifs of Indian History until Independence

1. If Congress Continued as a non-party organization as Gandhi wanted, so much degradation would not have happened because Gandhism was very ecological believing in balances and limiting development to resourcefulness. This alone would have ensued Sarvodaya, a long term and longer lasting destiny for India. This would have become the only NGO pervading the entire length and breadth of India as a vigilant on governments. Even capitalists would have kept at least a part of their wealth under Trusteeship.

2. If Indian Leadership did not hurry for independence and prep one it by 10 months, became conscious of what happened to Jewish ecology in Europe and the moves happening in Palestine, there would have been no partition, no ecological refugees of 1947 plaguing India and Pakistan even now. Left would have survived and help modernize Islam and balance with Hinduism. Habitats and Economics would have become more important than politics and conflicts.

3. Our concerns today are one of current living scenarios and assured futures. History therefore had to transcend from people, settlements, cultures, politics, interactions and consequences to what constitutes a successful life - resources - Soil, Water, Animals, minerals, Plants, Energy, Atmosphere, and Human Resourcefulness. Environmental concerns have become a global phenomena like the political and economic issues. Consequently environmental issues become jargonized, band-agonized and some scandalized as a natural corollary.

4. The ecology of natural resources is connected with the history of humankind. Ecoism and historicism combine to make futurology. History has

dealt with humans - past and present with their influences on future. But humans depend on natural resources which too changed and for worse. Worse, because, their qualities which depended on mutual interactions have steadily declined. In time ecology degraded. Thus history and ecology are interrelated and had to be studied if humans are to be safer. Historicism and ecoism hybridize here to build a new field environmental history.

History of every environmental component of today and interactions among these components are a prerequisite study if futures have to be sustainable and secure. History and ecology are thus an integrated and indispensable ingredient of all management strategies.

5. A question arises spontaneously. How one gets the history of natural resources and their ecologies. There are over a dozen sources-each suggesting some aspect of past of some human settlements. They are of different types- Literatures, Arts and Scriptures, Ancient People, Comparative linguistics, Museums and Monuments, Traditional agricultural and other practices cultural anthropology abandoned habitats, roads, temples, Archaeology, Archives, ancient historical treaties, travelogues. with the aid of the above one can build the history of most of the resources of the past. Not withstanding enormous blanks of information these investigations expose, the quantum and types of knowledge the historical approach digs up is very valuable, because there is no other way to know these. Without these, yesterday - today - tomorrow continuum will be an incomplete story. Consequently our understanding of the present ecology and managing the futures will be fraught with holes, sometimes gaping breaches.

F. History From an Ecological Perspective

1. The urge of human species has always been to obtain resources as per his perceptions and keep on upgrading them. Thus originated human habitats on soils near water sources, domestication of plants & animals, developing tools and thus civilizations.

Everything has a story and everything a connection to everything else. First is history and the second, ecology, two vital parts of human wisdoms. While ecology incorporates eight resources in about 50 self-evolved units called ecosystems, history too has its vicissitudes. This coupling, unfortunately, does not find recognition in the current knowledge systems, calling for inputs.

2. The long story of humans - their settlements in suitable ecosystems, civilizations, dependence on resources, searching for lands, trades, military campaigns, naval powers, industrial revolution, science and Technology, colonization policies of Europe in the rest of the world - reveal a single fact. The efforts of humankind are in the unquenchable quest for resources embedded in the ecosystems functioning according to ecological principles. Therefore

all conflicts are for them, whether or not fought with weaponry and are inherently global and ecological.

Ancient Times: 1. The humans were nomads, Hunters, Gatherers, herders, settlers, domesticators, enslavers, etc. They evolved for millennia on all continents, climates and ecosystems, maintaining harmony with nature, without which the people would not have survived. Resources of nature consequently assumed the status of a great force of life.

2. Thus human populations have gradually flourished in numbers and began to occupy considerable territories denuding the nature for habitat, agriculture and other uses. Yet the ecosystems are hardly affected owing to their immensity. Every settlement gradually learnt of other settlements and their resources, some of which they did not possess. Desire to acquire them must have initiated the practice of exchange and trading.

No wonder therefore that ecological understanding existed in **Vedic times** and scriptures! **Hinduism, Buddhism, Christianity and Islam** occupied significant areas of the Planet as kingdoms and empires involving in trades, wars, victories, defeats, expansions, conversions, etc. And all were rich in ecological knowledge.

3. During these times of direct dependence on nature, humankind developed insights into the structures, functions and behaviors of natural systems, which may be of use in future. This habit improved and modified the utility of nature and methods of discovering them. These constituted the science and technology of the ancient times. Very little of it is documented, but lot of it is in evidence. Attempts to annex such settlements for their resourcefulness have been a legion until the middle of 15th century. Enslavement of human resources have been very common during these times. This trend continued until the advent of industrial revolution, the giant leap into the modern times.

Medieval Conflicts: 1. The onset of industrial revolution accelerated resource appropriation by humans. This period is the fore-runner of the modern times, characterized by trade routes, usurpation of others' lands and using science & technology for efficient exploitation of resources. Migrations of entrepreneurs and enslavement of commoners are the other features of these times. Creation of First World and Third World countries and the beginnings of transition from agricultural feudalism to industrial capitalism followed.

2. The human settlements, whether Kingdoms, Empires or Principalities, desired to improve their quality of life through the resources of others. This was possible only by trading or by conquests. Thus, there were trade routes from Europe to Asia and Africa and vice versa. Some of these routes were blocked by the enemy Kingdoms. Consequently new trade routes have to be established. This is how that the Atlantic countries had to resort to sea-faring.

3. The Atlantic countries- England, France, Holland, Belgium, Spain and Portugal - had launched new trade routes towards the Asiatic lands of plenty. In this enterprise, they discovered new lands - South America, North America and Africa. These lands have been owned up eventually by these maritime powers, due to their alluring resources. New lands caused migrations of people from Europe and domination over the natives. This domination and ownership had different mechanisms depending upon the state of development of the local societies.

4. For example, in Africa, the people could be easily enslaved due to their very low level of development. In North America, Genocide had to be resorted to. In South America, the natives were deep in the forests leaving the shore lines to the colonizers. But in Asia where societies were advanced, they had to employ a cunning mixture of market, military and missionary techniques. Thus the European nations have taken over a large part of the planet for resource exploitation and banking them for future.

All these lead to the formation of the First and the (the dominant) Second World (the dominated) countries. Further the principle of self-reliance (Autarkism), and barter are eclipsed by the rise of capitalism in the European colonial nations.

5. Maritime advancements and the advent of Industrial revolution bestowed tremendous advantages on the human resourcefulness in Europe. Through these advantages, the resources of three continents were owned up by a few European Nations. The remaining nations had to get along or compete which lead to conflicts among them. Consequently Europe determined the destinies of the most of the World's nations.

World War- I: 1. The conflicts for resources lead to a major war among European countries one group longing for colonial resources while the others holding a monopoly over them. Disparities thus borne among nations, percolated to people including the ruling classes. This simmering discontent has lead to political strifes and ultimately to the first all-European war, popularly called, World War- I. This war therefore was essentially between two groups of colonialists spatially limited to Europe but motivated for control over the world's resources. In this process however, some of the colonialists dragged gullible colonies. 2. After the war, the victors and vanquished came to an understanding on the colonies in Africa and on the Arctic. But the oppressed classes in some European countries turned against their own masters, so that they could develop their own resources. These victims of capitalism and industrial revolution have organized into an entirely new economic forces. This is the origin of the second world, borne shatteringly from the first one. This Second World, popularly known as the communist world, has provided for nearly 70 years an alternate form of development where one is not allowed to exploit the other.

World War- II: 1. This war was also between two groups of colonialists- At stake are the same global resources. Colonized people are also cunningly drawn into this. Everything is a repetition of the First World War, except for the magnitude and spread. Once again, the desire to share the colonial wealth having failed, Antarctica, high seas, sea beds and outer space were globalized as a compensation for the eventual loss of colonies. The Second World War taught that it is better to share rather than fight and that they can no longer control the resources of their colonies while rebuilding themselves. As a sequel to sharing principle, High seas and Antarctica were internationalized for transport and resources!

2. Progress in S & T in converting resources into commodities and the consequent economic wealth happened not only in the two groups of colonialists but also in the allied economies such as Canada, USA & Australia and the communistic Second World. But all these countries diverted considerable wealth developed from nature's resources for building military machine to destroy the perceived competitor. This is exclusion principle of Population Ecology in action. This is beginning of life boat ethics.

3. The USA has aided both the warring groups in the form of Marshall plan, except USSR. The decade re-building of these countries was accomplished chiefly due to S & T. The colonies in Africa and Asia mindful of their exploitation obtained freedom but looked towards USSR and the Second World countries for help in building their nations. Consequently a competition between the First and Second World countries to influence the Third World began and thus the seeds of future wars - neo-colonialism, cold wars, Environmental - were sown.

4. At last the world was divided into a first capitalistic world, a second communistic world and a third world of former colonies and vassals of the first world. The first one has greatly progressed while the second was lagging behind. The third world was looking for help and obtained through scientific, technological and trade exchanges with the developed countries. This initiation of exchanges, related to resources ultimately, is the beginning of a protracted war of neo-colonialism, the Third World War. A cold war between the First and Second world countries has become a necessary ingredient of the Third World War because of their desires to wean the former colonies into their areas of influence.

First and Second World Wars were thus waged between two groups of developed countries for hegemony over world's resources i.e. between colony holders and non holders.

World War- III: (1950-70s): 1. The Third World War beginning from the Marshal Plan and tapering off by the Limits to Growth Paradigm of Club of Rome was waged by the former powers - the first world, on their former colonies - the Third world, for continuing their affluence at the latter's expense through science, technology, education, Trade, Aid, Loans, debts, Judases and Quislings, political arm twisting, fomenting internal squabbles and skirmishes at he borders. Many local wars wastefully consumed the products of their military- industrial complexes. This is bleeding national wealth through open veins. Disarmament and peace became urgent needs for the third world- but always remained a mirages even now! While we lost resources to the First World countries by the above mechanisms and war-related payments, we could not develop enough because of faulty development models which bred local bandicoots. Thus it was a 4-front (all front) loss with the addition of degradative pollutants- pentaple (5-fold) trouble. By the time we realized, it was too late.

2. Pollution of the global resources, depletions and resource crunches, over population and ill health, imbalance between and within countries, irrevocability of the bad impacts even on the exploiter, have all come to stay. Enormous development and consumerism on one hand, divisions, conflicts, hatreds and violence on the others, came to co-exist! Most current violences - like the historical ones - are thus basically ecological.

World War-IV (1972-92): 1. The realization stated above lead to the Stockholm Conference of all nations, in 1972, which was rechristened at Rio de Janeiro Conference in 1992. This period, called the World War- IV, is a chronicle of righteous battles against self -societies and governments for fully realizing the objectives of the third war (1950-72) before late. The battles against themselves in all countries (or most) were to self-regulate because all were getting endangered- vanquished as well as the victor, consumer as well as the hungry, the developed as well as the un-developing, knowledgeable as well as the ignorant.

2. Governments, Citizens, private and public agencies, are to make efforts to stop further degradation of the planet. It is however significant to note that the third war mechanisms continue side by side; Consumerism and over population are not diminishing despite the accent on recycling, alternative resources and materials, better science- better Technology- Awareness- Education- communication - public relations.

For want of the colonial fixed deposits, excessive spending on defense sector, unsustainable philanthropy towards the then third world and capitalistic blundering on ecology, USSR broke down. Thus it is ecology that felled Soviet Union, followed by others like Yugoslavia, and threatening others like India. Are there seeds of a final war here settling who is running where and how, for the rest of the planetary life ?

3. Meanwhile Capitalism continued its inexorable march from the village, to a locality, to a region, to a nation, to a country, to a group of countries, to a zone in the world and finally to the entire globe. And it has come to stay by privatizing everything and liberalizing regulations. And this was carried out

by all political forces irrespective of ideologies! The mechanisms and players here wage the next war for or against their own people- The World War- V.

World War -V (1992-until now): 1. The Fifth World War is ongoing within each country, each regional group, each economic group and among all these. Features of all the four earlier wars are making merciless occasional reappearances pitting the neighbour against the neighbour. Thus there are SEVEN NEIGHBOURS to contend with- The First World (G7+), Second World (the Nova Rich & part of erstwhile first world), Third World (formerly Marxist Europe), Fourth World (New emergences like China & ASEAN), Fifth World (G15 like India), Sixth World (LDC whose product equals debt) and Seventh World (whose product equals interest on the debt) countries. It is noteworthy that there are '***Souths in each North Country*** **(First, Second, and Third Worlds)** and ***Norths in each South Country*** (Fourth, Fifth, Sixth, and Seventh Worlds). A war of attrition, ultimately settling for a triaged existence.

2. Resources just do not exist for all. They can be amplified through Science and Technology for which good education is the foundation. Products of such Human resourcefulness can help a nation only if the interactions among people and nature are balanced, i.e. ecological and not consumeristic. If time-honoured ecology and human-developed education are uncorrected, where is the reason to hope. But efforts in this direction continue.

F. Historical Sources & Ecological Inferences

1. Archeology and other historical tools- Excavations of Indus, Maya, Inca, Aztec, Egypt, Mesopotamian, Chinese, Abyssinian and other civilizational sites, Museums, Monuments, Arts- Sculptures, Numismatics, Stamps, Abandoned habitats, Roads, Temples, Tanks, Forts- are useful informers of bygone Human cultures and their ecologies.

2. Human Settlements, their ecology, diversities, and traditions are discerned from an anthropological study of natives of Africa, Australia, South and North Americas, India, Philippines, etc, as they are vestiges of past. Professions such as agriculture and others reveal divisions of labours, Classes, castes, Traditions, dominions over the workers and women and are suggestive of their appropriation of Resources

3. Literatures in the Libraries and Archives contribute recorded history of all experiences, such as Religious literatures - their philosophies and perceptions on nature; Literary works from ancient times in Sanskrit, Tamil, Greek, Latin, Pali reveal the then environmental consciousness including in Vedas, Kalidasa, Sri Harsha, Shakespeare, etc; Arts Dance and Music indicate life in the human settlements; Travelogues record, environments long gone; Colonial History tells us the governance models and practices of altering and trading in the then resources - forests, water, agriculture and human settlements

4. Resistances to injustices : Revolts, Rebellions and Freedom Movements reveal the conflicts for resources between the natives and the rulers.

5. Military exercises : Wars and campaigns cause not only concentrations and migrations of people but also devastation of resources. Routes of Campaigns are a great source of clues for habitat restoration through Paleobotany.

6. Extant civilizations: While all civilizations fell and replaced newer ones, Indian and Chinese from antiquity- source of most ancient and reliable information

G. Compendium of Topics

1. From the above, it is evident that social sciences and humanities played greater positive roles than Sciences. Yet why are these subjects recognized as early as late Sixties and early Seventies still are nowhere in the Indian sub-continent touted as the Upcoming Knowledge Society? The leaders of Certificates and Degrees, were either ignorant of history of curricula, or did not bother about them or lazy to work or more serious with power than students' future. A list is offered here as a base work on further. It embraces the disciplines of Philosophy, Politics, Economics, Sociology, Social Work, Anthropology and constitute the bulk of Human Ecology and History. Due to overlap sequentiality could not be maintained always.

H. Conclusion

A Seminar to Review Curricula and make Recommendations to the Knowledge commission is already overdue. A fresh look at literatures in humanities and social sciences pertaining to ecology and environment, to revise curricula, will be in defense of knowledge society..

In science it often happens that scientists say, 'You know that's a really good argument; my position is mistaken,' and then they would actually change their minds and you never hear that old view from them again. They really do it. It doesn't happen as often as it should, because scientists are human and change is sometimes painful. But it happens every day. I cannot recall the last time something like that happened in politics or religion. (1987) – Carl Sagan

Table 35.1.
Student Assignments

1.	Ancient wisdoms on nature, Native people, Archives and scriptures, Archaeology,
2.	Religions on ecology and environment
3.	Scientific world view, Colonialism and World Wars I and II
4.	World Wars III, IV and the V - Neocolonialism and Neo-fascism - The Developed North and the Undeveloping South - Rainbow Warriors.
5.	Doomists, Boomists and the Existentialism, Triage and Philosophies on Man's
	existence and future - Noah's Ark, Spaceship, Gaia and Deep, Club of Rome.
6.	The prophets and their stands- Marx & Gandhi, Brown & Sartre, Hardin & Paddock, Ward, Simon & Kahn, Lovelock & Margulis, Huxley, Stevenson, Boulding, Ehrlich, Commoner.
7.	Green philosophies and People's movements
8.	Global common future vs. local tragedies of the commons, Judaism of the global north and Quislings of the local north.
9.	Religious world view vs. Scientific world view vs. Ecological world view - Religions Marx and Gandhi form the Environmentalism - Socio-ecology - Commonsense II?
10.	Philosophical Considerations of Ecology
11.	Historical Implications Ecology and Development : Past, Present and Futures
12.	Economic Aspects of Environment : W T O Regimes
13.	Sociological Considerations : Divisions and Conflict
14.	Legislative and Legal Aspects of Conserving Environment
15.	Political Policies and Mechanisms Pertaining to Environmental issues
16.	Science, Technology and Agriculture : Politics of Global biodiversity
17.	International Relations : perspective of Global and regional conflicts
18.	International Efforts: Ecology, Politics and violence : feedbacks and Backlashes
19.	Cr - Unche - Unep - Challenges And Responses -higher Education - Governments
20.	Regional & Local Environments, Environmental Resistances, Judicial Courts, Ngos

Table 35.2.
Student Assignments - Global Issues

Cuban Sugarcane	Cold War, Technology, Science
Ecological Espionage	Vietnam War
Ghanian Cocoa	Japan's Eco-fascism
Crop failure in USSR	North Sea oil
Brazilian forests	Politics of Climate
Oil resources in USA	Protectorates: Papua, Namibia
Zairean copper	Antarctic, Arctic, Seabed,
Metal Stocking in USA	Mineral Technology
Water stoppage by Turkey	Revolution and forests in S. America
Oil Technology - Trans Siberian Pipe line	Rhine -Maine -Danube
Water pollution by Germany	Vietnam forests - Imperialism
Consequences of Air Pollution by USA, Europe	Forest fires of Indonesia - Malaysia

Table 35.3.
Student Assignments - Regional Issues

Indian Cow - Islamic Meat!
GM Crops
GATT and Dunkel Draft
Civic Environmentalism : Local
People, Local Government,
Regional Industry, Regional
Academics,Regional Forestry,
Drainage, tanks and Rainfall loss
Ecological rights and Naxalism
Criminal tribes and forest cutting
Gharwal forests & chipko
Narmada and other river Valley projects
PL48 and Indian Agriculture
Regional Ecosystems

Presented at a Conference on Environment, Social sciences and History at CBR Foundation Chennai, 2006

APPENDICES

Chapter - 36

Forebodings of A Native American Chief

"End of Living and the 'Beginning of Survival

Prof. Krishna Murty, his wife and son, Calicut University

Protest and Prophecy
The First Voice

The Earth is our home....
the only one we will ever have,
the only one our children will.
Will they inherit a barren, hostile,
and dead planet on the morrow?
In parting, let's imbibe the splendour
that is the Earth, the magic that is our home .

In **1854**, the "**Great White Chief in Washington**" made an offer for a large area of Indian land and promised a **'reservation'** for the Indian people.

Chief Seattle's reply,
has been described as the most
beautiful and profound statement
on the environment ever made.
How can you buy or sell the sky, the warmth of the land?
The idea is strange to us.
If we do not own the freshness of the air
and the sparkle of the water,
how can you buy them?

All Sacred

Every part of this Earth is sacred to my people.
Every shining pine needle,
every sandy shore,
every mist in the dark woods,
every clearing and hum-ming insect is holy
in the memory and expe-rience of my people.
The sap which courses through the trees
car-ries the memories of the red man.
The white man's dead forget the country of their birth
when they go to walk among the stars.
Our dead never forget this beautiful Earth,

for it is the mother of the red man.
We are part of the Earth
and it is part of us.
The perfumed flowers are our sisters;
the deer, the horse, the great eagle,
these are our brothers.
The rocky crests,
the juices in the meadows,
the body heat of the pony and man
all belong to the same family.

Not Easy

So, when the Great Chief in Washington sends word that he wishes to buy our land, he asks much of us. The Great Chief sends word he will reserve us a place so that we can live comfortably to ourselves. He will be our fa-ther and we will be his children. So we will consider your offer to buy our land.

But it will not be easy. For this land is sacred to us.
This shining water that moves in the streams and rivers
is not just water but the blood of our ancestors.
If we sell you land, you must remember that it is sacred,
and you must teach your children that it is sa-cred and
that each ghostly reflection in the clear water of the lakes
tells of events and memories in the life of my people.
The water's mur-mur is the voice of my father's father.

Kindness

The rivers are our brothers, they quench our thirst.
The rivers carry our canoes, and feed our children.
If we sell you our land, you must remem-ber,
and teach your children, that the rivers are our brothers,
and yours, and you must henceforth give the rivers
the kindness you would give any brother.
We know that the white man does not under-stand our ways.
One portion of land is the same to him as the next,

for he is a stranger who comes in the night and
takes from the land whatever he needs.

The Earth is not his brother, but his enemy,
and when he has conquered it, he moves on.
He leaves his father's graves behind,
and he does not care.
He kidnaps the Earth from his children,
and he does not care.

His father's grave, and his children's birthright, are forgotten.

He treats his mother, the Earth,
and his brother, the sky, as things to be bought,
plundered, sold like sheep or bright beads
His appetite will devour the Earth
and leave behind only a desert.
I do not know.

Our ways are different from your ways
The sight of your cities pains the eyes of the red man.
But perhaps it is because
the red man is a savage and does not understand.

There is no quiet place in the white man's cities.
or the rustle of an insect's wings.
But perhaps it is because
I am a savage and do not understand.

The clatter only seems to insult the ears.
And what is there to life if a man cannot hear
the lonely cry of the whip-poor will or
the arguments of the frogs around a pond at night?
I am a red man and do not understand.

The Indian prefers the soft sound of the wind
darting over the face of a pond, and the smell of the wind itself,
cleaned by a midday rain, or scented with the Pinon pine

Precious

The air is precious to the red man,
for all things share the same breath ,
the beast, the tree, the man,
they all share the same breath.

The white man does not seem to notice the air he breathes.
Like a man dying for many days, he is numb to the stench.

But if we sell you our land,
you must remember that the air is precious to us,
that the air shares its spirit with all the life it supports.
The wind that gave our grandfather
his first breath also receives his last sigh.

And if we sell you land, you must keep it apart and sacred,
as a place where even the white man can go to taste the wind
that is sweetened by the meadow's flowers.

One Condition

So we will consider your offer to buy our land.
If we decide to accept, I will make one condition:
The white man must treat the beasts of this land as his brothers.

I am a savage and I do not understand any other way.
I have seen a thousand rotting buffaloes on the prairie,
left by the white man who shot them from a passing train.

I am a savage and I do not understand
how the smoking iron horse can be more important
than the buffalo that we kill only to stay alive.
What is man without the beasts?
If all the beasts were gone,
man would die from a great loneliness of spirit.
For whatever happens to the beasts,
soon happens to man.
All things are connected.

The Ashes

You must teach your children
that the ground beneath their feet
is the ashes of your grandfathers.
So that they will respect the land,
tell your children that the Earth
is rich with the lives of our kin.

Teach your children
what we have taught our children,
that the Earth is our mother.
Whatever befalls the Earth
befalls the sons of the Earth.
If men spit upon the ground,
they spit upon themselves.

This we know:
The Earth does not belong to man;
man belongs to the Earth.

This we know.
All things are connected like the blood .
which unites one family.

All things are connected.
Whatever befalls the Earth befalls the sons of the Earth.
Man did not weave the web of life; he is merely a strand in it.
Whatever he does to the web, he does to himself.
Even the white man, whose God walks and talks
with him as friend to friend, cannot be exempt from
the common destiny. We may be brothers after all. We shall see.

One thing we know,
which the white man may one day discover,
our God is the same God.
You may think now that you own Him
as you wish to own our land; but you cannot.

He is the God of man, and His compassion is equal
for the red man and the white.

This Earth is precious to Him,
and to harm the Earth
is to heap contempt on its Creator.
The whites too shall pass;
perhaps sooner than all other tribes.
Contaminate your bed and
you will one night suffocate
in your own waste.
But in your perishing
you will shine brightly.
Fired by the strength of the God
who brought you to this land and
for some special purpose gave you dominion
over this land and over the red man.

That destiny is a mystery to us, for we do not understand
when the buffalo are all slaughtered, the wild horses are tamed,
the secret corners of the forest heavy with scent of many men, and
the view of the ripe hills blotted by talking wires.

Where is the Thicket? Gone.
Where is the Eagle? Gone
The End of living and
the Beginning of Survival

1854

Chapter - 37

Concens of the United Nations Taking a Stand

Shareef, Pondichery University

Declaration of the United Nations Conference on the Human Environment 21st Plenary Meeting

The United Nations Conference on the Human Environment, having met at Stockholm from 5 to 16 June 1972, having considered the need for a common outlook and for common principles to inspire and guide the people of the world in the preservation and enhancement of the human environment.

Proclaims that: 1. Man is both creature and moulder of his environment, which gives him physical sustenance and affords him the opportunity for intellectual, moral, social and spiritual growth. In the long and tortuous evolution of the human race on this planet a stage has been reached when, through the rapid acceleration of science and technology, man has acquired the power to transform his environment in countless ways and on an unprecedented scale. Both aspects of man's environment, the natural and the man-made, are essential to his well-being and to the enjoyment of basic human rights the right to life itself.

2. The protection and improvement of the human environment is a major issue which affects the well-being of people and economic development throughout the world; it is the urgent desire of the people of the whole world and the duty of all Governments.

3. Man has constantly to sum up experience and go on discovering, inventing, creating and advancing. In our time, man's capability to transform his surroundings, if used wisely, can bring to all people the benefits of development and the opportunity to enhance the quality of life. Wrongly or heedlessly applied, the same power can do incalculable harm to human beings and the human environment. We see around us growing evidence of man-made harm in many regions of the Earth: dangerous levels of pollution in water, air, Earth and living beings; major and undesirable disturbances to the ecological balance of the biosphere; destruction and depletion of irreplaceable resources; and gross deficiencies, harmful to the physical, mental and social health of man, in the man-made environment, particularly in the living and working environment.

4. In the developing countries most of the environmental problems are caused by under-development. Millions continue to live far below the minimum levels required for a decent human existence, deprived of adequate food and clothing, shelter and education, health and sanitation. Therefore, the developing countries must direct their efforts to development, bearing in mind their priorities and the need to safeguard and improve the environment. For the same purpose, the industrialized countries should make efforts to reduce the gap themselves and the developing countries. In the industrialized countries, environmental problems are generally related to industrialization and technological development.

5. The natural growth of population continuously presents problems for the preservation of the environment, and adequate policies and measures should be adopted, as appropriate, to face these problems. Of all things in the world, people are the most precious. It is the people that propel social progress, create social wealth, develop science and technology and, through their hard work, continuously transform the human environment. Along with social progress and the advance of production, science and technology, the capability of man to improve the environment increases with each passing day.

6. A point has been reached in history when we must shape our actions throughout the world with a more prudent care for their environmental

consequences. Through ignorance or indifference we can do massive and irreversible harm to the earthly environment on which our life and well being depend. Conversely, through fuller knowledge and wiser action, we can achieve for ourselves and our posterity a better life in an environment more in keeping with human needs and hopes. There are broad vistas for the enhancement of environmental quality and the creation of a good life. What is needed is an enthusiastic but calm state of mind and intense but orderly work. For the purpose of attaining freedom in the world of nature, man must use knowledge to build, in collaboration with nature, a better environment. To defend and improve the human environment for present and future generations has become an imperative goal for humankind-a goal to be pursued together with, and in harmony with, the established and fundamental goals of peace and of worldwide economic and social development.

7. To achieve this environmental goal will demand the acceptance of responsibility by citizens and communities and by enterprises and institutions at every level, all sharing equitably in common efforts. Individuals in all walks of life as well as organizations in many fields, by their values and the sum of their actions, will shape the world environment of the future.

Local and national governments will bear the greatest burden for large-scale environmental policy and action within their jurisdictions. International cooperation is also needed in order to raise resources to support the developing countries in carrying out their responsibilities in this field. A growing class of environmental problems, because they are regional or global in extent or because they affect the common international realm, will require extensive cooperation among nations and action by international organizations in the common interest.

The Conference calls upon Governments and people to exert common efforts for the preservation and improvement of the human environment, for the benefit of all the people and for their posterity.

Principles

States the common conviction that:

Principle 1: Man has the fundamental right to freedom, equality and adequate conditions of life, in an environment of a quality that permits a life of dignity and well-being, and he bears a solemn responsibility to protect and improve the environment for present and future generations. In this respect, policies promoting or perpetuating apartheid, racial segregation, discrimination, colonial and other forms of oppression and foreign domination stand condemned and must be eliminated.

Principle 2: The natural resources of the Earth, including the air, water, land, flora and fauna and especially representative samples of natural ecosystems, must be safeguarded for the benefit of present and future generations through careful planning or management, as appropriate.

Principle 3: The capacity of the Earth to produce vital renewable resources must be maintained and, wherever practicable, restored or improved.

Principle 4: Man has a special responsibility to safeguard and wisely manage the heritage of wildlife and its habitat, which are now gravely imperiled by a combination of adverse factors. Nature conservation, including wildlife, must therefore receive importance in planning for economic development.

Principle 5: The non-renewable resources of the Earth must be employed in such a way as to guard against the danger of their future exhaustion and to ensure that benefits from such employment are shared by all humankind.

Principle 6: The discharge of toxic substances or of other substances and the release of heat, in such quantities or concentrations as to exceed the capacity of the environment to render them harmless, must be halted in order to ensure that serious or irreversible damage is not inflicted upon ecosystems. The just struggle of the people of ill countries against pollution should be supported.

Principle 7: States shall take all possible steps to prevent pollution of the seas by substances that are liable to create hazards to human health, to harm living resources and marine life, to damage amenities or to interfere with other legitimate uses of the sea.

Principle 8: Economic and social development is essential for ensuring a favorable living and working environment for man and for creating conditions on Earth that are necessary for the improvement of the quality of life.

Principle 9: Environmental deficiencies generated by the conditions of under-development and natural disasters pose grave problems and can best be remedied by accelerated development through the transfer of substantial quantities of financial and technological assistance as a supplement to the domestic effort of the developing countries and such timely assistance as may be required.

Principle 10: For the developing countries, stability of prices and adequate earnings for primary commodities and raw materials are essential to environmental management, since economic factors as well as ecological processes must be taken into account.

Principle 11: The environmental policies of all States should enhance and not adversely affect the present or future development potential of developing countries, nor should they hamper the attainment of better living conditions for all, and appropriate steps should be taken by States and international organizations with a view to reaching agreement on meeting the possible national and international economic consequences resulting from the application of environmental measures.

Principle 12: Resources should be made available to preserve and improve the environment, taking into account the circumstances and particular requirements of developing countries and any costs which may emanate- from their incorporating environmental safeguards into their development planning and the need for making available to them, upon their request, additional international technical and financial assistance for this purpose.

Principle 13: In order to achieve a more rational management of resources and thus to improve the environment, States should adopt an integrated and coordinated approach to their development planning so as to ensure that development is compatible with the need to protect and improve environment for the benefit of their population.

Principle 14: Rational planning constitutes an essential tool for reconciling any conflict between the needs of development and the need to protect and improve the environment.

Principle 15: Planning must be applied to human settlements and urbanization with a view to avoiding adverse effects on the environment and obtaining maximum social, economic and environmental benefits for all. In this respect projects which arc designed for colonialist and racist domination must be abandoned.

Principle 16: Demographic policies which are without prejudice to basic human rights and which are deemed appropriate by Governments concerned should be applied in those regions where the rate of population growth or excessive population concentrations are likely to have adverse effects on the environment of the human environment and impede development.

Principle 17: Appropriate national institutions must be entrusted with the task of planning, managing or controlling the nine environmental resources of States with a view to enhancing environmental quality.

Principle 18: Science and technology, as part of their contribution to economic and social development, must be applied to the identification, avoidance and control of environmental risks and the solution of environmental problems and for the common good of humankind.

Principle 19: Education in environmental matters, for the younger generation as well as adults, giving due consideration to the underprivileged, is essential in order to broaden the basis for an enlightened opinion and responsible conduct by individuals, enterprises and communities in protecting and improving the environment in its full human dimension. It is also essential that mass media of communications avoid contributing to the deterioration of the environment, but, on the contrary, disseminates information of an educational nature on the need to project and improve the environment in order to enable mal to develop in every respect.

Principle 20: Scientific research and development in the context of environmental problems, both national and multinational, must be promoted in all countries, especially the developing countries. In this connection, the free flow of up-to-date scientific information and transfer of experience must be supported and assisted, to facilitate the solution of environmental problems; environmental technologies should be made available to developing countries on terms which would encourage their wide dissemination without constituting an economic burden on the developing countries.

Principle 21: States have, in accordance with the Charter of the United Nations and the principles of international law, the sovereign right to exploit their own resources pursuant to their own environmental policies, and the responsibility to ensure that activities within their jurisdiction or control do not cause damage to the environment of other States or of areas beyond the limits of national jurisdiction.

Principle 22: States shall cooperate to develop further the international law regarding liability and compensation for the victims of pollution and other environmental damage caused by activities within the jurisdiction or control of such States to areas beyond their jurisdiction.

Principle 23: Without prejudice to such criteria as may be agreed upon by the international community, or to standards which will have to be determined nationally, it will be essential in all cases to consider the systems of values prevailing in each country, and the extent of the applicability of standards which are valid for the most advanced countries but which may be inappropriate and of unwarranted social cost for the developing countries.

Principle 24: International matters concerning the protection and improvement of the environment should be handled in a cooperative spirit by all countries, big and small, on an equal footing.

Cooperation through multilateral or bilateral arrangements or other appropriate means is essential to effectively control, prevent, reduce and eliminate adverse environmental effects resulting from activities conducted in all spheres, in such a way that due account is taken of the sovereignty and interests of all States.

Principle 25: States shall ensure that international organizations play a coordinated, efficient and dynamic role for the protection and improvement of the environment.

Principle 26: Man and his environment must be spared the effects of nuclear weapons and all other means of mass destruction. States must strive to reach prompt agreement, in the relevant international organs, on the elimination and complete destruction of such weapons.

16 June 1972

Chapter - 38

Vintage Wisdoms: A Bibliography 1990

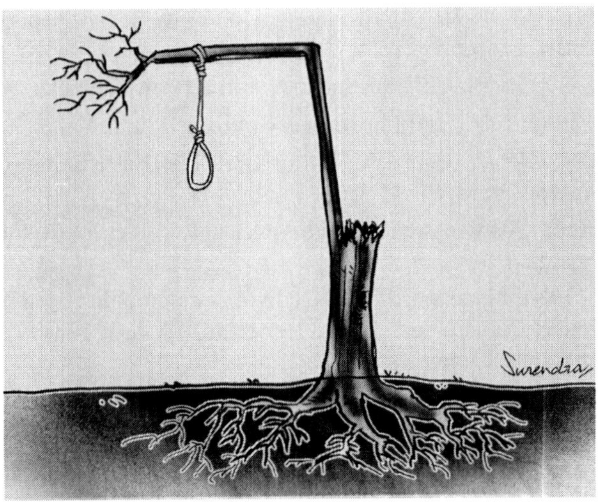

Where the mind is without FEAR
And the head is held high
Where the KNOWLEDGE is free
....
Where the words come out
From the depth of TRUTH
....
Into that heaven of FREEDOM,
Let my country AWAKE

Rabindanath Tagore

American Association for Advancement in Science 1976. Materials: Renewable and Nonrenewable Resources AAAS Washington, D.C.

Academy of Natural Sciences, 1977. The changing scene in Natural Sciences 1797-1976, Acad. Nat. Sci.

Agarwal, A 1981. Mud, Mud - the potential of earth based materials for third world housing, Earth Scan.

Agarwal, A, D. Monte D and Samarthu 1987. The Fight for Survival. CSE, New Delhi, p361

Agarwal, G 1986. Col hearths and barren slopes - the wood fuel crisis in the third world, Allied Publ.

Allsopp B, 1972. Ecological Morality, Muller

Ananichev K.V. 1976. Environment: International Aspects, Progress Publishers, Moscow

Anderson, E 1952. Plants, Man and Life. Little Brown and Co., Boston.

Ardery R, 1967. Territorial Imperative. Collins.

Arshur D.R. 1971. Survival: Man and his Environment. English Universities Press.

Borrelli, P(ed) 1988 Crossroads: Environmental Priorities for the Future.Island Press. p. 52.

Bouvier L, 1984. Planet Earth 1984-2034: A demographic vision Populat-ion Reference Bureau, Inc. Vol. 39 and No.1 F.A.O 1981. Agriculture: Toward 2000 Rome.

Budyko MI 1982. The Earth's Climate: Past and Future.International Geophysics Series Vol.29. pp.307. Academic Press London.

Box TW 1978. The Arid Lands Revisited. 100 years since John Wesley Powell. Honor Lecture, Utalh State Univ. pp. 30.

Bach R, 1977. Illusions: the adventures of a reluctant messiah, Pan, London.

Barbour, I G. 1972. Earth May be fair: Reflections on Ethics, Religion and Ecology. Prentice Hall, Englewoood cliffs, N.J.

Barnet, R, 1980. The lean years: Politics in the age of Scarcity. Simon and Schuster, New York.

Barney, GO (ed) 1982. The Global 2000 Report to the president.Penguin Books, Hammonds worth, U.K

Baker H, 1970. Plants and Civilization, Wadsworth Pub. Co. Belmont, California

Balfour EB 1975. The Living Soil, Faber and Faber, London.

Barbour IG 1972. Earth may be fair: Reflections on Ethics Religion and Ecology, Englewood Cliffs, USA

Barnet R, 1980. The lean years: Politics in the age of scarcity. Simon and Schuster, New York.

Barnett, HJL, Morse C 1963. Scarcity and Growth: The economics of Natural Resources Availability, Johns Hopkins Prss, Baltimore.

Barney GO (ed) 1982. The Global 2000 Report to the President Penguin Books Ltd, dsworth, UK.

Batchelder, RB, Hirt HF 1966. Fire in Tropical Forests and Grasslands, U.S. Army Natick Laboratories

Baumol and Dates WE 1975. The Theory of Environmental Policy, Prentice Hall, Englewood Cliff, NJ,

Behrman D 2973. In partnership with Nature - UNESCO and the Environment. UNESCO, Paris

Bergstrom G 1950. Too many, Macmillan, New York.

Bergstrom G 1967. The Hungry Planet, Collier Books, New York.

Bergstron G 1972. The Hungry Planet, Macmillan, New York.

Berry W 1977. Unsetting of America: Culture and Agriculture, Sierra Club, San Francisco

Bertell R, 1985. Immediate Danger: Prognosis for a Radioactive Earth, The Women's press, London

Behnke JA, 1972 Challenging Biological Problems, Oxford Univ. Press, New York, pp.502, $8.70

Birch C, 1976. Confronting the Future, Penguin, New York.

Bishnoi, R.S. 1987. Conservation as Creed, Jugal Kishore, Dehradun

Bolin B et al 1986. The Green House Effect: Climate Change and Ecosystems, John Wiley and Sons, Chichester, UK.

Bookchin, M. 1974 Our synthetic environment, Harper & Row

Bookchin M 1980. Towards an Ecological Society Black Rose Books, Montreal.

Box TW 1978. The Arid Lands Revisited. 100 years since John Wesley Powell. Honor Lecture, Utah State Univ. pp.30

Borrelli, P (ed) 1988 Crossroads: Environmental Priorities for the Future. Island Press. pp.352.

Boulding K 1965. The Meaning of the 20th century, Allen & Unwin.

Bouvier L, 1984. Planet Earth 1984-2034: A demographic vision. Population Reference Bureau, Inc. Vol.39 and No.1

Boyte H 1984. Community is Possible. Repairing America's Roots Harper & Row, New York

Braidwood, RJ 1967. The Preshistoric Man, Scolt & Foreman, Glenview, Illinois.

Bresh, D Hershner, CH, Milgram JH 1974. Oil Spills and the Marine Environment, Ballinger, Cambridge.

Bridges EL 1949 Uppermost part of the Earth, E.P. Dutton Co New York, pp.558.

Brooker R and Corder M. Environmental Economy, E & F.N. Spon Ltd., pp. 224.

Brockington F 1958. World Health, Pelican, pp. 405.

Brown LR, Binsterbasch GW 1972. Man and His Environment: Food, Harper and Row, New York.

Brown LR and Eckholm EP 1974 By Bread Alone, Praeger, New York.

Brown LR 1976. World Population Trends. Signs of Hope, Signs of stress, World Watch Institute, Washington.

Brown LR 1976. Twenty-two Dimensions of the population problem, World Watch Institute, Washington

Brown LR 1978. The Twenty-ninth Day, WW. Norton, New York.

Brown LR 1978. The world wide loss of cropland, Worldwatch Institute, Washington.

Brown LR 1978 The Global Economic Prospect: New Signs Economic Stress, World Watch Institute, Washington.

Brown LR 1981 Building a sustainable Society, W.W. Norton, New York.

Brown, MB 1976. Resources and the Environment: A Socialist Perspective, Spokesman Books, Nottingham

Brown M 1980. Laying Waste: The poisoning of America by toxic wastes Pantheon Books, New York.

Bryant J 1969. Health and the Developing world. Cornell, Ithaca, N.Y.

Bryson RA, Murray TJ 1977. Climates of Hunger, Univ. Wisconsin Press, Madison.

Burton I et al 1978. The Environment as hazard, Oxford Univ. Press, New York

Budyko MI 1982. The Earth's Climate : Past and Future. International Geophysics Series Vol.29. pp.307. Academic Press, London.

Burroughs J, The ways of Nature, Houghton, Mifflin Co Calder N (ed) 1964. The World in 1984 (Vol 1 & 2), Pelican Books

Calder N 1973. Nature in the Round: A guide to Environment Science, Weidenfeld & Nicholson, London.

Callenbach EJ. 1975. Ecotopia, Bantam, pp. 212.

Callenbach EJ Living poor with style.

Carson, RL 1962. Silent spring, Houghton and Mifflin Co. Greenwich, Connn.

Carmody J 1983. Ecology and Religion: Toward a New Christian Theology of Nature. Paulist Press, New York.

Carson RI 1941. Under the Sea Wind. Oxford Univ. Press.

Carson RL 1951. The Sea around us, Oxford Univ Press.

Carson RL 1962. Silent Spring, Houghton, Mifflin Co Fawcett Books, Greenwich, Conn. USA.

Caufield C 1982. Tropical Moist forests. The Resources, The People, The Threat, Earthscan.

Caufield C, 1986 In the Rain Forest, Picador, UK, pp.304.

Caudill HM 1863 Night comes to the Cumberlands: Biography of Depressed area, Little, Brown, Boston.

Chisholm A, 1972 Philosophers of the Earth, Sidgwick and Jackson, London pp. 201.

Clarke R and Timberlake W 1982. Stockholm plus Ten: Promises, Promises? The decade since the 1972 UN Environment Conference Earthscan.

Cobbett W 1930. Rural Rides, Penguin, Hammondsworth, UK

Cody ML, Diamond JM (ed) 1975. Ecology and Evolution of Communities, Harvard Univ Press, Cambridge, Mass

Cole H et al 1973. Thinking about the Future: a Critique of the Limits to growth, Sussex Univ. Press.

Cole J 1976. The Poor of the Earth. Westview Press, Boulder, Co, USA.

Coleman-Cooke J 1965. The Harvest that kills Odhams, London, pp.208.

Collis JS 1973. The Vision of Glory. The extrordinary nature of the ordinary, Braziller, New York.

Commoner B, 1971/1972. The Closing Circle. Confronting the Environmental Crisis, Jonathan Cape, London/Alfred Knopf, New York.

Commoner B 1979. The Politics of Energy. Alfred Knopf, New York.

Conservation Foundation 1986. Agriculture and the Environment in a changing World Economy. The Conservation Foundation, Washington, D.C.

Cook RC 1951. Human Fertility: the Modern dilemma, William Sloane Associates, New York.

Cotgrove S, 1982. Catastrophe or Cornucopia. The environment, Politics and Future Wiley.

Council on Environmental Quality, 1971 Toxic Substances, U.S. Govt Printing Office

Council on Environmental Quality 1981. Global Future: Time to Act, US govt. Printing office, Washington D.C.

Critchfield R 1981. Villages,Anchor Press/Doubel Day Garden city, N.Y.

Cuny FC 1983 Disasters and Development, Oxford Univ. Press

Dales JH 1968 Pollution, Property and Prices, Univ. of Toronto Press, Toronto, Canada Dale T and Carter VG 1974. Topsoil and civilisation, University of Oklahma Press, USA

Darwin C 1845. The voyage of the Beagle, Dent & Sons, pp. 486.

Darwin C 1859. On the origins of Species by means of Natural Selection or the preservation of Favored races in the struggle for life, John Munnay, London.

Dasmann RF 1972 Planet in Peril? Man and the Biosphere Today, Penguin.

Dasmann FF Milton JP and Freeman PH 1973. Ecological Principle for Economic Development, Wiley, New York.

Dasmann RF 1976. Environmental Conservation, Wiley, New York

Davies JC III, Davies B 1978. The Politics of Pollution Bobbs-Merill, Indianapolis, USA

Devall B and Sessions G 1984. Deep Ecology. Peregme Smith Press, Layton, Utah.

Disch R (ed) 1970. The Ecological Conscience: Values for Survival Prentice Hall , Engleword Cliffs, NJ.

Dorst J 1970 Before Nature Dies, Collins

Douglas W 1972 The Three hundred year war, New York.

Dubos R. Man Adapting.

Dubos R. So human an Animal

Dubos R. Reason Awake

Dumont R, Bernard R 1969. The Hungry Future, Preager, New York.

Eckholm EP & Harriman R 1971. How to be a Survivor Hallantine, London.

Eckholm E and Record F 1975. The two faces of Malnutrition World Watch Institute, Washington Eckholm EP 1975. The other energy crisis: Firewood Worldwatch Institute, New York.

Eckholm E 1976. The Picture of Health: Environmental Sources of Disease, W.W. Norton, New York.

Eckholm E.P. 1976. Loosing ground, Pergamon Press

Eckholm E 1978 Disappearing Species: The Social Challenge Worldwatch Institute, Washington

Eckholm E 1979. The Dispossessed of the Earth: Land Reform and the Sustainable Development, World Watch Institute, Washington

Eckholm E.P 1982 Down to Earth: Environment and Human Needs Pluto Press, London / W.W. Norton, New York.

Ehrlich PR 1968 Population Bomb, Ballantine Books

Ehrlich PR and Ehrlich AH 1972. Population, Resources and Environment W.H. Freman Ehrlich PR, Ehrlich AH, Holdren JP. 1973 Human Ecology: Problems and Solution, W.H. Freeman, San Fransciso

Ehrlich PR and Holden JP 1977. Ecoscience: Population, Resources, Environment, W.H. Freeman, San Francisco.

Ehrlich PR, Ehrlich AH 1981. Extinction, Random House, New York.

El-Hinnvai EE 1981. Environmental Impacts of Production and use of Energy, Tycooly Press, Dublin, pp.322.

El-Hinnawi E. 1955. Environmental Refugees, UNEP, Nairobi

Eiseley L, 1985. The Immense Journey, Random House

Eisenbud, M 1963. Environmental Radioactivity, McGraw Hill, New York.

Frakes LA 1979. Climates through out Geologic time pp.261. Elsevier, Amsterdam Ickes, HL, et al 1934. National Resources Board. U.S. Government Printing Office. Washington D.C. Kuch P. 1988. Natural Resources for the 21st Century. EPA department, Washington D.C. 125.pp..

Falk RA 1971. This Endangered Planet: Prospects and Proposals for Human Survival, Random House, New York.

F.A.O 1981. Agriculture: Toward 2000 Rome.

Farvar,MT & Milton JP(ed) 1972. The Careless Technology: Ecology and International Development. Natural History press, Garden City N.Y.

Farnworth EG, Golley FB 1974. Fragile Ecosystems: Evolution of Research and Application in the Neotropics, Springer-Verlag New York.

Farb P and Armelagos G 1980. Consuming Passions: The Anthropology of Eating Houghton Miffin, Boston.

Farvar MT, Milton JP (ed) 1972. The careless Technology: Ecology and International Development.

Natural History Press, Garden City, NY, pp.1030.

Fine RA 1972. The great drug deception, Stein and DAy, New York.

Fisher T, 1984. Place, Space and People, Macmillan Education, pp.122

Foley G, Moss P and Timberlake L, 1984. Stoves and Trees. Earthscan.

Forester JW 1977. World Dynamics, MIT Press, Cambridge USA

Frakes LA 1979. Climates through out Geologic time pp.261. Elsevier, Amsterdam Freeman C and Jahoda M (eds). World Futures: The Great Debate, Universe Books, New York.

Fuller JC. 1977. The poison that falls from the sky, Random House, New York.

Gaster TH (ed) 1964. Sir James Frazer's The New Golden Bough, Mentor, New York.

George S, 1976. How the other half dies, Penguin.

Georghiou G, Sato T (ed) 1983. Pest resistance to Pesticides, Plenum

Gibbons D 1986. The economic value of water, Resources for the Future, Washington D.C.

Goldberg ED 1976. The Health of the Oceans, UNESCO, Paris

Goodman GT. Edwards RW, Lambert JM. 1965. Ecology and the Industrial Society, Blackwell, pp.395.

Gopalakrishnan, TR 1975. Forests: 2000 A.D. A long range Perspective for India, Operations Research Group, Baroda.

GormalayWP 1976 Human Rights and Environment. The Need for Environmental cooperation. AW Sijthoff, Leyden.

Goulet D 1977. The uncertain promise: Value Conflicts in Technology Transfer, IDOC/North America, New York.

Grainger A, 1982. Desertification: How people make deserts, how people can stop and why they don't Earthscan. London.

Granberg-Michaeloon, W, 1984. A worldly spirituality: The call to take care of the Earth. Harper & Row. New York.

Gribbin J, 1988. The hole in the sky, Corge, London

Gray ED 1971.Green Paradise Lost, Round Table Press, Wellesley, MA,USA

Hagman G 1984. Prevention better than Cure, Swedish Red Cross, Stockholm Hall G 1974. Great Planning Disasters, Penguin Handler P (ed) 1970. Biology and the future of man, Oxford Unive press, London, pp.967.

Hardin G 1959. Nature and Man's fate Holt,Rhinehart & Winston, New York

Hardin G 1964. Population, Evolution and Birth Control: a college of controversial ideas, WH. Freeman

Hardin G. Baden J (eds) 1977. Managing the commons, Freeman, San Francisco

Harie V 1978. The Political Economy of Food, Saxon House, Herts

Harman WW 1976. An incomplete guide to the Future, Stanford Alumni Association, Stanford, Calif.

Hart J 1984 The Spirit of the Earth: A theology of the land Paulist Press, New York.

Harte J, Scolow RH. (ed) 1971. Patient Earth, Holt, Rienhart & Winston

Harrison H 1982 Make Room, Made Room, Penguin

Harrison F 1982. The Strange land: The countryside, myth and reality, Sidgwick & Jackson

Harvey D, 1973. Social Justice and the city, E. Arnold

Hayes D 1978. Repairs, Reuse, Recycling - First steps towards a Sustainable society, Worldwatch Institute, Washington

Heiser Jr. CB 1973. Seed to Civilisation, LH. Freeman, San Francisco, pp.243 $3.50

Heilbroner R, 1974.An Enquiry into the Human Prospect, WW.Norton, New York

Helfrich HW Jr (ed) 1970. The Environmental Crisis. Volume I & II Yale Unive Press, New Haven, Conn. USA.

Hewitt, K (ed)1983.Interpretations of Calamity, Allen & Unwin, Boston

Hes WN (ed) 1974. Weather and Climate modifications, John Wiley & Sons Hiatt HH et al 1977. Origins of human Cancer. Coldspring Habor Laboratory, Cold Spring Harbor, New York

Hillary E (ed) 1984. Ecology 2000 - The changing face of Earth Michael Joseph, London, pp.252.

Hopkins R, Paarlberg R and Wallerstein M, 1982. Food in the Global Arena: Actors, Values, Policies and Futures. Holt, Rinhart and Winston.

Hoult DP 1969. Oil on the Sea, Plenum, New York

Howard A, 1972. An Agricultural Testament, Oxford Univ. Press

Hughes, JD 1975 Ecology and Ancient civilization, U. New Mexico press, Albuquerque, USA

Huntington E 1907 The Pulse of Asia, Houghton, Mifflin

Huntington E 1915 Civilization and Climate. Yale Unive Press

Huntington E 1945 Mainsprings of civilization, Wiley

Hufschmidt H (ed) 1981. Economic approaches to Natural Resources and Environmental Quality Analysis. East West Centre.

Huxley A 1963 The Politics of Ecology - The Question of Survival. The Fund for the Republic Inc

Idris SM (ed) 1984. Environment Development, Natural Resources Crisis in Asia and Africa, SAM, Penang, pp.802, $30

Isard W et al 1972. Ecologic-economic analysis for regional develo-pment, Free Press, New York.

Ickes, HL, et al 1934. National Resources Board. U.S. Government Printing Office. Washington D.C. Frakes LA 1979. Climates through out Geologic time pp.261. Elsevier, Amsterdam

Jacks GV and White RO 1939. The Rape of the Earth: A world survey of Soil Erosion, Faber, London.

Jacobs J,1984. Cities and the Wealth of Nations, Random House, New York

Jevons WI. 1911. The theory of Political Economy Macmillan & Co, London

Joranson PN and Butigan K (ed) 1984. Cry of the Environment. Rebuilding the christian Creation tradition, Bearand Co Santa Fe, CA

Joseph LE 1991. GAIA, Arakana Books, London, $6.99

Kuch P. 1988. Natural Resources for the 21st Century. EPA department, Washington D.C. 125.pp.

Kahn H and Wiener A 1969. The year 2000. Macmillan, London

Kay DA Skolnikoff EB 1972. World Eco-crisis: International Organ- isations in Response, Unive Wisconsin Press, Madison Wi, USA.

Kennard BH (ed) 1972. The Water's Edge. MIT Press, Cambridge Madras.

Kennard, B. 1982. Nothing Can be done. Everything is Possible, Brick House.

Khoshoo TN 1986 Environmental Priorities In India and Sustainable Development. Indian Science Congress Association pp.224.

King J 1985. Troubled water. Rodale Press, Emmaus, USA

King FA 1911. The breakdown of Nations EP. Dutton.

Kovarick B 1982. Energy and Environment in a hungry world.

Kramer M 1980 Three farms; Making Milk, Meat and Money from the American soil. Little, Brown, Boston

Kuch P. 1988. Natural Resources for the 21st Century. EPA department, Washington D.C. 125.pp..

Lamb HH 1986 The Changing Climate. Matheun, London.

Laird, RD (ed) 1965. Soviet Agriculture: The Permanent Crisis. Praeger, New York.

Laeakey LSB and Goyder J 1983. Environmental groups in Politics GeorgeAllen and Unwin

Lappe FM and Collins J 1978. Food First: Beyond the Myth of scarcity. Ballantine Books, New York.

Lappe FM 1971. Diet for a small planet, Ballantine, New York

Legator MS, Harper BL and Scott MJ 1985. The Health Detectives Handbook. John Hopkins Press, Baltimore.

Leipziger DM, Nudge JL 1976. Sea bed Mineral Resources and the Economic interests of Developing countries. Ballinger, Cambridge Unive. Press Leith H, Wittaker RH (eds) 1975. Primary Productivity of the Biosphere, Springer, New York.

Lenihan J and Fletcher, WW 1979. Economics of the Environment - Environment and Man. Academic Press, pp.193.

Leontief W et al 1977. The future of the world Economy Oxford Univ. Press

Leopold A 1949. A Sandy Country Amanac, Oxford University Press

Levine A 1982 Love Canal: Science, Politics and People. Lexington Books, Lexington, Mass

Leucas G (ed) 1976. Life Boat Ethics. Harper & Row, New York

Logus C. Operation Noah

Lovins AB 1977. Soft Energy Paths: Towards a Durable Peace. Ballinger, Cambridge Ma. USA.

Lorenz KZ 1952. King Solomon's Ring. Methuen & Co. London

Love P and Overzend R 1976. Tree Power: An assessment of the Energy potential of forest Biomassin Canada. Min. Energy, Mines, Resources, Ottawa, Cannada.

Lovett JV (ed) 1973. The Environmental, Economic and Social significance of Drought. Angus & Robertson, Sydney.

Lowe, P. and Goyder J. 1983. Environment Groups in Politics, George Allen & Unwin

Malthus TR 1798. An Essay on the Principles of population, London.

Mamdami M 1972 The Myth of population control. Monthly Review Press, New York.

Marden PG. Hodgson, DG and Mc Coy TL 1982. Population in the Global arena: Actors, Values, Policies and Futures, Holt, Rinehart and Winston.

Mark VH, Ervin FR 1970. Violence and the Brain, Harper & Row, New York.

Marsh GP 1864. Man and Nature, Harvard University Press. Cambridge.

Marsh GP 1885 The EArth as modified by Human action, Scribner, New York.

Mason JA 1975. The Ancient Civilizations of Peru. Penguin. pp.137 Mathews WH 1971. Man's Impact on the Climate, MIT Press, Cambridge USA Maltboy E 1986 Waterlogged wealth - Why waste the wold's wet places? I.I.E.D.? Earthscan - London and Washington pp.200.

Mason J and Singer P, 1980. Animal Factories, Crown Publishers, New York Maddox 1972. The dooms day Syndrome. Macmillan.

Marx W 1971 Man and his Environment: Waste. Harper and Row New York.

May RM (ed0 1976. Theoretical Ecology: Principles and Application. Blackwell, Oxford McCormick J 1985 Acid Earth. The global threat of air pollution. Earthscan London.

McNamara RS 1973. One hundred countries: Two Billion People: The Dimensions of Development. Praeger, New York McHarg I 1969. Design with Nature. natural History Press Mc Neil WH 1963. The Rise of the West. Mentor, New York pp.455

Meadows DH, Meadows DL, Randers J, Behrens WW 1972. Limits to Growth. Universe Books, New York.

Meadows D et al 1982. Groping in the Dark: The first decade of global Modelling. John Wiley & Sons, Chichestor, U.K.

Meadows D and Meadows D (eds) 1973. Toward Global Equilibrium. Wright Allen Press, Cambridge Mass.

Meadows DL 1974. Dynamics of growth in a finite world Wright-Allen, Cambridge, Mass Medawar, P, 1986. The Limits of Science. Oxford University Press.

Medawar P 1986. The Future of Man. Oxford University Press.

Medawar P and Medawar J 1985. Aristotle to Zoos. Oxford University Press Means PA 1931. Ancient Civilizations of the Andes. Gordian Press, New York.

Matre SV and Weiler B (eds) 1983. The Earth Speaks. Institute for Earth Education, Warrenville, Ill.

Mellanby K 1967. Pesticides and Pollution. Collins, London pp.219.

Meolford D 1973. Environmenal Harassement of Technology assessment? Amsterdam.

Mesarovic, M. Pestel, E. 1974. Mankind at the Turning Point. Reader's Digest Press, New American Library. New York Myers N 1980. Pre-sinking Ark. Pergmon Press, New York.

Mill JS 1865. Principles of Political Economy. Augurtim M. Kelly.

Miller LH 1985. Global order: Values and Power in International Politics. West View Press, Boulder, Colorado.

Moore WG 1947 The Worlds Wealth. Penguin, pp.255.

Morris D 1967. Naked Ape. Cape.

Muir J 1894. The Mountain of California. Appleton - Century -Crofts.

Mumford, L. The Pentagon of Power.

Mumford L 1929. Golden Day. Cape.

Murdoch WW 1975. Resources, Pollution and Society. Sinauer Association. Sunderland, Mass.

Myers N 1979. The Sinking Ark. Pergamon, New York.

Myers N 1980. Pre-Sinking Ark. Pergamon Press, New York

Myers N 1983. A Wealth of Wild Species. Westview Press, Boulder, USA

Myers G 1984. The Primary Source. W.W. Norton, New York.

Myrdal G 1970. The Challenge of world poverty: A world Anti Poverty program in outline, Pantheon, New York.

National Academy of Sciences, 1969. Resources and Man. W.H. Freeman, San Franscisco.

N.A.S. 1969. Eutroplication: Cause, Consequences, Correctives. Nat. Acad. Sci., Washington.

N.A.S. 1972. Genetic Variability of Major Crops, N.A.S. Washington.

N.A.S. 1977. Drinking Water and Health. Nat. Acad. Sci. Wahsington.

N.A.S. 1978. Plant Species in the People's Republic of China. Nat. Acad. Sci. Washington.

Naess A and Sessions G 1984. Ecophilosophy. Sierra College, Rocklin, CA 95677.

Neuhas R. 1971. In defence of People: Ecology and the reduction of Radicalism. Mac Millan.

Nash H (ed) 1977. Progress as if Survival Mattered. Friends of the Earth, San Francisco.

Newell FH 1920. Water Resources: Present and Future Resources. Yale Univ. Press, New Haven, Conn. USA.

Nicholson M 1969-70. The Environmental Revolution. Hodder/Penguin.

Nicholson Lord D 1987. The greening of the Cities. Routledge & Kegan Paul, pp.270.

Nijkamp P. 1977. Theory and Application of Environmental Economics. North-Holland, Amsterdam.

Norman C. 1978. Soft Technologies, Hard Choices. Worldwatch Institute, Washington.

Noyes CR 1948. Economic Man. Columbia Unive. Press, New York.

O'Riordan T. 1981. Environmentalism. Pion, London.

Opie J (ed) 1971. Americans and the Environment: The controversy over Ecology. D.C. Heath.

O.E.C.D. 1982. Economic and Ecological Interdependence. O.E.C.D.,Paris.

Osborn F. 1948. Our Plundered Planet. Little Brown and Co. Boston.

Osborn HF 1953. Limits of the Earth. Little Brown & Co. Boston.p238

Ophuls W. 1977. Ecology and the Politics of Scarcity. W.H. Freeman, San Francisco.

Portney P (ed) 1982. Current Issues in Natural Resource Policy. Resources for the Future, Washington, Pearson FS and Rochester JM 1984. International Relations: The global condition in the Late 20th century.

Paddock W and Paddock P 1967. Famine 1975 Little Brown, Boston.

Pearce DW 1976. Environmental Economics. Longman, New York.

Pepper D 1984. The roots of modern environmentalism. Croomholm, p242 Pimental D (ed) 1978. World Food, Pest losses, and the Environment. West View Press, Boulder Co. USA.

Pinchot G 1970. The fight for conservation. Harcourt Brace.

Pirages D 1978. The new context for international Relations: Global Ecopolitics. Dixbury Press, North Scituate, Mass.

Pregel B 1975. Environment and Society in transition:World Priorities. New York Acad. Sci. New York Prescott, A and Robert C 1982. What is wildlife worth Earthscan.

Prescott A and Robert C 1982. Genes from the wild-Using wild genetic resources for food and raw materials. Earthscan.

Price D de S 1961. Science since Babylon. Yale Univ. Press, New Haven, Conn.

Ratchiffe F 1948. Flying Fox and Drifting sand, Argus and Robertson.

Reece R 1979. The Sun Betrayed: A report on the Corporate seizure of U.S. Solar Energy Development. South End Press, Boston.

Reich C 1970. The greening of America. Random House.

Repetto R. 1985. The Global possible: Resources, Development and the new century, Yaee University Press, New Heaven, Conn. USA.

Rick J 1963. The Time has come. A.A. Khop's Inc., New York.

Ricker WE 1969. Resources and Man. W.H. Freeman, San Francisco.

Ridgeway J 1971. The Politics of Ecology. E.P. Dutton & Co. New York.

Roszak K 1970. The Making of a Counter Culture. Faber.

Russett, B and Starr H 1985. World Politics: The Menu for choice. W.H. Freeman, New York.

Rifkin J 1980. Entropy: A new world view. Viking Press, New York.

Ridley HN 1930. The dispersal of plants throughout the world. L.Reeve & Co. Ashford, U.K.

Rhorin, GS 1979. The Biosphere and Politics. Progress Publi. Moscow.

Rosenberg NJ Easterling WE Crosson P and Darmstadter J (eds) 1989 Greenhouse Warming: Abatement and Adaptation. Resources for the Future Press.

Rothchild E 1973. Paradise Lost: The decline of the Auto-indusrial Age. Random House: New York.

Ruddle K, Manshard W 1981. Renewable Natural Resources and Environment. United Nations University. East West centre pp.404.

Rudd. RC 1964. Pesticides and the Living Landscape. Univ. Wisconsin Press, Madison.

Rudloe, J. 1968. The sea brings forth. Alfred A. Knoff. Hall, New York.

Repetto R. 1985. The Global possible: Resources, Developmentand the new century, Yate University Press, New Heaven, Conn. USA.

Sale K 1982. Human Scale. Patnam Books. New York.

Sanbach F. 1980. Environment, Ideology and Policy, Blackwell.

Sant R. Carhard SC et al 1981. Eight Great Energy Myths: The Least-cost Energy Strategy - 1978-2000. Energy Productivity Center of the Mellon Institute. Arlington. Va.

Sarala Devi 1982. Revive our Dying Planet. Gyanodya Prakashan Nainital pp. 475. Rs.150.

Sartre JP 1948. Existentialism and Humanism. Mathuen, London.

Sauer GO 1969. Seeds, spades, Hearths and Herbs: The Domestication of Animal and Foodstuffs, M.I.T. press, Cambridge.

Schell J 1982. The Fate of the Earth. Avon Books, New York.

Shapely H (ed). Climatic Change: Evidence, Causes and Effects. Harvard Univ. Press, Cambridge.

Shirikov Y 1975. Man, Society and Environment. Progress Publi.,Moscow

Schmidt A 1971. The Concept of Nature in Marx. New Left Books.

Schnailberg A 1980. The Environment: From Surplus to Scarcity. Oxford University Press.

Schumacher EF 1973. Small is Beautiful: Economics as if People Mattered. Harper and Row, New York.

Sears PB. The Living Landscape. Lancer, New York, pp.222.

Sears PB, 1935. Deserts on the March. University Oklahama Press, Norman.

Seton ET. 1927. Lives of game animals. Vol I, II, III Doubleday and Co. Inc. New York.

Seward AC 1941. Plant life through the ages. Cambridge Univ. Press, London pp.607.

Simpson GG 1934. Attending Marvels. Macmillan.

Simon J 1981. The Ultimate Resource. Princeton University Press, Princeton NJ.

Simon J and Kahn, H.(ed) 1984. The Resourceful Earth, Blackwell, Oxford
Singh N 1976. Economics and the Crisis of Ecology. Oxford University Press, New Delhi.

Smith VK (ed) 1981. Ecology and Field Biology. Harper and Row. New York.

Smith FE and Goldberg ED 1971. Man's Impact on Terrestrial and Oceanic Ecosystems. The M.I.T. Press, Cambridge. M.A.

Snyder, DP and Edwards G 1984 Future Forces Amer. Soc. Assoc. Executives, Washington, p.109.

Solkoff J 1985. The Politics of Food. Sierra Club Books, Scan Francisco.

Soule ME (ed) 1986. Conservation Biology: Sciences of Scarcity and Diversity. Sinauer Associates Sunderland, Mass, U.S.A.

Sprout H and Sprout M 1978. The Context of Environmental Politics. University Press of Kenhucky, Lexington, KY.

Spretnak, C and Capra F. 1985. Green Politics, Paladin, London, pp. 262.

Stamp LD 1955. Man and the Land. Collins, London, pp. 272

Strausz-Hope R (ed) 1968. Society and Ecology. Sage Publ. Beverley Hills, Celif.

Steinback J, Ricketts EF 1951. Log from the Sea of Cortez. Viking Press.

Shoard, M 1980. The Theft of the countryside. Temple Smith.

Sternberg, HO 1975. The Amazon River of Brazil. Franz-Steiner Verlag, Wiesbalen.

Steward, JH (ed) 1950. Handbook of South American Indians, Smithsonian Institution. Bureau of American Ethnology, Washington.

Stretton H 1976. Capitalism, socialism and the Environment. Cambridge Unive. Press.

Struever S (ed) 1970. Prehistoric Agriculture. Natural History press, Double Day, Garden city, New York.

Sverdrup HV, Johnson MW, and Flemming RH 1942. The Oceans. Prentice-Hall, New York.

Seton ET. 1927. Lives of game animals. Vol I, II, III Doubleday and Co. Inc. New York.

Snyder, DP and Edwards G 1984 Future Forces Amer. Soc. Assoc. Executives, Washington, p.109.

Titus JG (ed) 1988. Greenhouse effect, sea level rise and coastal wetlands. EPA Office of Policy, Planning and Evaluation, Washington, D.C.

The Conservation Foundation 1986. Agriculture and the Environment in a changing World Economy. The Conservation Foundation, Washington, D.C.

Thomas WL Jr (ed) 1956. Man's Role in Changing the Face of the Earth Vol.I and II. U. Chicago, p. 1193.

Taylor G 1937. Environment, Race and Migration. University of Toronto Press. Thomas MF, Whittington BW 1969. Environment and Land Use in Africa, Methuen, London.

Thoreau HD (1854) 1942. Walden. New American Library, pp. 255.

Thorhacu A (ed) 1980. Botany in China. Stanford Univ. Press, Stanford.

Thurow L 1980. The Zerosum society. Basic Books: New York.

Tinbergen J 1976. Reshaping the International Order. Hutchinson.

Timbergen N. Social Behaviour in Animals. Wiley. Random House, New York.

Timberlake L 1985. Africa in Crisis. I.I.E.D., Earthscan, London.

Timberlake L 1987. Only One Earth: Living for the future. BBC/Earthscan London.

Tofler A 1970. Future Shock. Random House, New York.

Tolba MK. 1983. Earth Matters. U.N.E.P. pp. 163.

Trainer FE 1985. Abandon Affluence. Zeb Books. London.

Treshow M 1976. The Human Environment. McGraw Hill, New York.

Turnbull CM 1962. The Forest People: A study of the pygmies of the Congo. Simon and Schuster, New York.

Turnbull CM 1972. The Mountain People. Simon and Schuster, New York

Titus JG (ed) 1988. Greenhouse effect, sea level rise and coastal wetlands. EPA Office of Policy, Planning and Evaluation, Washington, D.C.

Ucko PJ 1969. Dimbleby GW (eds) 1969. The domestication and Exploitation of plants and animals. Aldine, Chicago.

UNEP, 1986. The State of the Environment: Environment and health. UNEP, Nairobi.

Urquhart J, Heilmann, K. 1984. Rish Watch: The odds of life. Facts on File, Bicestor, U.K.

Van Dyne GM 1966. Ecosystems, Systems Ecology and Systems Ecologists (ORNL - 3957). ORNL. Oak Ridge.

Van Tassel, AJ (ed) 1975. The Environmental Price of Energy. D.C. Heath & Co. Lexington.

Verrett J, Carper J. 1974. Eating May be hazardous to your Health. How your government fails to protect you from the dangers in your food. Simon and Schuster. New York.

Vichit-Vadakan, (ed). 1979. The Environmental Dimensions in Development. U.N.A.P. D.I. Bangkok, pp.304.

Vichit-Vadakan, V, Rogov AI (ed) 1980. Environment (Resources) Management and Development. UNAPDI, Bangkok, pp.521.

Vogt W 1948. Road to Survival. William Sloane, New York Ward B 1976. The Home of Man. W.W. Norton, New York.

Ward B, Dubos R. 1972. Only One Earth: The care and Maintenance of a Small Planet. W.W. Norton, New York

Ward B 1979. Progress for a small planet. W.W. Norton, New York.

Warnock M 1979. Existentialism. Oxford University Press, Oxford.

Watt, KEF 1968. Ecology and Resource Management. McGraw Hill, New York Westing AH 1986. Global Resources and International Conflict. Oxford Unive. Press pp. 280.

Webster CC, Wilson, PN 1966. Agriculture in the tropics. Longmans, London.

Weir D and Shapiro, M 1981. Circle of Poison. Institute for Food and Development Policy, San Francisco.

Whicher GF 1945. Walden Revisited. Farron, Straus & Co.

Whitton J 1980 Disasters: The Anatomy of Environmental Hazards. Penguin Books Ltd.

Wijkman A, Timberlake, L. 1984. Natural Disasters: Acts of God and Acts of Man? I.I.E.D./ Swedish Red Cross/Earthscan, London/ Washington pp.145.

Wilson CL, Mathews WH (eds) 1970. Man's Impact on the Global Environment. MIT Pres, Cambridge, Mass.

Wolstenholme G 1963. Man and his Future. Little, Brown and Co. Boston.

Woytinsky WS and Woytinsky ES 1953. World Population and Production, New York World Commission on Environment and Devlopment (WCED), 1987. Our Common Future. Oxford University Press. pp. 383. Rs.110.

Yergin D and Hillenbrand (eds) 1982. Global Insecurity: A Strategy for Energy and Economic Renewal. Houghton Mifflin, Boston.

PEOPLE WOULD RATHER BELIEVE THAN KNOW.

EDWARD O. WILSON

Index

A

Acred grove movement, 140
Activists, 351
Adirampattinam, 216
Africa, 45, 51, 55, 56, 64
Afro-Asia, 322
Agharkar, 341
Agricultural Wastes, 309, 310
Agriculture, 352, 355
Agriculture and Aquaculture, 226
Agriculture and Public Health, 292
Agroclimatic zones, 324
Agroecosystems, 254, 291, 296
Agroindustrial wastes, 309
Agronomists, 30
Agronomy, 203
Ahmedabad, 285
Air pollution, 187
Aksai Chin, 184
Alcohol Production, 307
Aldous Huxley, 74
All East India Companies, 185
All India Coordinated Project on Genotoxicity, 265
Alternate Nobel Prizes, 131
Altic States, 32
Alwar, 91
American and Europian efforts, 267
American Association for Advancement in Science, 119
American Solar Energy Society, 319
Amphibian Frogs, 292
Amritsar, 285
Anandavan, 91
Ancient Hindu materialism, 153
Andaman Sea, 131
Andhra Pradesh, 92, 93, 138, 157, 223, 235
Andhra University, 351
Animal husbandry, 90
Anthropology Philosophy, 353
Aquaculture, 90
Aquatic habitats, 263
Aquatic plants, 309
Aquatic Pollution, 291
Arab, 62
Arabidopsis, 284
Aromatic hydrocarbons, 109
Asia, 45, 51, 56, 64
Asia-Pacific Peoples Environment Network, 150
Atheist Movement, 176
Atomic Energy Regulation Board, 263, 268
Atomic Power Plants, 252
Auditor or Accountant Generals, 158
Auduborn Society, 127
Aurobindo Philosophy of action, 289
Auroville, 104
Auroville Movement, 139
Australia, 51, 56, 61, 73, 216, 261
Automatic chemical alarms, 116
Available technology, 314

B

Baba Amte, 135
Bangladesh, 62, 72

Bedthi Campaign, 133
Belgaum, 91
Belgium, 55, 185
Bengal tiger, 132
Benzapyrene, 109
Bhabha Atomic Research Centre, 283
Bhagavatula Charitable Trust, 150
Bhagavatula Trust, 105
Bharatpur, 296
Bharucha, 341
Bhoodan Movement, 219
Bhopal Gas leak accident, 45
Bhopal Gas Tragedy, 139
Bhopalpatnam-inchampal Dams, 133
Bihar, 157, 223, 235
Biochemists, 30
Biodiversity Park, 331
Biological Program, 351
Biomass experts, 318
Biotechnologists, 318
Biotechnology, 100
Bird Migrations, 187
Bird Sanctuary, 231
Birth defects, 267
Bishnoi tradition, 140
Bishnoi women, 165
Black holes, 169
Boards of Plant and Animal, 99
Boat Ethics, 193
Bolivia, 62, 94
Bombay, 285
Bombay Environmental Action Group, 133
Botanists, 30
Botany, 94, 346, 352
Brazil, 30, 181
British Colonialists, 83
British history, 87
British India, 83
British navy, 30
British Power, 137
British Times, 169
Buddhism and Hinduism, 53
Burma, 94, 187, 323
Butterfly species and birds, 225

C

Calcutta, 285
Californian Coast, 297
Cambodia, 62
Canada, 61, 73
Canadian soil, 32
Captive Female, 168
Carbohydrate, 315
Carbon containing compounds, 306
Carbon tetrachloride, 284
Carbondioxide, 312
Castro, 80
Cauvery River Dispute, 203
Center for science and environment, 140
Central Asian Republics, 187
Central Bureau of Investigation, 158
Central Forestry Commission data, 305
Central Pollution Control Board, 245
Centre for Environment Education, 150
Centre for Science and Environment, 132
Ceylon, 93, 187
Chad, 46
Chairmanship, 333
Chanakya, 34
Chandi Prasad Bhat, 132
Chandigarh, 285
Characteristics of Biogas, 308
Chattisgarh Movement of Sankar Guha Neogi, 139
Chemical effects, 287
Chemical industry, 250
Chemical spread, 154
Chemical weapon designers, 117
Chemistry, 94, 346
Chennai, 208

Cherapunji, 91
Chernobyl disaster, 108
Chico Movement of Philippines, 129
Chico river, 129
Chile, 62
Chilka Bachao Andolan, 136
Chilka Lake Ecosystem, 136
China, 73, 77, 85, 86, 323
Chinese Martin Thong, 130
Chipko activists, 132
Chipko Andolan, 150
Chipko Movement, 132
Chipko or Narmada Activism, 150
Chitrakoot, 91
Chloroacetophenone, 112
Climatologists, 202
Climatology, 203
Club of Rome, 77, 124, 154, 162, 181, 339
Coastal Aquaculture, 208
Coastal areas, 205
Coastal plain, 206
Coastal ponds of Orissa, 235
Coastal waters, 187
Coastal wetlands of Tamil nadu, 207
Columbia, 62, 93
Communicable / Infectious Diseases, 254, 289
Computer Modeling, 80
Concerns for Ecology and Environment, 132
Conference at Nairobi, 339
Confucius, 66
Congo, 185
Contaminated water, 109
Contract Farming, 94
Cooperatives and Panchayats, 232
Corchorus, 284
Corporate agriculture, 94
Cotton leaf worm, 322
Council for Scientific and Industrial Research, 158
Cuba, 62, 94
Cuddalore, 208
Cycles, 153

D

D. Photosynthetic Productivity, 301
Dams, 355
Dancing Shiva, 155, 353, 355
Decreasing taxonomic diversity, 297
Deep Ecology, 162
Delhi, 285, 332, 336
Deltaic plains, 206
Department of Atomic Energy, 110
Department of Environment and Forests of the Gover, 265
Department of Science and Technology, 158, 288
Deputy Registrar, 335
Dialogues, 73
Dichloroformoxime, 112
Dichlorvos, 277
Digha, 91
Director of Plant Science Research at Port Detrick, 119
Doon Mining, 133
Dr Chapekar, 332
Dr Davidar, 332
Dr Gnanam, 333, 335, 336
Dr Homji, 334
Dr Kannabiran, 331, 334, 335
Dr Parthasarathy, 332
Dr Poyyamoli, 332, 333
Dr Ramaswamy, 333
Dravidan Movement, 171
Drosophila, 284
Dunkel Draft, 100

E

E.V.R. Naicker, 171
Earth Summit in Brazil, 153

East Aztec Region of Mexico, 129
East Coast of India, 240
East Coast of Tamilnadu, 209
East Coast Road, 136
East Germany, 32
East India, 157
East India Company, 313
East India Company's Biologists, 183
East-Coastal regions, 83
Eastern Europe, 73, 77, 182
Eco-Agriculture, 333
Ecoagriculture, 230
Ecological Growth Centres, 95
Ecological Marxism, 177
Ecological World View, 66
Ecology, 203, 346
Ecology and environment, 132
Ecology and Politics, 74
Ecology School, 332
Economic and Political leaderships, 149
Economic Plants, 238
Economic viability, 314
Economics, 203
Economics and History, 353
Ecosystem pollution and Genetic Health, 259
Ecosystems, 125
Effects on Crop Systems, 276
Effects on Genetic Systems, 258
Efforts of Ecology School, 332
Electro Magnetic Fields, 261
Energy Crisis, 301
Energy From Phytomass, 301
Engineering, 352
Engineering College, 334
England, 55
Environment Departments, 94, 352
Environment Protection Act, 224
Environmental and ecological processes, 126
Environmental Education, 95
Environmental Genotoxicology, 297
Environmental History, 313, 356
Environmental Management Plan, 215
Environmental Mutagen Information Centre, 262
Environmental Mutagen Society of India, 264, 283
Environmental Sciences, 269
Environmental studies, 349
Equador, 62
Eritrea, 131
Ethiopia, 46, 62, 131, 185
Ethylene glycol, 284
Eucalyptus, 308
Europe, 51, 73, 323
European Community, 182
European countries, 57
European nations, 55
European Union, 92
Even Environmental economics, 355
Expansionism of development syndrome, 149
Extinct Phytomass, 301

F

Fauna, 153
Feminism, 356
Fifth World War, 70, 192
First World War, 57, 58, 191
Fisheries, 352
Five major elements of nature, 153
Flood plains, 206
Flora, 153
Floriculture, 232
Fodder and fertilizer, 226
Food chain hazards, 263
Forest Conservation Act, 224
Forest sources, 309
Forest Survival in Brazil, 128
Forestry, 203, 352

Forestry, 90
Fossil fuel combustion during transport, 252
Fourth World War, 64
France, 55
Francis, 332
French Owners, 118
Friends of the Earth Malaysia, 130
Fungicides, 278

G

Gajendragadkar commission, 158
Gandhamardhan Bauxite mining, 134
Gandhi, 356
Gandhi as an Ecological Mahatma, 132
Gandhigram, 91
General Agreement on Trade and Tariffs, 78
Genetic ailments, 109
Genetic Resources, 99
Genetic syndromes, 264
Genetic Toxicology, 290
Genetically Modified Crops, 138
Geneva Protocol, 120
Genotoxic agents, 264
Geographical and political boundaries, 124
Geographical Geological, 352
Geographical International, 127
Geography, 94, 203, 352
Geology, 94, 352
Gerhard Schrader, 113
German Scientist, 113
Gir lion, 132
Global Commons, 162
Global Resources, 77
Gnanam, 335
God and Religion, 54
Gods and Faiths, 52
Government Departments of Environment, 141
Government Global Model, 20
Great Indian Bustard, 132
Great Karma, 161
Greek, 168
Green movements, 130
Green Peace, 150
Green Peace and Friends of the Earth Movements, 128
Green Peace Movement, 130
Green Philosophers, 351
Green Politics, 154
Greenhouse gases, 40
Greens of Europe, 150
Greens of Germany, 129
Gross national product, 101
Gross natural/ecological product, 101
Guerilla war, 124
Guerrilla warfare, 129
Gujarat, 134
Gulf of Mannar, 208
Gurgaon, 91

H

Half female, 289
Half male, 289
Harmony and peace, 194
Harvard University, 119
Hawai island, 305
Health care systems, 263
Heaven on Earth, 169
Hebrew, 168
Hemoglobin of human blood, 110
Herbicide treatment, 118
Herbicides, 278
Heritable diseases, 267
heritable diseases, sperm abnormalities, 267
Himalayan Glaciers, 187
Hindu Mythology, 193
Historicism in Plant Selection, 313
Hochi Minh, 80

Holland, 55
Hoshangabad, 104
Hosur, 91
Human Ecosystems, 296
Human Environment, 351
Human evolution and development, 180
Human genetic Health, 107, 254
Human Resource Development, 20, 80, 269
Human Settlements, 52
Hungarian Statistics, 266
Hybrid Maize, 219
Hydrocarbon, 312
Hydrocarbon production, 319
Hydrocarbon rich oil, 312
Hydrology, 203
Hymn to Mother Earth, 153

I

Ideal Firewood Species, 305
Ideal Project Site, 334
Identification of hazard sources, 287
Indian Council of Medical Research, 263
Indian Institute of Chemical technology, 319
Indian Institute of Meteorology at Poona, 201
Indian National Congress, 83
Indian Ocean Rim States, 188
Indian Rural Ecosystesms, 33
Indian Territories, 186
Indian Universities, 156, 245, 340
Indian Waters, 355
Indira Gandhi, 132
Indira Gandhi Sanctuary, 139
Indo - Gangetic plains, 187
Indo-German Binational Conference, 110
Indonesia, 182, 187
Indus Waters, 184
Industrial complexes, 158
Industrial revolution, 124
Industrialization and standard of living, 252
Industrialized countries, 302
Industry, 203
Insecticides, 278
Insecticides Urethane, 284
International Agency for Research on Cancer, 268
International Association of the Environmental Mut, 284
International Boards of Plant and Animal, 99
International Committee for Protection against Env, 262, 268
International Programme for Chemical Safety, 262
International Registry of Potentially Toxic Chemic, 262
International Union for Conservation of Nature, 99, 224
Iran, 323
Irish Protestants, 126
Israel, 62

J

Jafna islands, 216
Jammu, 285
Japan, 86, 130, 261
Japanese invasion of East Indies, 119
Java Island, 182
Jawaharlal Nehru University, 351
Jhansi, 91
Jharkhand Movement, 139
John Bunyan's Pilgrims Progress, 150

K

Kaiga campaign, 134
Kalapet region, 244
Kalinga Prize Winner, 120
Kaliveli, 224, 230
Kaliyuga, 156

Kalki of Bhagavatham, 157
Kalyani, 285
Kambikudi, 224
Kanyakumari, 135, 136, 208
Karaikal, 91, 209, 245
Karimganz, 91
Karl Marx, 78
Karnataka, 92, 133, 138, 223, 235
Karnataka raita sangha, 140
Karnataka's joint sector projects on degraded fore, 140
Karve and Phule in Maharashtra, 170
Kaveri Festival at Erode, 203
KAVERI water sharing, 201
Kaveri waters, 137
Kaveripakkam, 224
Ken Sarowivo, 129
Kenya, 62
Kerala Sastra Sahitya Parishat, 150
Khurdistan, 93
Kilakkarai, 216
Kishore Bharati, 104
Konkan controversy, 135
Koodangulam, 107
Korea, 62
Krishna Chaitanya, 127
Kumaon Hills, 132
Kunrakudi, 91
Kutch in Gujarat, 134

L

Land grabbing, 230
Laos, 62
Latin, 168
League of Nations, 77
Life Boat Ethics, Sinking Ark, 162
Lignocellulose, 315
Loss of a gene pool, 101
Loss of an ecological link, 101
Love Canal, 267
low-level ambient exposures, 109
Lucknow, 285
Lysergic acid diethylamide, 113

M

Madhok principle, 187
Madhura Refinery and Tajmahal, 157
Madhya Pradesh, 134
Madras, 136, 285
Madras with Kanyakumari, 136
Madurai, 285
Madurai University, 351
Mahabharata, 161, 168
Maharashtra, 92, 134, 223, 235
Mahe, 245
Malaccas, 131
Malathion, 277
Malaya, 94
Malaysia, 62, 73, 128, 182, 187
Malda, 91
Man Baiting, 169
Man Eaters, 169
Man Eating, 169
Mangamma Salai, 229
Mani, 341
Manmohan Rao Dunkel, 203
Marshall plan, 59
Marxist Politics in Europe, 192
Medha Patkar, 135, 141
Medical and Environmental Education, 269
Medical college, 264
Medical Practitioner, 264
Medicine, 352
Medico-social surveys, 261
Methanogenic organisms, 310
Methyl parathion, 277
Mexico, 129
Microbes, 318
Middle East, 81

Ministry of Agriculture, 314
Mirattunilai, 224
Misra, 341
MN Roy, 80
Molecular formula, 305, 316
Molluscan Mussels, 292
Morocco, 118
Mother Earth, 193
Mother Earth Concept, 34
Movement of Tree Planting in China, 128
Movements in China, 128
Mr.Subramaniam, 331
Multinational Asiatic Corporations, 188
Municipal solid wastes, 309
Murshidabad, 91
Muzaffar Ahmed, 80
Myopia, 333

N

Nagarjuna Sagar Project, 123
Nairobi, 47, 181
Nalaban Island, 136
Narmada Bachao Andolan, 134, 150
Narmada river, 134
Narmada Sagar Projects, 134
National Botanical Research Institute at Lucknow, 319
National Commission on Wetlands, 224
National Environmental Engineering Research Instit, 265
National Fishermen's Forum, 135
National Geographic Society, 128
National Health Survey, 264
National Institute of Occupational Safety and Heal, 268
National Institutes of Health, 268
National Productivity Council, 319
National Tank Authority, 233, 235
National Toxicology Programme, 262
Natural vegetation, 322
Naxalism, 124
Naxalism, Srilanka problem, 124
Nehruvian era, 184
Nepal, 72, 187
Nepal rivers, 187
New Delhi, 94
New Education Policy, 267
New Guinea, 185
New trade objects, 57
Nicaragua, 94
Nigeria, 62
Nitrogen fixing plants, 310
Nitropyrenes, 109
Nobel Laureate From Kenya, 131
Nobel Prize, 182
Noida, 91
Non-Governmental Agencies, 232
Non-reproductive tissues, 290
North -Eastern India, 93
North America, 51, 55, 56, 87, 126
North Sea, 32
North-East, 240
North-South, 73
Northern Malaya, 130
Northern Philippines, 129
Norway, 32
Nursing home, 264
Nutritionists, 30

O

Oil Pollution, 32
Oils and Fats Division, 319
Okinawa, 120
Ooty Hills, 139
Operation Ranch Head, 118
Organic content, 304
Orissa High Court, 136
Orissa Revenue Minister Banka Behari Das, 136
Orthochlorobenzalme-lanonitrile, 112
Osmania University, 284

Ousteri, 224
Oxydemeton methyl, 277

P

Pakistan, 72, 187
Palestine, 81, 93
Palestinian violence, 124
Panchayati Raj, 233
Patriotic Society, 149
Patronage, 156
Pentagon, 120
Perestroika and Glasnost, 79, 80
Peru, 62, 93
Peruvian era, 183
Pesticide Genotoxicity Trends, 254
Pesticide Index, 273
Pesticide Task Group, 264
Pesticides-fertilizer interactions, 323
Petro-plants, 306
Petroleum and natural gas, 109
Petroleum plantations, 316
Pharmacologists, 30
Pharmacology and Toxicology, 269
Phosphamidon, 277
Phosphonoshiolates, 113
Photosynthetic products, 305
Physics, 94, 346
Pillaipakkam, 224
Planet Earth, 127
Poland, 32
Politics, 353
Polluted Ecosystems, 254, 296
Pollution, 153
Pollution Control Boards, 141
Polycyclic Aromatic, 262
Polycyclic Aromatic Hydrocarbons, 262
Pondicherry, 82, 91, 92, 104, 141
Pondicherry Gazetteer, 245
Pondicherry region, 245
Pondicherry University, 244
Poona, 105, 285
Population Bomb, 154
Post-Marshall Plan Europeanization, 81
Post-Project Monitoring Mechanism, 215
Prabhupada Bhakti Vedanta Swami, 127
Prehistoric Scenario, 167
Prithvi Suktam, 194
Privatization through Liberalization, 77
Professor Wangeri Maathai, 131
Pumpset Facility, 202
Punjab, 93, 157

Q

Quantum Of Phytosources, 317
Quinolines oxides of nitrogen, 109

R

R.V.Nagar, 331
Radha Kamal Mukherjee, 127
Rain Bow Warriors, 127, 142
Rain water harvesting, 242
Rainbow Warriors fashion, 145
Rajasthan, 134, 184, 223
Ramanathapuram, 208
Ramsar Convention, 224
Ramsar international treaty on wetland preservation, 136
Rayon factory pollution in Kerala, 134
Red Indian Chief Seattle, 127
Red Indians, 126
Regional Ecology Agreements, 188
Regional Environment Agreements, 188
Regional Issues, 83
Regional Wars, 62
Religious View of Life, 52
Resistance in Mexico, 129
Rhine river, 32
Right Livelihood Awards, 131
River Connection, 244
Rodent populations, 322

Rodenticides, 296
Role of plants, 260
Rubber plants, 306
Rural and urban systems, 90
Russia, 85, 323
Russian pollution of Scandinavian forests, 187

S

S.P. Pattinam, 208
Salim Ali, 334, 341
Sand Dunes, 238
Sanskrit, 168
Santana Dharma, 153
Santapau, 341
Sarala Devi, 166
Sarawak Tribal Movement, 128
Sardar Sarovar, 134
Scandinavia, 73
Scientific Culture, 318
Screening smokes, 115
SE Asia, 77
Second World countries, 59, 61
Second World War, 58, 249
Self Regulation Properties Of Earth, 66
Sericulture, 232
Sethusamudram Channel Project, 356
Sethusamudram Shipping Canal, 138
Sevagram, 91
Seveso, 267
Sexual abnormalities, 289
Shell Company, 129
Sholavandan, 91
Siamese fighting fish, 114
Sierra Club, 127
Sierra Club, 128
Silent Spring, 154
Silent Valley Campaign, 133
Sinking Ark, 154
Sisir Kumar Ghose, 127
Skolimowski, 355
Small Town Development, 94
Social-inequalities, 126
Sociology, 94, 353
Soil Campaign, 133
Somalia, 46, 131, 185
South African flora and fauna, 188
South America, 44, 45, 51, 55, 63, 64, 73
South American countries, 31
South Arcot district, 230
South Asia, 72
South East Asia, 118
South-South Co-operations, 73
South-West Monsoon, 240
Southern Nigeria's Resistance Movement, 129
Soviet Russia, 59
Soviet Union, 192, 249
Spaceship Earth, 154
Spaceship Earth Concept, 79
Spain, 55
Sperm abnormalities, 267, 289
Srilankan coast, 216
St.Francis of Assissi, 66
Stockholm, 181
Stockholm International Peace Research Institute, 119
Stone Age, 51
Sugadha Kumari, 166
Sundarayya, 80
Sunderlal Bahuguna, 132, 135
Surface water resources, 40
Sustainable Development, 355
Swadeshi problems, 95
Sweden, 32
Swedish Premier Earlander, 119

T

Taj Campaign, 133
Tale of Two Cities, 50
Tamil civilization, 138
Tamil Country, 319

Tamil Literatures, 195
Tamil Nadu, 92, 107, 141, 223, 235, 238, 239
Tamilnadu Green Movement, 138
Tamils and Sinhalese in Srilanka, 185
Tank ecosystem, 229
Tank irrigation, 233
Tata Project with Orissa Government, 136
Techno-Economic Feasibility Report, 215
Technology Transfers, 73
Tehri Dam Campaign, 135
Terrestrial weeds, 310
Terrestrial plants, 309
Thai Vaishet Campaign, 133
Thailand, 187
Thar Desert Agreement on desert research, 184
The atmosphere and above, 153
The Rain forest Protection in Brazil, 128
Thermal decomposition, 306
Third Global War, 60
Third World, 148
Third World countries, 61
Third World War, 62, 63, 144
Three Mile Island, 267
Timber Politics, 130
Tourism, 90
Tourism and traffic, 289
Toxic environment, 115
Toxicologists, 30
Traces of nitrogen, 312
Trade, 203
Trade Practices, 73
Traditional cultivars, 225
Traditional values, 126
Trans-Boundary Ecosystems, 296
Travancore Princely State, 137
Triage Principle, 162
Trimethoxy-amphetamine, 113
Tropical Pest Management, 274
Tuticorin, 216

U

UN World Model, 79
United Nations, 51
United Nations Committee, 351
United Nations Conference, 162
United Nations Conference on Environment and Development, 161
United Nations Conference on Human Environment, 162
United Nations Environment Program, 216
United Nations Environment Programme, 339
Universities Handbook, 340
University Departments of Environment, 141
University Grants Commission, 158
Upland plains, 206
Urban Krakatopia, 154
Uttar Pradesh, 132, 235
Uttarpradesh, 223

V

Vande Mataram, 150
Vanrai, 104
Varanasi, 285
Vedaranyam, 216
Vedaranyam to Tuticorin, 216
Vedic Cultures, 181
Vedic heritage, 153
Vedic literature, 194
Vedic scholarships of Ancient India, 154
Vedic-Hindu Ecocentrism, 153
Veeresalingam in Andhra, 170
Vegetable oils, 284
Vermiculture, 232
Veterinarians, 30

Vice Chancellors, 158, 336
Vidyasagar in Bengal, 170
Vietnam, 62
Vietnam War, 117
Vietnamese environment, 119
Vietnamese wildlife, 118
Violence in Kashmir, 157
Visakapatnam, 105
Vital environmental issue, 196
Voluntary Activity, 149
Voluntary Agencies, 147, 220
Vyasa Maharishi, 30

W

Waltair, 94
Water Act, 224
Water Hyacinth, 292
Water Movements In Maharashtra, Palamau and Sukhmo, 140
Watershed, 130
Watershed management, 139
Waxes, 307, 317
West African Nations, 71
West Asia, 62, 63
West Coast, 297
West German, 32
Western Ghats March, 135
Wetland Forestry, 226, 230
White Flag Armed Resurrection, 130
Wood Sources, 310
World Bank, 73
World Trade Organization, 181
World watch, 150
World Wide, 150
World Wide Fund, 224
World Wide Fund for Nature, 150

Y

Yanam, 245
Yelamanchili, 91
Yelamanchili coastal region, 221
Yugoslavia, 81, 323

Z

Zamindars, 231
Zeolite Catalyst, 305
Zoological, 352
Zoology, 94, 346, 352